Sedimentation and Thickening

T0210935

MATHEMATICAL MODELLING:
Theory and Applications

VOLUME 8

This series is aimed at publishing work dealing with the definition, development and application of fundamental theory and methodology, computational and algorithmic implementations and comprehensive empirical studies in mathematical modelling. Work on new mathematics inspired by the construction of mathematical models, combining theory and experiment and furthering the understanding of the systems being modelled are particularly welcomed.

Manuscripts to be considered for publication lie within the following, non-exhaustive list of areas: mathematical modelling in engineering, industrial mathematics, control theory, operations research, decision theory, economic modelling, mathematical programmering, mathematical system theory, geophysical sciences, climate modelling, environmental processes, mathematical modelling in psychology, political science, sociology and behavioural sciences, mathematical biology, mathematical ecology, image processing, computer vision, artificial intelligence, fuzzy systems, and approximate reasoning, genetic algorithms, neural networks, expert systems, pattern recognition, clustering, chaos and fractals.

Original monographs, comprehensive surveys as well as edited collections will be considered for publication.

Editors:
R. Lowen (*Antwerp, Belgium*)

Editorial Board:
E. Jouini (*University of Paris 1 and ENSAE, France*)
G.J. Klir (*New York, U.S.A.*)
J.-L. Lions (*Paris, France*)
P.G. Mezey (*Saskatchewan, Canada*)
F. Pfeiffer (*München, Germany*)
H.-J. Zimmerman (*Aachen, Germany*)

The titles published in this series are listed at the end of this volume.

Sedimentation and Thickening˙

Phenomenological Foundation and Mathematical Theory

by

María Cristina Bustos
Department of Mathematical Engineering,
University of Concepción, Chile

Fernando Concha
Department of Metallurgical Engineering,
University of Concepción, Chile

Raimund Bürger
Institute of Mathematics A,
University of Stuttgart, Germany

and

Elmer M. Tory
Department of Mathematics and Computer Science,
Mount Allison University, Sackville,
New Brunswick, Canada

KLUWER ACADEMIC PUBLISHERS
DORDRECHT / BOSTON / LONDON

A C.I.P. Catalogue record for this book is available from the Library of Congress.

ISBN 978-90-481-5316-9

Published by Kluwer Academic Publishers,
P.O. Box 17, 3300 AA Dordrecht, The Netherlands.

Sold and distributed in North, Central and South America
by Kluwer Academic Publishers,
101 Philip Drive, Norwell, MA 02061, U.S.A.

In all other countries, sold and distributed
by Kluwer Academic Publishers,
P.O. Box 322, 3300 AH Dordrecht, The Netherlands.

Printed on acid-free paper

All Rights Reserved
© 1999 Kluwer Academic Publishers
Softcover reprint of the hardcover 1st edition 1999
No part of the material protected by this copyright notice may be reproduced or
utilized in any form or by any means, electronic or mechanical,
including photocopying, recording or by any information storage and
retrieval system, without written permission from the copyright owner.

This book is dedicated to

PROFESSOR DR.-ING. WOLFGANG L. WENDLAND,

Professor of Mathematics at the University of Stuttgart.

*His constant support of research on sedimentation
for nearly two decades has made this book possible.*

Table of Contents

Preface

The aim of this book is to present a rigorous phenomenological and mathematical formulation of sedimentation processes and to show how this theory can be applied to the design and control of continuous thickeners. The book is directed to students and researchers in applied mathematics and engineering sciences, especially in metallurgical, chemical, mechanical and civil engineering, and to practicing engineers in the process industries. Such a vast and diverse audience should read this book differently. For this reason we have organized the chapters in such a way that the book can be read in two ways. Engineers and engineering students will find a rigorous formulation of the mathematical model of sedimentation and the exact and approximate solutions for the most important problems encountered in the laboratory and in industry in Chapters 1 to 3, 7 and 8, and 10 to 12, which form a self-contained subject. They can skip Chapters 4 to 6 and 9, which are most important to applied mathematicians, without losing the main features of sedimentation processes. On the other hand, applied mathematicians will find special interest in Chapters 4 to 6 and 9 which show some known but many recent results in the field of conservation laws of quasilinear hyperbolic and degenerate parabolic equations of great interest today. These two approaches to the theory keep their own styles: the mathematical approach with theorems and proofs, and the phenomenological approach with its deductive technique.

A great part of the theory, model formulation and results shown in this book is original and based on the publications of the authors from the sixties to the present day. The most distinctive feature of the book is its combination of a rigorous axiomatic phenomenological approach with a concern for mathematical correctness and its application to the most important practical industrial problems.

In Chapter 1 we give a rigorous, but limited, presentation of the theory of mixtures. We feel that, since this is a book on mechanics of particle-fluid motion, there is no necessity to develop the thermodynamics of mixtures. This chapter gives the framework for developing the concepts of a sedimentation process and the tools to study the two models, the kinematical and the dynamical sedimentation processes, which are presented in Chapters 2 and 3. An introduction discusses the condition required for a multiphase particle system to be regarded as a continuum. Next, the concept of mixture, components and configuration are laid down and the ideas of mass and deformation function for each component and for the mixture are discussed. The measures of deformation and motion lead to the macroscopic mass

xi

and momentum balances. For regions where the variables are continuous, these balances yield the local mass and momentum balances while at discontinuities mass and momentum jump balances are obtained. A dynamical process is defined and the necessity of formulating constitutive equations for the stresses and the interaction forces is established.

Chapter 2 analyses the mixture of fine particles and a fluid regarded as superimposed continua. The concepts of an ideal suspension and of an ideal thickener are presented and the field equations for the mass of the solid component and of the mixture are established. The models for batch and for continuous sedimentation are formulated and the kinematical or Kynch sedimentation processes are defined.

Chapter 3 studies the sedimentation of flocculated suspensions. The assumptions are established for those suspensions to be regarded as superimposed continua. Macroscopic and local balances are established and constitutive equations are formulated. Dimensional analysis shows that the convective terms are several orders of magnitude smaller than the other terms in the equation and can be eliminated from the local balances. Experimental variables are replaced for the solid and liquid component pressures and the dynamical sedimentation process is defined.

Chapters 4, 5 and 6 are formulated without appealing to sedimentation problems. Chapter 4 collects a variety of well-known results for initial value problems or Cauchy problems of scalar hyperbolic conservation laws. Weak solutions are defined and the method of characteristics is used to obtain the solution. At the intersection point of the characteristic lines, the solution is no longer unique and entropy conditions are imposed. Riemann problems of scalar conservation laws are solved. Chapter 5 studies all possible forms of solutions of Riemann problems with flux density functions with up to two inflection points. In Chapter 6 the theoretical framework for hyperbolic conservation laws is extended to initial-boundary value problems. Initial data are given only on a closed interval rather than on the whole real axis and, at the boundaries of the computational domain corresponding to the endpoints of that interval, solution values are prescribed as time-dependent functions. However, these boundary data are not always assumed in a pointwise sense by entropy weak solutions. Entropy boundary conditions are derived and characterized, and existence and uniqueness of entropy weak solutions of the initial-boundary value problem are shown by the vanishing viscosity method.

Chapters 7 and 8 apply the previous results to problems of batch and continuous sedimentation of ideal suspensions. Batch sedimentation is described by two adjacent Riemann problems in Chapter 7. Here these are solved for flux density functions with up to two inflection points giving seven types of solutions, the so-called modes of sedimentation. Finally Dafermos' polygonal approximate method is used to solve batch sedimentation problems, allowing an a priori error estimate. In Chapter 8, one Riemann problem located in the interior of the ideal continuous thickener is considered as an initial state and the upper and lower constant are extended to the boundaries. All possible weak solutions of this configuration for flux density functions with up to two inflection points are determined, giving rise to five modes of continuous sedimentation.

Chapter 9 formulates batch and continuous sedimentation of flocculated suspensions as an initial-boundary value problem for the governing quasilinear degenerate parabolic equation. An appropriate definition of weak or generalized solutions of this initial-boundary problem is formulated, from which jump and entropy boundary conditions are derived. Existence and uniqueness results for this problem and special properties of its solutions are summarized.

Chapter 10 presents a numerical algorithm together with a selection of examples of simulated batch and continuous sedimentation processes illustrating the predictive power of the mathematical model of sedimentation with compression.

In the light of the discussion in the previous chapters, the different methods of thickener design that have been proposed in the literature are described and analyzed in Chapter 11. Three types of methods are distinguished: those based on macroscopic balances, those based on kinematical models and those based on dynamical models. All, from Mishler's and Coe and Clevenger's methods to Kynch's, Talmage and Fitch's, Oltman's, Yoshioka and Hassett's, Wilhelm and Naide's and Adorján's methods fall into these categories. This classification permits the analysis of thickener design procedures with a clear perspective of their applicability and limitations.

Finally in Chapter 12 alternative treatments to sedimentation than those based on continuum mechanics are looked at, in particular cases where Kynch's theory is a useful approximation although his assumptions do not hold. Open problems are also examined and we try to assess which can be treated within the theory of mixtures and which may require other treatments.

We wish to acknowledge financial support of Dr. Elmer Tory as Visiting Professor at the University of Concepción by the Iberoamerican Program of Science and Technology for Development (CYTED) and of Dr. Raimund Bürger as Postdoctoral Fellow by Fundación Andes/Alexander-von-Humboldt-Stiftung, project C-13131, and by FONDEF project D97-I2042.

We thank the University of Concepción, its Engineering Faculty, its Faculty of Physics and Mathematics, its Research Council and its Graduate School for the permanent support they have given us during the last 25 years of research in the area of sedimentation. Our thanks also go to FONDECYT through the many projects they have financed during this period. We cannot forget to thank our many graduate and undergraduate students who have contributed to our work.

The research cooperation between the University of Concepción and first the Technical University of Darmstadt and then the University of Stuttgart was made possible by the German Academic Exchange Service (DAAD) and by the German Research Foundation (DFG), which continues to support research on sedimentation within the Sonderforschungsbereich 404 at the University of Stuttgart. The preparation of the final version of this book has also benefitted from the research cooperation of the University of Stuttgart with the University of Bergen, Norway, supported by the recently initiated 'Applied Mathematics for Industrial Flow Problems (AMIF)' program of the European Science Foundation (ESF).

Introduction

Sedimentation processes in history

Sedimentation is the process of deposition of solid material from a fluid, usually air or water, from a state of suspension. It is widely observed in nature in the formation of rocks and ore deposits. One important branch of Geology, Sedimentology or Sedimentary Petrology, studies the origin of rocks by this method. Sedimentation has also a great scientific and industrial importance, especially in the mining and chemical industry.

The physics of the most common sedimentation process, the settling of solid particles in a fluid medium, has long been known. The settling velocity equation formulated by G.G. Stokes in 1851 is the classic starting point for any discussion of the sedimentation process. Stokes showed that the terminal velocity of a sphere in a fluid is directly proportional to the density difference between the solid and the fluid, to the square of the radius of the sphere, to the force of gravity and inversely proportional to the fluid viscosity. This equation is valid only for very slow motions, so more elaborate equations have been developed for faster-moving particles.

When the concentration of a suspension is low, the distance between particles is large compared to the size of the particles and the effects of mutual interference are often disregarded. The rate of settling of a constituent particle under these conditions, called *free settling*, is then calculated from a single particle motion. This settling behavior of particles in fluids is the basis for the most frequent methods of particle size determination. However, *cluster settling* in dilute dispersions can distort these results (Tory *et al.* 1992, Tory and Kamel 1997). At higher concentration, conditions within the suspension are considerably modified, particularly in that the upward velocity of the fluid displaced by the settling particles is much greater and the flow patterns are appreciably altered. The process is then known as *hindered settling* and is commonly encountered industrially in solid-liquid separation of suspensions by sedimentation in a thickener.

In the Americas, the invention of the Dorr thickener in 1915 (Dorr, 1915) can be mentioned as the starting point of the modern thickening era. John V.N. Dorr, D.Sc., a chemist, cyanide mill owner and operator and a consulting engineer and plant designer, tells (Dorr, 1936): *"The first mill I built and operated was turned*

1

into a profitable undertaking by my invention of the Dorr classifier in 1904 and, in remodeling another mill in the same district, the Dorr thickener was born in 1905. [...] *Its recognition of the importance of mechanical control and continuous operation in fine solid-liquid mixtures and the size of its units have opened the way for advances in sewage and water purification and made wet chemical processes and industrial mineral processes possible.*"

Thickening or dewatering may be defined as the removal of a portion of the liquid from a suspension, also known as pulp or slime, made up of a mixture of finely divided solids and liquids. The early methods of thickening employed plain, flat-bottomed tanks into which the pulp was fed until the tank was full. The solid were then allowed to settle as long as required, the top liquid decanted, the settled solids were discharged and the operation was repeated. Such settling was usually carried out in a number of tanks so that regular cycles of filling, settling and discharging could be maintained.

The invention of the Dorr thickener made possible the continuous dewatering of a dilute pulp whereby a regular discharge of a thick pulp of uniform density took place concurrently with overflow of clarified solution. Scraper blades or rakes, driven by a suitable mechanism, rotating slowly over the bottom of the tank, which usually slopes gently toward the center, move the material as fast as it settles without enough agitation to interfere with the settling. The standard construction of the thickener mechanism is of iron and steel. The tanks are usually made of steel or wood for medium sized machines, but in the larger sizes they are often constructed of concrete.

Thickeners are used in the metallurgical field to thicken prior to agitation and filtration, in the countercurrent washing of cyanide slimes, for thickening ahead of flotation, for thickening concentrates and for dewatering tailings to recover the water for reuse in the mill.

Thickening is not a modern undertaking and it was certainly not discovered as a process in the Americas. Whenever an ore has been dressed to obtain a concentrate, two processes have been utilized in an inseparable form: *crushing* and *washing*. There is evidence that in the IV Egyptian dynasty, some 4000 years BC, the ancient Egyptians dug for and washed gold. There is also evidence of gold washing in Sudan in the XII dynasty. Nevertheless, the earliest written reference for crushing and washing processes in Egypt is that of Agatharchides, a Greek geographer who lived 200 years before Christ. Ardaillon, author of the book "Les mines du Laurion dans l'antiquité", describes in 1897 the processes used in the extensive installation for crushing and washing ores in Greece between the V to the III century BC: *"The ore was first hand-picked and afterwards it was apparently crushed in stone mortars some 16 to 24 inches in diameter and thence passed to the mill. These mills, which crushed dry, were of the upper and lower millstone order, like the old-fashioned flour mills and were turned by hand. The stones were capable of adjustment in such a way as to yield different sizes. The sand was sifted and the oversize returned to the mill. From the mill it was taken to the washing plants, which consisted essentially of an inclined area, below which a canal, sometimes with riffles, lead through a series of basins, ultimately returning the water again to near the head of the area.*"

Agricola, in his book "De re metallica" (1556), describes several methods of washing gold, silver, tin and other metallic ores. He describes settling tanks used as classifiers, jigs and thickeners and settling ponds used as thickeners or clarifiers. These devices operated in a batch or semi-continuous manner. Typical descriptions are the following: "*The large settling pit which is outside the building is generally made of joined flooring, and is eight feet in length, breadth and depth. When a large quantity of mud mixed with very fine tin-stone has settled in it, first all water is let out by withdrawing a plug, then the mud which is taken out is washed outside the house on the canvas strakes and, afterwards, the concentrates are washed on the strake which is inside the building. By these methods the very finest tin-stones are made clean. The mud mixed with the very fine tin-stone, which has neither settled in the large settling-pit nor in the transverse launder, which is outside the room and below the canvas strakes, flows away and settles in the bed of the stream or river*" and "*To concentrate copper at the Neusohl in the Carpathians the ore is crushed and washed and passed through three consecutive washer-sifters. The fine particles are washed through a sieve in a tub full of water, where the undersize settles to the bottom of the tub. At a certain stage of filling of the tub with sediment, the plug is drawn to let the water run away. Then the mud is removed with a shovel and taken to a second tub and then to a third tub where the whole process is repeated. The copper concentrate which has settled in the last tub is taken out and smelted.*"

It is evident from these references that, by using the *washing* and *sifting* processes, the ancient Egyptians and Greeks and medieval Germans knew the practical effect of the difference in specific gravity of the various components of an ore and used sedimentation in operations that can now be identified as classification, clarification and thickening. There is also evidence that in the early days no clear distinction was made between these three operations (Agricola 1556, Richards and Locke 1909).

Classification, clarification and thickening all involve the settling of one substance in solid particulate form through a second substance in liquid form, but the development of each one of these operations has followed different paths. While clarification deals with very dilute suspensions, classification and thickening are forced to use more concentrated pulps. Maybe this is reason for which clarification was the first of these operations amenable to a mathematical description. The work by Hazen in 1904 was the first analysis of factors affecting the settling of solid particles from dilute suspensions in water. It shows that detention time was not a factor in the design of settling tanks, but rather that the portion of solid removed was proportional to the surface area of the tank and to the settling properties of the solid matter, and inversely proportional to the flow through the tank. Depth of the tank had no effect, so that a shallow tank would produce a result as good as a deep tank for the same flow. Several other works continued this line of research introducing the effect of turbulence (Dobbins 1943, Camp 1945, 1946).

Classification used, at the beginning, equipment that mimicked the clarification settling tanks, adding devices to discharge the settled sediment. Therefore the first theories of mechanical classification were based on Hazen's theory of settling basins. New developments led to the restriction of clarification to its proper place. The introduction of hydraulic classifiers and hydrocyclones led to theories of hindered settling in gravitational and centrifugal fields, respectively.

Modern thickening research

Thickening, the main concern in this book, began in the Americas with the introduction of the Dorr thickener in the mining concentrators of South Dakota, as we have already mentioned. The first references to batch thickening are those of Nichols (1908) and Ashley (1909) on the effect of temperature and viscosity on the settlement of slimes. On thickener performance, the first reference is that of Forbes (1912), where he proposes certain standards for conducting batch settling tests which should enable operators to better predict thickener performance, and manufacturers to meet settling requirements of equipment. Experiments with four types of slimes gave results tending to show that settling efficiency decreases as depth of the settling column is increased, and that lime, cyanide and other electrolytes are accelerators for the settling of slimes. Mishler (1912), an engineer and superintendent at the Tigre Mining Company concentrator in the Sonora desert of Mexico, was the first to show by experiments that the rate of settling of slimes is different for dilute and for concentrated suspensions. While the settling speed of dilute slimes is usually independent of the depth of the settling column, extremely thick slimes were governed by a different law. He devised a formula that allowed data from laboratory tests to be used as the for sizing thickeners. However, the milestone in this early American history of thickening is the paper by Coe and Clevenger in 1916, where they established firmly the method for thickener design used until today. This paper, produced at the University of Stanford and at the U.S. Bureau of Mines, is possibly the most cited paper in the thickening literature worldwide. Coe and Clevenger presented a design similar to Mishler's, but advised to find, for each discharge concentration, the concentration that limits the thickener capacity. In doing so they defined handling capacities of different zones in the thickener.

After the important contributions made in the development of thickening technology in the first two decades of this century, the invention of the industrial continuous thickener and the development of a method for its design, the next decade was surely one of expansion of this technology. No further important contribution was made until the beginning of the thirties. Stewart and Roberts (1933) give, in a review paper, a good idea of the state of the art of thickening in the twenties. They say: *"The basic theory is old but limitations and modifications are still but partially developed. Especially in the realm of flocculent suspensions is the underlying theory incomplete. Practical testing methods for determining the size of machines to be used are available, but the invention and development of new machines will no doubt be greatly stimulated by further investigation of the many interesting phenomena observed in practice and as fresh problems are uncovered."*

The University of Illinois became very active in thickening research in the thirties and fourties. At least nine B.Sc. theses were made, mainly on the effect of operating variables in continuous thickening. The mechanism of continuous sedimentation was investigated in the laboratory in order to explain behavior in continuous thickeners. Comings describes these findings in an important paper (1940) and later in a joint paper (Comings et al. 1954). Four zones were described in descending order: clarification, settling, upper compression and rake action zones. Most interesting was their observation that the concentration in the settling zone is nearly constant for a thickener at steady state and that it depends on the rate at which the solids are introduced in the feed. At low feed rate, the solids settled rapidly at a very low concentration. When the feed rate was increased and approached the settling handling capacity of this zone, the concentration rose to a definite value and this concentration was maintained as the feed rate was increased. Solids fed in excess of the settling handling capacity left the thickener by the overflow. The value of concentration in these zones was unknown. It was shown that for the same feed rate, the underflow rate could be adjusted by increasing the sediment depth, resulting in a correspondingly higher sediment residence time. Another observation of those experiments was that the feed to a continuous thickener was diluted on entering the thickener. Their findings were presented by Comings in 1954 (Comings et al. 1954).

Kynch wrote his celebrated paper "A theory of sedimentation" in 1952. Kynch was a mathematician at the University of Birmingham in Great Britain, and his is the first mathematical theory of sedimentation. He proposed a kinematical theory of sedimentation based on the propagation of concentration waves in the suspension. This paper had the greatest influence in the development of thickening thereafter. When Comings moved from Illinois to Purdue, the research on thickening continued there for another ten years. Although Comings soon moved on to Delaware, work continued at Purdue under the direction of P.T. Shannon. A PhD thesis by Tory and M.Sc. theses by De Haas and Stroupe analyzed Kynch's theory and proved its validity by experiments with glass beads. Their results were published in a series of joint papers by these authors in 1963, 1964, 1965 and 1966. Batch and continuous thickening was regarded as the process of propagating concentration changes upwards from the bottom of the settling vessel as a result of the downward movement of the solids. Equations were derived and experimental results for the batch settling of rigid spheres in water were found to be in excellent agreement with Kynch's theory.

Kynch's paper also motivated industry to explore the possibilities of this new theory in thickener design. Again the Dorr Co. went a step further in their contribution to thickening by devising the Talmage and Fitch method of thickener design (Talmage and Fitch 1955). Several authors (for example Yoshioka et al. 1957, Michaels and Bolger 1962, Hassett 1958, 1968, Fitch 1967, 1983, Scott 1968, Tiller 1981) followed, especially modifying Kynch's theory to fit the sedimentation of industrial pulps, because Kynch's theory is valid only for suspensions of equally-sized dispersed rigid spheres and industrial slimes are usually compressible.

The next important contribution to thickening theory was the recognition that

Kynch's theory could only describe a part of the settling process and that a different theory was necessary for testing the compression of the sediment. Ad-hoc theories were developed by several authors, among them by Fitch (1979), Tiller (1981), Kos (1977a,b, 1980), Adorján (1975, 1977) and Shirato *et al.* (1970). These theories treated compression either as a filtration or as a consolidation phenomenon.

Strong and important research on thickening that unfortunately rarely was published in mainstream international journals (but is well documented in local journals and conference volumes) was going on in Brazil in the decade of the 1970s. At the Graduate School of the Universidade Federal de Rio de Janeiro, COPPE, several researchers and graduate students, such as G. Massarani, A. Silva Telles, R. Sampaio, I. Liu, J. Freire, L. Kay and J. D'Avila (Telles 1977, D'Avila 1976a,b, Sampaio and D'Avila 1977a,b, D'Avila 1978), to mention only a few, were involved in the application of a newly developed mathematical tool, the theory of mixtures of continuum mechanics, to particulate systems. They were especially interested in the flow in porous media and in sedimentation. This theory gave the sedimentation process a proper phenomenological framework. At about the same time, and strongly related with the Brazilians, the University of Concepción in Chile started to work in the same direction and findings were presented in B.Sc. theses by Bascur (1976) and Barrientos (1978), at the Engineering Foundation Conference on Particle Technology in New Hampshire in 1980, and recently in Chapter 3 of Tory's book "Sedimentation of particles in a viscous fluid" (Tory 1996). Between 1976 and 1979 a few other papers using the theory of mixtures appeared in the international literature (Bedford and Hill 1976, Hill and Bedford 1979, Thacker and Lavelle 1977, 1978). In spite of the fact that the theory of mixtures did a great job in unifying the sedimentation of dispersed and of flocculated suspensions and gave rise to a robust framework in which any suspension could be simulated once appropriate constitutive equations were formulated, the mathematical analysis of these models did not exist. The University of Concepción in cooperation with, first, the Technische Hochschule Darmstadt, and then with the University of Stuttgart in Germany studied, in the last ten years, the mathematical properties of the solutions of sedimentation processes for ideal and flocculated suspensions. These results were reported in a series of papers by M.C. Bustos, F. Concha, A. Barrientos, W.L. Wendland, M. Kunik, R. Bürger and others, which are widely cited in this book.

Chapter 1

Theory of mixtures

1.1 Introduction

In developing the balance equations for a multicomponent flow, there are two approaches available. Some authors (Whitaker 1967, Slattery 1967) start from the equations of fluid mechanics, valid on the particle scale, and then integrate these equations in regions sufficiently large to contain a representative mass of all components. In case of multiphase particulate flow, those regions must be much greater than the size of the particles contained in the system. The spatially averaged properties then become field variables and the new balance equations constitute a set of local equations describing the flow of a multicomponent mixture. The second approach is the theory of mixtures which uses the concepts of continuum mechanics, considering all components as superimposed interacting continua (Bowen 1976, Truesdell 1984, Dobran 1985). The macroscopic balances are established as the fundamental equations and, from them, the local balances and jump conditions are deduced. The field variables in the continuum approach are equivalent to the averaged variables in the first approach, so that both methods give the same results (Drew 1983). In both cases the local variables cannot be experimentally measured and are not to be confused with the experimental variables of fluid mechanics. In the work that follows, we use the continuum approach of the theory of mixtures.

1.2 Theory of mixtures

The theory of mixtures, put forward originally by Fick (1855) and Stefan (1871) but developed more recently by Truesdell and Toupin (1957), postulates that each point in space is simultaneously occupied by a finite number of particles, one for each component of the mixture. Then, the mixture may be represented as a superposition of continuous media, each following its own movement with the only restriction imposed by the interaction between components. This means that each component will obey the laws of conservation of mass and momentum, incorporating terms to account for the interchange of mass and momentum between components. To obtain a rational theory we must require that the properties of the mixture are a consequence of the properties of the components and that the mixture follows the same laws as a body of a single component, that is, that the mixture behaves as a single component body. Treatments similar or alternative to this may be found in many articles or books such as Bowen (1976), Bedford and Drumheller (1983), Drew (1983), Truesdell (1984), Ungarish (1993), Rajagopal and Tao (1995), and Drew and Passman (1999).

1.2.1 Kinematics

Body, configuration and type of mixture

We denote by a mixture a body B formed by n components $B_\alpha \subset B$, $\alpha = 1, 2, \ldots, n$. The elements of B_α are called *particles* and are denoted by p_α. Each body B_α occupies a determined region of the Euclidean three-dimensional space E_3 called the *configuration* of the body. The elements of the configurations are *points* $X_\alpha \subset E_3$ whose positions are given by $\mathbf{r}$. Then, the position of a particle $p_\alpha \subset B_\alpha$ is given by

$$\mathbf{r} = \chi_\alpha(p_\alpha), \quad \alpha = 1, 2, \ldots, n. \tag{1.1}$$

For the mathematical properties of χ_α see Bowen (1976). The configuration $\chi(B)$ of the mixture is

$$\chi(B) = \bigcup_{\alpha=1}^{n} \chi_\alpha(B_\alpha) \tag{1.2}$$

The volume of $\chi(B)$ is called the *material volume* and is denoted by $V_m = V(\chi(B))$. To every body B_α we can assign a positive, continuous and additive function m_α that measures the amount of matter it contains, such that

$$m(B) = \sum_{\alpha=1}^{n} m_\alpha(B_\alpha), \tag{1.3}$$

where $m_\alpha(B_\alpha)$ is the *mass* of the α component and $m(B)$ is the mass of the mixture. Due to the continuous nature of the mass, we can define a *mass density* $\bar{\rho}_\alpha$ at the point $\mathbf{r}$ at time t in the form

$$\bar{\rho}_\alpha(\mathbf{r}, t) = \lim_{k \to \infty} \frac{m_\alpha(P_k)}{V_m(P_k)}, \tag{1.4}$$

where $P_{k+1} \subset P_k$ are parts of the mixture having the position $\mathbf{r}$ in common at time t. Due to the hypothesis that mass is an absolutely continuous function of volume for a continuum, the function $\bar{\rho}_\alpha$ exists almost everywhere in B, see Drew and Passman (1999). This mass density is called the *apparent density* of B_α. The total mass of B_α is given in terms of $\bar{\rho}_\alpha$ by

$$m_\alpha = \int_{V_m} \bar{\rho}_\alpha(\mathbf{r}, t) dV, \tag{1.5}$$

where t denotes time. For each body B_α we can select a reference configuration $\chi_{\alpha\kappa}$ such that in that configuration it is the only component of the mixture (pure state). Let $\rho_{\alpha\kappa}$ be the mass density of the α component in this configuration. Then we can write

$$m_\alpha = \int_{V_m} \bar{\rho}_\alpha dV = \int_{V_{\alpha\kappa}} \rho_{\alpha\kappa} dV. \tag{1.6}$$

The density of B_α in the reference configuration (pure) is denoted by $\rho_{\alpha\kappa}$ and called the *material density*. Denote the material density of B_α in the actual configuration by ρ_α and define the function

$$\varphi_\alpha(\mathbf{r}, t) = \bar{\rho}_\alpha(\mathbf{r}, t)/\rho_\alpha(\mathbf{r}, t). \tag{1.7}$$

Substituting into equation (1.5) yields

$$m_\alpha = \int_{V_m} \bar{\rho}_\alpha dV = \int_{V_m} \rho_\alpha \varphi_\alpha dV. \tag{1.8}$$

Define a new element of volume $dV_\alpha = \varphi_\alpha dV$ such that

$$m_\alpha = \int_{V_m} \bar{\rho}_\alpha dV = \int_{V_\alpha} \rho_\alpha dV_\alpha. \tag{1.9}$$

The volume V_α in the actual configuration of B_α is called the *partial volume*, and the function $\varphi_\alpha(\mathbf{r}, t)$ the *volume fraction* of B_α. Since the sum of the partial volumes should give the total volume, φ should obey the restriction

$$\sum_{\alpha=1}^{n} \varphi_\alpha(\mathbf{r}, t) = 1. \tag{1.10}$$

We can distinguish two types of mixtures: homogeneous and heterogeneous. Homogeneous mixtures fulfill completely the condition of continuity for the material because the mixing between components occurs at the molecular level. Those mixtures are frequently called *solutions*. For a homogeneous mixture, $\bar{\rho}_\alpha$ is the concentration of the component B_α. In a heterogeneous mixture, the mixing of components is at the macroscopic level, and for them to be considered as a continuum, the size of the integration volume V_m in the previous equations must be greater than that of the mixing level. These mixtures are also called *multiphase mixtures* because each component can be identified as a different phase. In these types of mixtures, $\varphi(\mathbf{r}, t)$ is a measure of the local structure of the mixture, and $\bar{\rho}_\alpha$ is called the *bulk density*.

It is sometimes convenient to define other reference configurations, such as χ_0 or χ_c with material volume V_0 and V_c, respectively, that may correspond or not to certain instants in the motion of the mixture. The mass density of the component α in these new configurations is denoted by $\bar{\rho}_{a0}$ or $\bar{\rho}_{ac}$ which are related to $\rho_{a\kappa}$ in the following way:

$$m_\alpha = \int_{V_0} \bar{\rho}_{a0} dV = \int_{V_c} \bar{\rho}_{ac} dV = \int_{V_{a\kappa}} \rho_{a\kappa} dV. \qquad (1.11)$$

We denote by *material point* the position of the particle p_α in reference configuration $\chi_{a\kappa}$:

$$\mathbf{R}_\alpha = \chi_{a\kappa}(p_\alpha). \qquad (1.12)$$

We assume that (1.12) has an inverse, $p_\alpha = \chi_{a\kappa}^{-1}(\mathbf{R}_\alpha)$.

Configuration and motion

The motion of $p_\alpha \in B_\alpha$ is a continous sequence of configurations:

$$\mathbf{r} = \chi_\alpha(p_\alpha, t), \qquad \alpha = 1, 2, ..., n. \qquad (1.13)$$

Substituting (1.12) into (1.13) yields $\mathbf{r} = \mathbf{f}_\alpha(\mathbf{R}_\alpha, t)$, where $\mathbf{f}_\alpha$ is the α *deformation function* given by

$$\mathbf{f}_\alpha = \chi_\alpha \circ \chi_{a\kappa}^{-1}. \qquad (1.14)$$

We require $\mathbf{f}_\alpha$ to be twice differentiable and to have an inverse, such that

$$\mathbf{R}_\alpha = \mathbf{f}_\alpha^{-1}(\mathbf{r}, t). \qquad (1.15)$$

The cartesian components of $\mathbf{r}$ and $\mathbf{R}_\alpha$ are the *spatial* and *material* coordinates of p_α,

$$\mathbf{r} = x_i \mathbf{e}_i \quad \text{and} \quad \mathbf{R}_\alpha = X_{\alpha i} \mathbf{e}_i. \qquad (1.16)$$

Any property G_α of the body B_α can be described in terms of material or spatial coordinates. From $G(p_\alpha, t)$ we can write either

$$G_\alpha = G_\alpha\left(\chi_{a\kappa}^{-1}(\mathbf{R}_\alpha), t\right) = g_{\alpha 1}(\mathbf{R}_\alpha, t) \qquad (1.17)$$

or

$$G_\alpha = G_\alpha\left(\chi_\alpha^{-1}(\mathbf{r}), t\right) = g_{\alpha 2}(\mathbf{r}, t). \qquad (1.18)$$

Of course, the properties $g_{\alpha 1}(\mathbf{R}_\alpha, t)$ and $g_{\alpha 2}(\mathbf{r}, t)$ are equivalent. We refer to the first notation as the *material property* and to the second as the *spatial property* G of the body B_α.

Partial material and spatial derivatives

Since the property G_α is a function of two variables $(\mathbf{R}_\alpha, t)$ or $(\mathbf{r}, t)$ it is possible to obtain the partial derivative of G with respect to each of these variables. The gradient of G_α is the partial derivative of G_α with respect to the space variable. Since there are two such variables, there will be two gradients, the material gradient

$$\frac{\partial G_\alpha}{\partial \mathbf{R}_\alpha} = \frac{\partial g_{\alpha 1}(\mathbf{R}_\alpha, t)}{\partial \mathbf{R}_\alpha} = \frac{\partial g_{\alpha 1}}{\partial X_{\alpha i}} \mathbf{e}_i \equiv \mathrm{Grad}\, G_\alpha \qquad (1.19)$$

and the spatial gradient

$$\frac{\partial G_\alpha}{\partial \mathbf{r}} = \frac{\partial g_{\alpha 2}(\mathbf{r}, t)}{\partial \mathbf{r}} = \frac{\partial g_{\alpha 2}}{\partial X_i} \mathbf{e}_i \equiv \mathrm{grad}\, G_\alpha. \qquad (1.20)$$

In the same way we can define material and spatial time derivatives of G_α: the material derivative

$$\dot{G}_\alpha = \frac{D_\alpha G_\alpha}{Dt} = \frac{\partial g_{\alpha 1}(\mathbf{R}_\alpha, t)}{\partial t} = \frac{\partial g_{\alpha 2}(\mathbf{r}, t)}{\partial t}\bigg|_{\mathbf{R}_\alpha} \qquad (1.21)$$

represents the derivatives of G_α with respect to time holding the material point $\mathbf{R}_\alpha$ fixed, while the spatial time derivative

$$\frac{\partial G_\alpha}{\partial t} = \frac{\partial g_{\alpha 2}(\mathbf{r}, t)}{\partial t} = \frac{\partial g_{\alpha 1}(\mathbf{R}_\alpha, t)}{\partial t}\bigg|_{\mathbf{r}} \qquad (1.22)$$

represents the derivative of G_α with respect to time holding the place $\mathbf{r}$ fixed. The relationship between the material and spatial derivatives is obtained by applying the chain rule of differentiation to (1.21):

$$\dot{G}_\alpha = \frac{\partial g_{\alpha 2}(\mathbf{r}, t)}{\partial t}\bigg|_{\mathbf{R}_\alpha} = \frac{\partial g_{\alpha 2}(\mathbf{r}, t)}{\partial t} + \frac{\partial g_{\alpha 2}(\mathbf{r}, t)}{\partial \mathbf{r}} \frac{\partial \mathbf{r}}{\partial t}\bigg|_{\mathbf{R}_\alpha},$$

$$\dot{G}_\alpha = \frac{\partial g_{\alpha 2}}{\partial t} + \mathrm{grad}\, g_{\alpha 2} \cdot \dot{\mathbf{r}},$$

$$\dot{G}_\alpha = \frac{\partial G_\alpha}{\partial t} + \mathrm{grad}\, G_\alpha \cdot \dot{\mathbf{r}}, \qquad (1.23)$$

where $\dot{\mathbf{r}}$ is the material derivative of the deformation function $\mathbf{r} = \mathbf{f}_\alpha(\mathbf{R}_\alpha, t)$.

Gradient of deformation tensor

Consider two particles p_α, $q_\alpha \in B_\alpha$ having the positions $\mathbf{R}_\alpha$ and $\mathbf{R}_\alpha + d\mathbf{R}_\alpha$ in the reference configuration; see Figure 1.1. The positions of p_α and q_α are given by

$$\mathbf{r} = \mathbf{f}_\alpha(\mathbf{R}_\alpha, t), \quad \mathbf{r} + \Delta \mathbf{r} = \mathbf{f}_\alpha(\mathbf{R}_\alpha + d\mathbf{R}_\alpha, t). \qquad (1.24)$$

The position of q_α in time t can be approximated in the vicinity of $\mathbf{r}$ by a linear function of $d\mathbf{R}_\alpha$,

$$\mathbf{f}_\alpha(\mathbf{R}_\alpha + d\mathbf{R}_\alpha, t) \approx \mathbf{f}_\alpha(\mathbf{R}_\alpha, t) + \frac{\partial \mathbf{f}_\alpha}{\partial \mathbf{R}_\alpha} \cdot d\mathbf{R}_\alpha. \qquad (1.25)$$

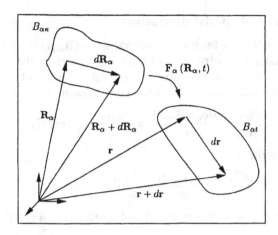

Figure 1.1: Deformation of a body from the reference configuration $B_{\alpha\kappa}$ to the present configuration $B_{\alpha t}$.

From (1.24) and (1.25) we deduce

$$\Delta \mathbf{r} \approx d\mathbf{r} = \frac{\partial \mathbf{f}_\alpha}{\partial \mathbf{R}_\alpha} \cdot d\mathbf{R}_\alpha. \qquad (1.26)$$

The tensor $\partial \mathbf{f}_\alpha / \partial \mathbf{R}_\alpha$ that approximates the deformation function of q_α in the neighborhood of $\mathbf{r}$ is called the *gradient of deformation tensor of the α component* and is denoted by $\mathbf{F}_\alpha$,

$$\mathbf{F}_\alpha(\mathbf{R}_\alpha, t) = \frac{\partial \mathbf{f}_\alpha(\mathbf{R}_\alpha, t)}{\partial \mathbf{R}_\alpha} = \text{Grad}\, \mathbf{r}. \qquad (1.27)$$

To ensure the existence of an inverse, $\det \mathbf{F}_\alpha \neq 0$. In cartesian and matrix notation the deformation tensor can be written in the form

$$\mathbf{F}_\alpha(\mathbf{R}_\alpha, t) = \frac{\partial \mathbf{r}}{\partial \mathbf{R}_\alpha} = \mathbf{B}^{\mathrm{T}} \begin{pmatrix} \dfrac{\partial x_1}{\partial X_{\alpha 1}} & \dfrac{\partial x_1}{\partial X_{\alpha 2}} & \dfrac{\partial x_1}{\partial X_{\alpha 3}} \\[2mm] \dfrac{\partial x_2}{\partial X_{\alpha 1}} & \dfrac{\partial x_2}{\partial X_{\alpha 2}} & \dfrac{\partial x_2}{\partial X_{\alpha 3}} \\[2mm] \dfrac{\partial x_3}{\partial X_{\alpha 1}} & \dfrac{\partial x_3}{\partial X_{\alpha 2}} & \dfrac{\partial x_3}{\partial X_{\alpha 3}} \end{pmatrix} \mathbf{B}, \qquad (1.28)$$

where $\mathbf{B}$ is a basis of orthogonal unit vectors. The transformation of a line element $d\mathbf{R}_\alpha$ from the reference configuration to the actual configuration $d\mathbf{r}$ is then

$$d\mathbf{r} = \mathbf{F}_\alpha(\mathbf{R}_\alpha, t) \cdot d\mathbf{R}_\alpha. \qquad (1.29)$$

A deformation is called *homogeneous* if $\mathbf{F}_\alpha$ is independent of $\mathbf{R}_\alpha$.

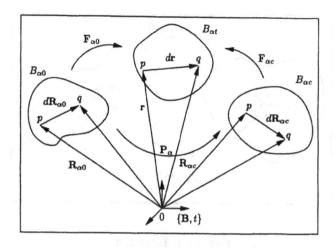

Figure 1.2: Change of reference configuration.

Change of reference configuration

The deformation quantified by $\mathbf{F}_\alpha(\mathbf{R}_\alpha, t)$ depends on the reference configuration chosen. Since this reference configuration is arbitrary it is convenient to know how a change of reference configuration affects $\mathbf{F}_\alpha$.

Consider two reference configurations $B_{\alpha 0}$ and $B_{\alpha c}$ and the actual configuration $B_{\alpha t}$ of a body. Call $\mathbf{F}_{\alpha 0}$, $\mathbf{F}_{\alpha c}$ and $\mathbf{P}_\alpha$ the gradient of deformation tensor to go from $B_{\alpha 0}$ to $B_{\alpha t}$, from $B_{\alpha c}$ to $B_{\alpha t}$ and from $B_{\alpha 0}$ to $B_{\alpha c}$ respectively; see Figure 1.2. We can write

$$dr = \mathbf{F}_{\alpha 0} \cdot d\mathbf{R}_{\alpha 0} = \mathbf{F}_{\alpha c} \cdot d\mathbf{R}_{\alpha c}, \quad d\mathbf{R}_{\alpha c} = \mathbf{P}_\alpha \cdot d\mathbf{R}_{\alpha 0}.$$

Then $dr = \mathbf{F}_{\alpha 0} \cdot d\mathbf{R}_{\alpha 0} = \mathbf{F}_{\alpha c} \cdot \mathbf{P}_\alpha \cdot d\mathbf{R}_{\alpha 0}$, and therefore

$$\mathbf{F}_{\alpha 0} = \mathbf{F}_{\alpha c} \cdot \mathbf{P}_\alpha. \qquad (1.30)$$

Dilatation

Consider an element of material volume $dV_{\alpha \kappa}$ in the reference configuration in the form of a parallelepiped, then

$$dV_{\alpha \kappa} = d\mathbf{R}_{\alpha 1} \cdot d\mathbf{R}_{\alpha 2} \times d\mathbf{R}_{\alpha 3} \equiv [d\mathbf{R}_{\alpha 1}, d\mathbf{R}_{\alpha 2}, d\mathbf{R}_{\alpha 3}]. \qquad (1.31)$$

After the deformation, the volume (see Figure 1.3) will be

$$dV = dr_1 \cdot dr_2 \times dr_3 \equiv [dr_1, dr_2, dr_3]. \qquad (1.32)$$

Using (1.29) yields

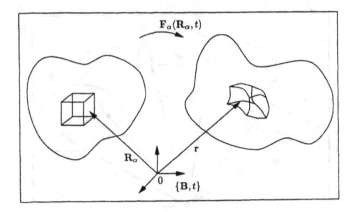

Figure 1.3: Dilatation.

$$dV = [\mathbf{F}_\alpha \cdot d\mathbf{R}_{\alpha 1}, \mathbf{F}_\alpha \cdot d\mathbf{R}_{\alpha 2}, \mathbf{F}_\alpha \cdot d\mathbf{R}_{\alpha 3}]$$
$$= \det \mathbf{F}_\alpha [d\mathbf{R}_{\alpha 1}, d\mathbf{R}_{\alpha 2}, d\mathbf{R}_{\alpha 3}] = (\det \mathbf{F}_\alpha) dV_{\alpha \kappa}. \quad (1.33)$$

Since $\det \mathbf{F}_\alpha = dV/dV_{\alpha \kappa}$, it is called the *dilatation*

$$J_\alpha = \det \mathbf{F}_\alpha. \quad (1.34)$$

Equation (1.33) shows that $\det \mathbf{F}_\alpha > 0$.

Rigid deformation

A special type of deformation is the *rigid deformation* defined as a deformation in which the distances between the particles do not change. Consider two particles $p_\alpha, q_\alpha \in B_\alpha$ which during the motion maintain their distance. Call $d\mathbf{R}_\alpha = ds_{0\alpha}\mathbf{e}_0$ and $d\mathbf{r} = ds\mathbf{e}$. Refer to Figure 1.1 and write

$$ds^2 - ds_0^2 = 0,$$
$$d\mathbf{r} \cdot d\mathbf{r} - d\mathbf{R}_\alpha \cdot d\mathbf{R}_\alpha = 0,$$
$$(\mathbf{F}_\alpha \cdot d\mathbf{R}_\alpha) \cdot (\mathbf{F}_\alpha \cdot d\mathbf{R}_\alpha) - d\mathbf{R}_\alpha \cdot d\mathbf{R}_\alpha = 0,$$
$$d\mathbf{R}_\alpha \cdot \mathbf{F}_\alpha^T \cdot \mathbf{F}_\alpha \cdot d\mathbf{R}_\alpha - d\mathbf{R}_\alpha \cdot \mathbf{I} \cdot d\mathbf{R}_\alpha = 0,$$
$$d\mathbf{R}_\alpha \cdot (\mathbf{F}_\alpha^T \cdot \mathbf{F}_\alpha - \mathbf{I}) \cdot d\mathbf{R}_\alpha = 0. \quad (1.35)$$

Equation (1.35) shows that for a rigid deformation

$$\mathbf{F}_\alpha^T \cdot \mathbf{F}_\alpha = \mathbf{I} \quad (1.36)$$

should be valid. There are two cases for which (1.36) is valid: if $\mathbf{F}_\alpha = \mathbf{I}$ and if $\mathbf{F}_\alpha = \mathbf{Q}_\alpha$ is a rotation. The first case is called *translation*.

Stretching

Since the deformation function $\mathbf{f}_\alpha(\mathbf{R}_\alpha, t)$ implies $\det \mathbf{F}_\alpha > 0$, the polar decomposition (Gurtin 1981) may be applied to $\mathbf{F}_\alpha$,

$$\mathbf{F}_\alpha = \mathbf{Q}_\alpha \cdot \mathbf{U}_\alpha = \mathbf{V}_\alpha \cdot \mathbf{Q}_\alpha. \tag{1.37}$$

Then

$$\mathbf{U}_\alpha^2 = \mathbf{F}_\alpha^T \cdot \mathbf{F}_\alpha = \mathbf{C}_\alpha, \quad \mathbf{V}_\alpha^2 = \mathbf{F}_\alpha \cdot \mathbf{F}_\alpha^T = \mathbf{B}_\alpha. \tag{1.38}$$

Since $\mathbf{U}_\alpha$ and $\mathbf{V}_\alpha$ are symmetric and positive definite tensors and $\mathbf{Q}_\alpha$ is an orthogonal tensor, it can be shown that the characteristic values of $\mathbf{U}_\alpha$ and $\mathbf{V}_\alpha$ are the same and that their characteristic vectors are related by a rotation $\mathbf{Q}_\alpha$:

$$\mathbf{U}_\alpha = \sum_k \lambda_{\alpha k} \mathbf{u}_k \mathbf{u}_k, \quad \mathbf{V}_\alpha = \sum_k \lambda_{\alpha k} \mathbf{v}_k \mathbf{v}_k, \quad \mathbf{v}_k = \mathbf{Q}_\alpha \cdot \mathbf{u}_k. \tag{1.39}$$

Velocity and acceleration

The velocity and acceleration of a particle $p_\alpha \in B_\alpha$ are the first and second material derivatives of the deformation function $\mathbf{F}_\alpha$:

$$\mathbf{v}_\alpha = \frac{D_\alpha \mathbf{f}_\alpha(\mathbf{R}_\alpha, t)}{Dt} = \frac{D_\alpha \mathbf{r}}{Dt} = \dot{\mathbf{r}}(\mathbf{R}_\alpha, t), \tag{1.40}$$

$$\mathbf{a}_\alpha = \dot{\mathbf{v}}_\alpha = \frac{D_\alpha \mathbf{v}_\alpha}{Dt} = \frac{D_\alpha^2 \mathbf{r}}{Dt^2}. \tag{1.41}$$

If the flow field is expressed in spatial coordinates, the acceleration is

$$\mathbf{a}_\alpha = \frac{D_\alpha \mathbf{v}_\alpha}{Dt} = \left.\frac{\partial \mathbf{v}_\alpha(\mathbf{R}_\alpha, t)}{\partial t}\right|_{\mathbf{r}} + \frac{\partial \mathbf{v}_\alpha(\mathbf{R}_\alpha, t)}{\partial \mathbf{r}} \cdot \dot{\mathbf{r}}(\mathbf{R}_\alpha, t) = \frac{\partial \mathbf{v}_\alpha}{\partial t} + \operatorname{grad} \mathbf{v}_\alpha \cdot \mathbf{v}_\alpha. \tag{1.42}$$

Velocity gradient

Consider two neighboring particles $p_\alpha, q_\alpha \in B_\alpha$. If p_α has a velocity $\mathbf{v}_\alpha$, the velocity $\mathbf{v}_\alpha + \Delta \mathbf{v}_\alpha$ of q_α can be approximated by a linear function of $d\mathbf{r}$ (see Figure 1.4),

$$\mathbf{v}_\alpha(\mathbf{r} + d\mathbf{r}) \approx \mathbf{v}_\alpha(\mathbf{r}) + \frac{\partial \mathbf{v}_\alpha}{\partial \mathbf{r}} \cdot d\mathbf{r}.$$

Since $\mathbf{v}_\alpha(\mathbf{r} + \Delta \mathbf{r}) = \mathbf{v}_\alpha(\mathbf{r}) + \Delta \mathbf{v}_\alpha$, we can write $\Delta \mathbf{v}_\alpha \approx d\mathbf{v}_\alpha = \operatorname{grad} \mathbf{v}_\alpha \cdot d\mathbf{r}$. The velocity gradient $\operatorname{grad} \mathbf{v}_\alpha$ is called the *velocity gradient tensor* and is denoted by $\mathbf{L}_\alpha(\mathbf{r}, t)$. In cartesian and matrix notation, $\mathbf{L}_\alpha$ is given by

$$\mathbf{L}_\alpha = \frac{\partial v_{\alpha i}}{\partial x_j} \mathbf{e}_i \mathbf{e}_j = \mathbf{B}^T \begin{pmatrix} \dfrac{\partial v_{\alpha 1}}{\partial x_1} & \dfrac{\partial v_{\alpha 1}}{\partial x_2} & \dfrac{\partial v_{\alpha 1}}{\partial x_3} \\[2ex] \dfrac{\partial v_{\alpha 2}}{\partial x_1} & \dfrac{\partial v_{\alpha 2}}{\partial x_2} & \dfrac{\partial v_{\alpha 2}}{\partial x_3} \\[2ex] \dfrac{\partial v_{\alpha 3}}{\partial x_1} & \dfrac{\partial v_{\alpha 3}}{\partial x_2} & \dfrac{\partial v_{\alpha 3}}{\partial x_3} \end{pmatrix} \mathbf{B}. \tag{1.43}$$

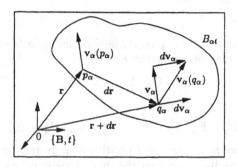

Figure 1.4: Velocity gradient.

The relationship between $\mathbf{L}_\alpha$ and $\mathbf{F}_\alpha$ can be obtained by calculating $d\mathbf{v} = d\dot{\mathbf{r}}$ from (1.29):

$$d\mathbf{v} = \dot{\mathbf{F}}_\alpha \cdot d\mathbf{R}_\alpha = \dot{\mathbf{F}}_\alpha \cdot \mathbf{F}_\alpha^{-1} d\mathbf{r}, \qquad (1.44)$$

and therefore $\mathbf{L}_\alpha = \dot{\mathbf{F}}_\alpha \cdot \mathbf{F}_\alpha^{-1}$.

Rate of dilatation

Taking the material derivative of the dilatation we obtain the rate of dilatation,

$$\dot{J}_\alpha(\mathbf{R}_\alpha, t) = \frac{D}{Dt}(\det \mathbf{F}_\alpha) = \det \mathbf{F}_\alpha \mathrm{tr}(\dot{\mathbf{F}}_\alpha \cdot \mathbf{F}_\alpha^{-1}) = \det \mathbf{F}_\alpha(\mathrm{tr}\mathbf{L}_\alpha). \qquad (1.45)$$

It is clear from equation (1.43) that $\mathrm{tr}\mathbf{L}_\alpha = \mathrm{div}\,\mathbf{v}_\alpha$, so that:

$$\dot{J}_\alpha(\mathbf{R}_\alpha, t) = J_\alpha \mathrm{div}\,\mathbf{v}_\alpha \qquad (1.46)$$

Expansion rate, stretching and spin

The velocity gradient tensor can be separated into three irreducible parts which are mutually orthogonal:

$$\mathbf{L}_\alpha = \underbrace{\frac{1}{3}(\mathrm{tr}\mathbf{L}_\alpha)\mathbf{I}}_{\substack{\text{rate of dilatation}\\\text{tensor or rate of}\\\text{expansion}}} + \underbrace{\left\{\frac{1}{2}\left(\mathbf{L}_\alpha + \mathbf{L}_\alpha^{\mathrm{T}}\right) - \frac{1}{3}(\mathrm{tr}\mathbf{L}_\alpha)\mathbf{I}\right\}}_{\substack{\text{rate of shear tensor or}\\\text{stretching}}} + \underbrace{\frac{1}{2}\left(\mathbf{L}_\alpha - \mathbf{L}_\alpha^{\mathrm{T}}\right)}_{\substack{\text{rate of rotation}\\\text{tensor or spin}}}.$$

$$\qquad (1.47)$$

Defining the symmetric part and the skew part of $\mathbf{L}_\alpha$, respectively, by

$$\mathbf{D}_\alpha = \frac{1}{2}\left(\mathbf{L}_\alpha + \mathbf{L}_\alpha^{\mathrm{T}}\right), \quad \mathbf{W}_\alpha = \frac{1}{2}\left(\mathbf{L}_\alpha - \mathbf{L}_\alpha^{\mathrm{T}}\right) \qquad (1.48)$$

and noting that $\mathrm{tr}\mathbf{L}_\alpha = \mathrm{tr}\mathbf{D}_\alpha$, we can write

$$\mathbf{L}_\alpha = \frac{1}{3}(\mathrm{tr}\mathbf{D}_\alpha)\mathbf{I} + \left\{\mathbf{D}_\alpha - \frac{1}{3}(\mathrm{tr}\mathbf{D}_\alpha)\mathbf{I}\right\} + \mathbf{W}_\alpha \qquad (1.49)$$

or

$$\mathbf{L}_\alpha = \frac{1}{3}(\mathrm{div}\,\mathbf{v})\mathbf{I} + \left\{\frac{1}{2}\left(\mathrm{grad}\,\mathbf{v} + \mathrm{grad}\,\mathbf{v}^T\right) - \frac{1}{3}(\mathrm{div}\,\mathbf{v})\mathbf{I}\right\} + \frac{1}{2}\left(\mathrm{grad}\,\mathbf{v} - \mathrm{grad}\,\mathbf{v}^T\right). \qquad (1.50)$$

1.2.2 Mass balance

Let the components B_α interchange mass between them and call $\bar{g}_\alpha(\mathbf{r}, t)$ the rate of mass transfer, *per unit volume*, to the component α from all other components. Another name for $\bar{g}_\alpha$ is the *mass growth rate*. Then, the rate of change of mass of the α component is

$$\frac{d}{dt}\int_{V_m} \bar{\rho}_\alpha dV = \int_{V_m} \bar{g}_\alpha dV, \qquad (1.51)$$

where dV is an element of the material volume B_α. There is another common way to express the macroscopic mass balance for a mixture. To obtain it, go to the reference configuration in the left hand side of (1.51) by using equation (1.33)

$$\int_{V_m} \bar{g}_\alpha dV = \frac{d}{dt}\int_{V_{\alpha\kappa}} \bar{\rho}_\alpha J_\alpha dV_{\alpha\kappa} = \int_{V_{\alpha\kappa}} \overline{\dot{\bar{\rho}_\alpha J_\alpha}}\, dV_{\alpha\kappa}$$

$$= \int_{V_{\alpha\kappa}} (\dot{\bar{\rho}}_\alpha J_\alpha + \bar{\rho}_\alpha \dot{J}_\alpha)dV_{\alpha\kappa}$$

$$= \int_{V_{\alpha\kappa}} \left\{\left(\frac{\partial\bar{\rho}_\alpha}{\partial t} + \mathrm{grad}\,\bar{\rho}_\alpha \cdot \mathbf{v}_\alpha\right)J_\alpha + \bar{\rho}_\alpha(\mathrm{div}\,\mathbf{v}_\alpha)J_\alpha\right\}dV_{\alpha\kappa}$$

$$= \int_{V_{\alpha\kappa}} \left(\frac{\partial\bar{\rho}_\alpha}{\partial t} + \mathrm{div}\,(\bar{\rho}_\alpha\mathbf{v}_\alpha)\right)J_\alpha dV_{\alpha\kappa}$$

$$= \int_{V_m} \left(\frac{\partial\bar{\rho}_\alpha}{\partial t} + \mathrm{div}\,(\bar{\rho}_\alpha\mathbf{v}_\alpha)\right)dV.$$

Using the Green-Gauss-Ostrogradsky theorem, we obtain

$$\int_{V_m} \frac{\partial\bar{\rho}_\alpha}{\partial t}dV + \int_{S_m} \bar{\rho}_\alpha\mathbf{v}_\alpha \cdot \mathbf{n}dS = \int_{V_m} \bar{g}_\alpha dV. \qquad (1.52)$$

Equation (1.52) is the other form of the macroscopic mass balance. Substituting the volume element dV in the actual configuration $B_{\alpha t}$ for an element $dV_{\alpha\kappa}$ in the reference configuration $B_{\alpha\kappa}$ yields

$$\frac{d}{dt}\int_{V_{\alpha\kappa}} \bar{\rho}_\alpha J_\alpha dV = \int_{V_{\alpha\kappa}} \bar{g}_\alpha J_\alpha dV.$$

Since the volume $V_{\alpha\kappa}$ is constant, the derivative can be introduced inside the integral,

$$\int_{V_{\alpha\kappa}} \left(\frac{D}{Dt}(\bar{\rho}_\alpha J_\alpha) - \bar{g}_\alpha J_\alpha \right) dV_{\alpha\kappa}. \tag{1.53}$$

Now, using the localization theorem (Gurtin 1981), we get

$$\frac{D}{Dt}(\bar{\rho}_\alpha J_\alpha) - \bar{g}_\alpha J_\alpha = 0. \tag{1.54}$$

Dividing both terms by $\bar{\rho}_\alpha J_\alpha$ and denoting the rate of growth *per unit mass* of the α component by $\hat{g}_\alpha = \bar{g}_\alpha / \bar{\rho}_\alpha$, we can write

$$\frac{D}{Dt} \ln(\bar{\rho}_\alpha J_\alpha) = \hat{g}_\alpha. \tag{1.55}$$

Integrating this last equation with boundary conditions $\bar{\rho}_\alpha = \bar{\rho}_{\alpha\kappa}$ and noting that $J_\alpha = 1$ in the reference configuration, we obtain

$$\bar{\rho}_\alpha J_\alpha = \bar{\rho}_{\alpha\kappa} \exp \left(\int_{t_\kappa}^{T} \hat{g}_\alpha dt \right). \tag{1.56}$$

In those cases in which there is no mass interchange between the components of the mixture, $\hat{g}_\alpha = 0$ and equation (1.56) reduces to

$$\bar{\rho}_\alpha J_\alpha = \bar{\rho}_{\alpha\kappa}. \tag{1.57}$$

Equation (1.57) is the local mass balance for B_α. There is another convenient way to express the local mass balance. Perform the operation $\dot{\bar{\rho}}_\alpha J_\alpha + \bar{\rho}_\alpha \dot{J}_\alpha = \bar{g}_\alpha J_\alpha$ in (1.54) and use (1.46) to obtain

$$\dot{\bar{\rho}}_\alpha + \bar{\rho}_\alpha \text{div } \mathbf{v}_\alpha = \bar{g}_\alpha, \tag{1.58}$$

or, writing out the material derivative,

$$\frac{\partial \bar{\rho}_\alpha}{\partial t} + \text{div }(\bar{\rho}_\alpha \mathbf{v}_\alpha) = \bar{g}_\alpha. \tag{1.59}$$

Equation (1.58) or (1.59) is known as the *continuity equation* for B_α.

Mass jump conditions

For bodies having discontinuities, the local balance equations are not valid. In these cases it is necessary to analyze the macroscopic mass balance further. Consider a body $B_\alpha \in B$ having a surface of discontinuity S_I that separates the body into two parts $B_{\alpha t}^+$ and $B_{\alpha t}^-$ in the actual configuration; see Figure 1.5. The following conditions hold:

$$V_m = V^+ + V^-, \quad S_m = S^+ + S^-, \quad S_I = B_{\alpha t}^+ \cap B_{\alpha t}^-. \tag{1.60}$$

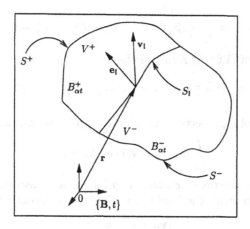

Figure 1.5: Surface of discontinuity.

Applying the macroscopic balance (1.51) to each side of the body yields

$$\frac{d}{dt}\int_{V+}\bar{\rho}_\alpha dV + \frac{d}{dt}\int_{V-}\bar{\rho}_\alpha dV = \int_{V+}\bar{g}_\alpha dV + \int_{V-}\bar{g}_\alpha dV.$$

Going to the reference configuration in the left side of this equation:

$$\int_{V+}\bar{g}_\alpha dV + \int_{V-}\bar{g}_\alpha dV = \int_{V_{\alpha\kappa}^+}\frac{D_\alpha}{Dt}(\bar{\rho}_\alpha J_\alpha)dV_{\alpha\kappa} + \int_{V_{\alpha\kappa}^-}\frac{D_\alpha}{Dt}(\bar{\rho}_\alpha J_\alpha)dV_{\alpha\kappa}$$

$$= \int_{V_{\alpha\kappa}^+}(\dot{\bar{\rho}}_\alpha + \bar{\rho}_\alpha \operatorname{div}\mathbf{v}_\alpha)J_\alpha dV_{\alpha\kappa} + \int_{V_{\alpha\kappa}^-}(\dot{\bar{\rho}}_\alpha + \bar{\rho}_\alpha \operatorname{div}\mathbf{v}_\alpha)J_\alpha dV_{\alpha\kappa}$$

$$= \int_{V+}\left(\frac{\partial\bar{\rho}_\alpha}{\partial t} + \operatorname{div}(\bar{\rho}_\alpha\mathbf{v}_\alpha)\right)dV + \int_{V-}\left(\frac{\partial\bar{\rho}_\alpha}{\partial t} + \operatorname{div}(\bar{\rho}_\alpha\mathbf{v}_\alpha)\right)dV.$$

Since the surface of discontinuity is not a material surface, using the Green-Gauss-Ostrogradsky theorem we obtain

$$\int_{V+}\frac{\partial\bar{\rho}_\alpha}{\partial t}dV + \int_{S+}\bar{\rho}_\alpha\mathbf{v}_\alpha\cdot\mathbf{n}dS + \int_{S_I}(\bar{\rho}_\alpha)^+(\mathbf{v}_\alpha^+ - \mathbf{v}_I)\cdot(-\mathbf{e}_I)dS + \int_{V-}\frac{\partial\bar{\rho}_\alpha}{\partial t}dV$$

$$+ \int_{S-}\bar{\rho}_\alpha\mathbf{v}_\alpha\cdot\mathbf{n}dS + \int_{S_I}(\bar{\rho}_\alpha)^-(\mathbf{v}_\alpha^- - \mathbf{v}_I)\cdot(\mathbf{e}_I)dS = \int_{V+}\bar{g}_\alpha dV + \int_{V-}\bar{g}_\alpha dV,$$

where $\mathbf{v}_I$ is the velocity of the surface discontinuity and $\mathbf{e}_I$ is a unit vector normal to that surface. Adding the volume integrals and the surface integrals and using (1.60) results in

$$\int_{V_m}\frac{\partial\bar{\rho}_\alpha}{\partial t}dV + \int_{S_m}\bar{\rho}_\alpha\mathbf{v}_\alpha\cdot\mathbf{n}dS - \int_{V_m}\bar{g}_\alpha dV$$

$$= \int_{S_I} \left\{ (\bar{\rho}_\alpha)^+ (v_\alpha^+ - v_I) - (\bar{\rho}_\alpha)^- (v_\alpha^- - v_I) \right\} \cdot e_I dS. \quad (1.61)$$

The left-hand side of (1.61) is zero, therefore

$$\int_{S_I} \left\{ ((\bar{\rho}_\alpha)^+ (v_\alpha^+ - v_I) - (\bar{\rho}_\alpha)^- (v_\alpha^- - v_I)) \right\} \cdot e_I dS = 0. \quad (1.62)$$

Defining the jump of a property G as $[G] = G^+ - G^-$, we can write

$$\int_{S_I} [\bar{\rho}_\alpha (v_\alpha - v_I) \cdot e_I] dS = 0. \quad (1.63)$$

This equation is the *macroscopic mass jump balance* at a discontinuity. Using the localization theorem yields the *local mass balance* at a discontinuity,

$$[\bar{\rho}_\alpha (v_\alpha - v_I) \cdot e_I] = 0 \quad (1.64)$$

or

$$\sigma = [\bar{\rho}_\alpha v_\alpha \cdot e_I]/[\bar{\rho}_\alpha], \quad (1.65)$$

where $\sigma = v_I \cdot e_I$ is the displacement velocity of the discontinuity.

Density and average velocity of the mixture

To obtain the continuity equation and the jump condition for the mixture we must add the equations of the components. If we choose the continuity equation (1.59) and the jump condition (1.65), we obtain

$$\frac{\partial}{\partial t} \left(\sum_\alpha \bar{\rho}_\alpha \right) + \operatorname{div} \left(\sum_\alpha \bar{\rho}_\alpha v_\alpha \right) = \sum_\alpha \bar{g}_\alpha, \quad (1.66)$$

$$\left[\sum_\alpha \bar{\rho}_\alpha (v_\alpha \cdot e_I) \right] = \sigma \left[\sum_\alpha \bar{\rho}_\alpha \right]. \quad (1.67)$$

According to the initial postulates, the mixture should follow the laws of a pure material; therefore the continuity equation and the jump mass balance for the mixture should be

$$\frac{\partial \rho}{\partial t} + \operatorname{div}(\rho v) = 0, \quad (1.68)$$

$$[\rho(v \cdot e_I)] = [\rho]\sigma, \quad (1.69)$$

where ρ and v are the density and the velocity of the mixture. From equations (1.66) and (1.68) we obtain

$$\rho = \sum_\alpha \bar{\rho}_\alpha, \quad \rho v = \sum_\alpha \bar{\rho}_\alpha v_\alpha, \quad \sum_\alpha \bar{g}_\alpha = 0. \quad (1.70)$$

The velocity v of the mixture is called the *mass average velocity*. The restriction represented by the third equation of (1.70) indicates that no net production of mass occurs.

Convective diffusion equation

Sometimes it is of interest to write the continuity equation of each component in terms of the *convective mass flux* of that component. A convective mass flux is that associated with the motion of the mixture. Since this motion is described by the mass average velocity $\mathbf{v}$, the convective flux density of the α component is

$$\mathbf{j}_{c\alpha} = \overline{\rho}_\alpha \mathbf{v}. \tag{1.71}$$

Adding and subtracting the convective term $\operatorname{div}(\overline{\rho}_\alpha \mathbf{v})$ to equation (1.59) yields

$$\frac{\partial \overline{\rho}_\alpha}{\partial t} + \operatorname{div}(\overline{\rho}_\alpha \mathbf{v}) = -\operatorname{div}(\overline{\rho}_\alpha(\mathbf{v}_\alpha - \mathbf{v})) + \overline{g}_\alpha. \tag{1.72}$$

The first term on the right hand side of equation (1.72) represents the difference between the real mass flux-density of the α component and the convective mass flux-density of the same component. This difference is called the *diffusive mass flux-density* $\mathbf{j}_{D\alpha}$ of the α component, such that

$$\mathbf{j}_{D\alpha} = \overline{\rho}_\alpha \mathbf{u}_\alpha, \tag{1.73}$$

where $\mathbf{u}_\alpha = \mathbf{v}_\alpha - \mathbf{v}$. In terms of the diffusive flux density, the continuity equation for the α component can be written in the form

$$\frac{\partial \overline{\rho}_\alpha}{\partial t} + \operatorname{div}(\overline{\rho}_\alpha \mathbf{v}) = -\operatorname{div}\mathbf{j}_{D\alpha} + \overline{g}_\alpha. \tag{1.74}$$

This equation is known as the *convective diffusion equation*. Summing equation (1.74) for all values of α, we obtain

$$\frac{\partial}{\partial t}\left\{\sum_\alpha \overline{\rho}_\alpha\right\} + \operatorname{div}\left\{\left(\sum_\alpha \overline{\rho}_\alpha\right)\mathbf{v}\right\} = -\operatorname{div}\left\{\sum_\alpha \mathbf{j}_{D\alpha}\right\} + \sum_\alpha \overline{g}_\alpha. \tag{1.75}$$

Using equations (1.68) and (1.70) we conclude that

$$\sum_\alpha \mathbf{j}_{D\alpha} = 0, \tag{1.76}$$

and therefore

$$\sum_\alpha \overline{\rho}_\alpha \mathbf{u}_\alpha = 0. \tag{1.77}$$

Continuity equation and jump condition for mixtures of incompressible components

Some mixtures have components with constant material densities. In these cases we talk of *incompressible components*, realizing that the mixture itself can be compressible due to the term φ_α. Using this definition of volume fraction given by equation (1.7), equations (1.59) and (1.65) become

$$\frac{\partial}{\partial t}(\rho_\alpha \varphi_\alpha) + \text{div}\,(\rho_\alpha \varphi_\alpha \mathbf{v}_\alpha) = \bar{g}_\alpha, \qquad (1.78)$$

$$[\rho_\alpha \varphi_\alpha (\mathbf{v}_\alpha \cdot \mathbf{e}_I)] = \sigma[\rho_\alpha \varphi_\alpha]. \qquad (1.79)$$

Since ρ_α is a constant, we can divide both equations by ρ_α obtaining the *continuity equation and jump condition for incompressible components*,

$$\frac{\partial \varphi_\alpha}{\partial t} + \text{div}\,(\varphi_\alpha \mathbf{v}_\alpha) = \hat{g}_\alpha \qquad (1.80)$$

where $\hat{g}_\alpha = \bar{g}/\rho_\alpha$, and $[\varphi_\alpha (\mathbf{v}_\alpha \cdot \mathbf{e}_I)] = \sigma[\varphi_\alpha]$. Summing all α components yields

$$\frac{\partial}{\partial t}\left\{\sum_\alpha \varphi_\alpha\right\} + \text{div}\left\{\sum_\alpha \varphi_\alpha \mathbf{v}_\alpha\right\} = \sum_\alpha \hat{g}_\alpha, \qquad (1.81)$$

$$\sum_\alpha \varphi_\alpha (\mathbf{v}_\alpha \cdot \mathbf{e}_I) = \sigma \sum_\alpha \varphi_\alpha. \qquad (1.82)$$

Using equation (1.10) and defining the *volume average velocity*

$$\mathbf{q} = \sum_\alpha \varphi_\alpha \mathbf{v}_\alpha, \qquad (1.83)$$

we see that the equation of continuity and the mass jump balance for the mixture with incompressible components are given by

$$\text{div}\,\mathbf{q} = \sum_\alpha \hat{g}_\alpha, \quad [\mathbf{q} \cdot \mathbf{e}_I] = 0.$$

The volume average velocity suffers no jump across the surface of discontinuity.

1.2.3   Linear momentum balance

Applying the axiom of linear momentum and the Cauchy stress principle to each body B_α, we arrive at the *macroscopic balance of linear momentum*

$$\frac{d}{dt}\int_{V_m} \bar{\rho}_\alpha \mathbf{v}_\alpha dV = \int_{S_m} \mathbf{T}_\alpha \cdot \mathbf{n} dS + \int_{V_m} (\mathbf{b}_\alpha + \mathbf{m}_\alpha + \bar{g}_\alpha \mathbf{v}_\alpha) dV, \qquad (1.84)$$

where $\mathbf{T}_\alpha$ is the stress field in B_α, called the *partial stress*, $\mathbf{b}_\alpha$ is the body force on B_α, $\mathbf{m}_\alpha$ is the interaction force between components, that is, the force per unit volume exerted on B_α by all other components of the mixture and $\bar{g}_\alpha$ is the rate of mass growth as defined previously. The first term on the right-hand side of equation (1.84) represents the diffusive flux of linear momentum in B_α and the last term, the source of momentum per unit volume in B_α due to the body forces, interaction forces between components and to the mass generation. Using the Green-Gauss-Ostrogradsky theorem on the surface integral yields

$$\frac{d}{dt}\int_{V_m}\bar{\rho}_\alpha\mathbf{v}_\alpha dV = \int_{V_m}(\operatorname{div}\mathbf{T}_\alpha + \mathbf{b}_\alpha + \mathbf{m}_\alpha + \bar{g}_\alpha\mathbf{v}_\alpha)dV.$$

Changing the left hand side to the reference configuration and taking the material derivative, we obtain

$$\int_{V_{\alpha\kappa}}(\overline{\bar{\rho}_\alpha\mathbf{v}_\alpha}J_\alpha + \bar{\rho}_\alpha\mathbf{v}_\alpha\dot{J}_\alpha)dV_{\alpha\kappa} = \int_{V_m}(\operatorname{div}\mathbf{T}_\alpha + \mathbf{b}_\alpha + \mathbf{m}_\alpha + \bar{g}_\alpha\mathbf{v}_\alpha)dV.$$

Using equation (1.46) and changing back to the actual configuration,

$$\int_{V_m}(\overline{\bar{\rho}_\alpha\mathbf{v}_\alpha} + \bar{\rho}_\alpha\mathbf{v}_\alpha\operatorname{div}\mathbf{v}_\alpha - \operatorname{div}\mathbf{T}_\alpha - \mathbf{b}_\alpha - \mathbf{m}_\alpha - \bar{g}_\alpha\mathbf{v}_\alpha)dV = 0.$$

Changing the material derivative into a spatial derivative and convective term yields

$$\int_{V_m}\left\{\frac{\partial}{\partial t}(\bar{\rho}_\alpha\mathbf{v}_\alpha) + \operatorname{div}(\bar{\rho}_\alpha\mathbf{v}_\alpha\mathbf{v}_\alpha) - \operatorname{div}\mathbf{T}_\alpha - (\mathbf{b}_\alpha + \mathbf{m}_\alpha + \bar{g}_\alpha\mathbf{v}_\alpha)\right\}dV = 0. \quad (1.85)$$

When the field variables in equation (1.85) are smooth and continuous, the use of the localization theorem (Gurtin 1981) leads to the *local linear momentum balance equation*

$$\frac{\partial}{\partial t}(\bar{\rho}_\alpha\mathbf{v}_\alpha) + \operatorname{div}(\bar{\rho}_\alpha\mathbf{v}_\alpha\mathbf{v}_\alpha) = \operatorname{div}\mathbf{T}_\alpha + \mathbf{b}_\alpha + \mathbf{m}_\alpha + \bar{g}_\alpha\mathbf{v}_\alpha. \quad (1.86)$$

This equation is said to have a *conservation form*. Introducing the continuity equation (1.59) yields

$$\bar{\rho}_\alpha\dot{\mathbf{v}}_\alpha = \operatorname{div}\mathbf{T}_\alpha + \mathbf{b}_\alpha + \mathbf{m}_\alpha. \quad (1.87)$$

Linear momentum jump conditions

In regions having discontinuities, equations (1.86) and (1.87) are still valid on each side of the discontinuity, but they are not valid at the discontinuity. Following a procedure similar to that used in the previous section for the mass jump balance, we write the macroscopic momentum balance in the form

$$\int_{V_m}\left\{\frac{\partial}{\partial t}(\bar{\rho}_\alpha\mathbf{v}_\alpha) - (\mathbf{b}_\alpha + \mathbf{m}_\alpha + \bar{g}_\alpha\mathbf{v}_\alpha)\right\}dV =$$
$$-\int_{S_m}\bar{\rho}_\alpha\mathbf{v}_\alpha\mathbf{v}_\alpha \cdot \mathbf{n}dS + \int_{S_m}\mathbf{T}_\alpha \cdot \mathbf{n}dS. \quad (1.88)$$

Applying this equation at each side of the discontinuity yields

$$\int_{V+}\left\{\frac{\partial}{\partial t}(\bar{\rho}_\alpha\mathbf{v}_\alpha) - (\mathbf{b}_\alpha + \mathbf{m}_\alpha + \bar{g}_\alpha\mathbf{v}_\alpha)\right\}dV = -\int_{S+}\bar{\rho}_\alpha\mathbf{v}_\alpha\mathbf{v}_\alpha \cdot \mathbf{n}dS$$

$$- \int_{S_I} (\bar{\rho}_\alpha)^+ \mathbf{v}_\alpha^+ (\mathbf{v}_\alpha^+ - \mathbf{v}_I) \cdot (-\mathbf{e}_I) dS + \int_{S^+} \mathbf{T}_\alpha \cdot \mathbf{n} dS + \int_{S_I} \mathbf{T}_\alpha^+ \cdot (-\mathbf{e}_I) dS, \quad (1.89)$$

$$\int_{V-} \left\{ \frac{\partial}{\partial t} (\bar{\rho}_\alpha \mathbf{v}_\alpha) - (\mathbf{b}_\alpha + \mathbf{m}_\alpha + \bar{g}_\alpha \mathbf{v}_\alpha) \right\} dV = - \int_{S-} \bar{\rho}_\alpha \mathbf{v}_\alpha \mathbf{v}_\alpha \cdot \mathbf{n} dS$$

$$- \int_{S_I} (\bar{\rho}_\alpha)^- \mathbf{v}_\alpha^- (\mathbf{v}_\alpha^- - \mathbf{v}_I) \cdot \mathbf{e}_I) dS + \int_{S-} \mathbf{T}_\alpha \cdot \mathbf{n} dS + \int_{S_I} \mathbf{T}_\alpha^- \cdot \mathbf{e}_I dS. \quad (1.90)$$

Adding equations (1.89) and (1.90) yields

$$\int_{V_m} \left\{ \frac{\partial}{\partial t} (\bar{\rho}_\alpha \mathbf{v}_\alpha) - (\mathbf{b}_\alpha + \mathbf{m}_\alpha + \bar{g}_\alpha \mathbf{v}_\alpha) \right\} dV \quad (1.91)$$

$$= - \int_{S_m} \bar{\rho}_\alpha \mathbf{v}_\alpha \mathbf{v}_\alpha \cdot \mathbf{n} dS + \int_{S_m} \mathbf{T}_\alpha \cdot \mathbf{n} dS + \int_{S_I} (\bar{\rho}_\alpha)^+ \mathbf{v}_\alpha^+ (\mathbf{v}_\alpha^+ - \mathbf{v}_I) \cdot \mathbf{e}_I dS$$

$$- \int_{S_I} \mathbf{T}_\alpha^+ \cdot \mathbf{e}_I dS - \int_{S_I} (\bar{\rho}_\alpha)^- \mathbf{v}_\alpha^- (\mathbf{v}_\alpha^- - \mathbf{v}_I) \cdot \mathbf{e}_I dS + \int_{S_I} \mathbf{T}_\alpha^- \cdot \mathbf{e}_I dS.$$

Comparing equations (1.88) and (1.91) yields the *macroscopic momentum jump balance*

$$\int_{S_I} \{ [\bar{\rho}_\alpha \mathbf{v}_\alpha (\mathbf{v}_\alpha - \mathbf{v}_I) \cdot \mathbf{e}_I] - [\mathbf{T}_\alpha \cdot \mathbf{e}_I] \} dS = 0. \quad (1.92)$$

The local equation is then $[\bar{\rho}_\alpha \mathbf{v}_\alpha (\mathbf{v}_\alpha - \mathbf{v}_I) \cdot \mathbf{e}_I)] = [\mathbf{T}_\alpha \cdot \mathbf{e}_I]$, or, using $\sigma = \mathbf{v}_I \cdot \mathbf{e}_I$,

$$[\bar{\rho}_\alpha \mathbf{v}_\alpha (\mathbf{v}_\alpha \cdot \mathbf{e}_I)] = [\bar{\rho}_\alpha \mathbf{v}_\alpha] \sigma + [\mathbf{T}_\alpha \cdot \mathbf{e}_I]. \quad (1.93)$$

Linear momentum balance for the mixture

Summing equations (1.86) yields

$$\frac{\partial}{\partial t} \left\{ \sum_\alpha \bar{\rho}_\alpha \mathbf{v}_\alpha \right\} + \operatorname{div} \left\{ \sum_\alpha \bar{\rho}_\alpha \mathbf{v}_\alpha \mathbf{v}_\alpha \right\} = \operatorname{div} \left\{ \sum_\alpha \mathbf{T}_\alpha \right\} + \sum_\alpha (\mathbf{b}_\alpha + \mathbf{m}_\alpha + \bar{g}_\alpha \mathbf{v}_\alpha).$$

$$(1.94)$$

Via the second equation of (1.70), defining the *diffusion velocity*

$$\mathbf{u}_\alpha = \mathbf{v}_\alpha - \mathbf{v} \quad (1.95)$$

and using (1.77), the above expression becomes

$$\frac{\partial}{\partial t} (\rho \mathbf{v}) + \operatorname{div} (\rho \mathbf{v} \mathbf{v}) = \operatorname{div} \left\{ \sum_\alpha \mathbf{T}_\alpha - \sum_\alpha \bar{\rho}_\alpha \mathbf{u}_\alpha \mathbf{u}_\alpha \right\} + \sum_\alpha \mathbf{b}_\alpha + \sum_\alpha (\mathbf{m}_\alpha + \bar{g}_\alpha \mathbf{v}_\alpha).$$

$$(1.96)$$

Comparing this equation with the linear momentum for a single material,

$$\frac{\partial}{\partial t} (\rho \mathbf{v}) + \operatorname{div} (\rho \mathbf{v} \mathbf{v}) = \operatorname{div} \mathbf{T} + \mathbf{b}, \quad (1.97)$$

we conclude that it is necessary that

$$T_I = \sum_\alpha T_\alpha, \tag{1.98}$$

$$T = T_I - \sum_\alpha \bar{\rho}_\alpha u_\alpha u_\alpha, \quad b = \sum_\alpha b_\alpha, \quad \sum_\alpha (m_\alpha + \bar{g}_\alpha v_\alpha) = 0, \tag{1.99}$$

where T_I is called the *inner part* of the stress tensor (Truesdell 1984). The restriction represented by the last equation of (1.99) indicates that no net production of linear momentum exists, and that the growth in one component is done at the expense of the linear momentum of other components.

Angular momentum balance

The application of the axiom of angular momentum, also known as Euler's second law, and the Cauchy stress principle to each of the bodies B_α gives the *macroscopic angular momentum balance*:

$$\frac{d}{dt} \int_{V_m} ((\mathbf{r} - \mathbf{r}_q) \times \bar{\rho}_\alpha v_\alpha) dV = \int_{S_m} ((\mathbf{r} - \mathbf{r}_q) \times T_\alpha \cdot \mathbf{n}) dS$$

$$+ \int_{V_m} ((\mathbf{r} - \mathbf{r}_q) \times (b_\alpha + m_\alpha + \bar{g}_\alpha v_\alpha)) dV + \int_{V_m} \bar{a}_{\alpha q} dV, \tag{1.100}$$

where the first term corresponds to the angular momentum of the α component, the second term is the torque of the contact forces, the third term the torque of the body forces and interaction forces between components, and the last term represents the source of angular momentum due to the interaction between components. Here $\mathbf{r}_q$ is the position of the fixed point Q with respect to which the torques and angular momentum are calculated. When the field variables are smooth and continuous, a procedure similar to that of the previous section leads to

$$T_\alpha - T_\alpha^T = A_{\alpha q}, \tag{1.101}$$

where $A_{\alpha q}$ is the skew tensor corresponding to the axial vector $\bar{a}_{\alpha q}$. If we assume that there is no interchange of angular momentum between components, $\bar{a}_{\alpha q} = 0$ and $A_{\alpha q} = 0$, then $T_\alpha = T_\alpha^T$ and $T_I = T_I^T$. The angular momentum balance at a discontinuity does not provide information beyond that obtained from the linear momentum balance.

Dynamical processes

Consider a mixture B formed by the components $B_\alpha \subset B$, with $\alpha = 1, 2, ..., n$. We have defined the following field variables: $\mathbf{r} = f_\alpha(R_\alpha, t)$, $\bar{\rho}_\alpha = \bar{\rho}_\alpha(\mathbf{r}, t)$, $T_\alpha = T_\alpha(\mathbf{r}, t)$, $b_\alpha = b_\alpha(\mathbf{r}, t)$, $\bar{g}_\alpha = \bar{g}_\alpha(\mathbf{r}, t)$, $m_\alpha = m_\alpha(\mathbf{r}, t)$. We say that these six field variables constitute a *dynamical process* if, in those regions where they are smooth and continuous, they satisfy the following two field equations

$$\frac{\partial \bar{\rho}_\alpha}{\partial t} + \text{div}(\bar{\rho}_\alpha v_\alpha) = \bar{g}_\alpha \quad \text{or} \quad \bar{\rho}_\alpha \det F_\alpha = \bar{\rho}_{\alpha \kappa}, \tag{1.102}$$

$$\frac{\partial}{\partial t}(\bar{\rho}_\alpha \mathbf{v}_\alpha) + \operatorname{div}(\bar{\rho}_\alpha \mathbf{v}_\alpha \mathbf{v}_\alpha) = \operatorname{div} \mathbf{T}_\alpha + \mathbf{m}_\alpha + \bar{g}_\alpha \mathbf{v}_\alpha. \tag{1.103}$$

for all $\alpha = 1, \ldots, n$. At discontinuities the following jump conditions substitute for the field equations:

$$[\bar{\rho}_\alpha(\mathbf{v}_\alpha \cdot \mathbf{e}_I)] = \sigma[\bar{\rho}_\alpha], \quad [\bar{\rho}_\alpha \mathbf{v}_\alpha(\mathbf{v}_\alpha \cdot \mathbf{e}_I)] = \sigma[\bar{\rho}_\alpha(\mathbf{v}_\alpha \cdot \mathbf{e}_I)] + [\mathbf{T}_\alpha \cdot \mathbf{e}_I]. \tag{1.104}$$

For this dynamical process to be complete, we have to postulate four *constitutive equations* that relate the dynamical variables to the motion, that is, $(\mathbf{T}_\alpha, \mathbf{r})$, $(\mathbf{b}_\alpha, \mathbf{r})$, $(\mathbf{m}_\alpha, \mathbf{r})$ and $(\bar{g}_\alpha, \mathbf{r})$. A dynamical process for these six field variables $\mathbf{r}, \bar{\rho}_\alpha, \mathbf{T}_\alpha, \mathbf{b}_\alpha, \bar{g}_\alpha$ and $\mathbf{m}_\alpha$ is admissible when the six equations are satisfied.

Chapter 2

Theory of sedimentation of ideal suspensions

2.1 Introduction

Particle sedimentation is analyzed in this chapter within the framework of the continuum theory of mixtures by regarding the particles and fluid as superimposed continua. We consider the problem of sedimentation of ideal suspensions in a thickener under gravity only. For *batch* operation, the thickener is called a *settling column*, and is filled with the suspension which settles, producing at the bottom a blanket of thick mud and a supernatant liquid. In *continuous* operation, the suspension is fed at some point of a thickener and two discharges are provided, a thickened slurry underflow at the bottom and a clear liquid overflowing at the top of the tank. In this manner of operation, the feed and discharge of the thickener are continuous.

A theory of sedimentation which is more general than that found in this chapter could be based on the three-dimensional treatment of the theory of mixtures given in Chapter 1 (see Bürger *et al.* 1998c).

2.2 Kinematical sedimentation process

Consider a mixture of solid particles in a fluid that satisfies the properties stated by Kynch (1952). Denote by $\mathbf{v_s}$ and $\mathbf{v_f}$ the velocities of the solid and the fluid components of the mixture respectively. These properties are the following:

(1) *All the solid particles are small and of the same size, shape and density.*

(2) *The solid and the fluid in the mixture are incompressible. There is no mass transfer between components.*

27

(3) *The relative solid-fluid velocity*

$$\mathbf{u} = \mathbf{v_s} - \mathbf{v_f} \tag{2.1}$$

in the mixture is a function only of the local solid concentration φ.

Property (1) was intended to imply equal velocities for all particles, but closer observation of suspensions of very closely sized spheres as well as theoretical and numerical studies of identical spheres show that there is actually a distribution of velocities. These studies are summarized in Tory *et al.* (1992) and Tory and Hesse (1996). Without knowing the positions of all particles, it is impossible to predict the velocity of any particle. Thus, the assumption of equal velocities is not unreasonable. However, the assumption that settling is a Markov process yields additional information at the cost of increased complexity. The current status of this work and references to earlier studies can be found in Hesse and Tory (1996) and Tory and Hesse (1996). Implicit in property (3) is the assumption that inertial and transient forces can be neglected for small particles. This is justified formally in Chapter 3.

A mixture which satisfies properties (1)–(3) is called an *ideal suspension* (Shannon and Tory 1966, Bustos *et al.* 1990, Concha and Bustos 1992) and it may be regarded as a superimposed continuous medium with two incompressible components. The concentration $\varphi(\mathbf{r}, t)$ of the suspension is in general a function of the three space variables and of the time. The respective continuity equations for the solid and the fluid are obtained from (1.80) with $\hat{g}_\alpha = 0$ for those regions where the field variables are smooth,

$$\frac{\partial \varphi}{\partial t} + \mathrm{div}\,(\varphi \mathbf{v_s}) = 0, \quad \frac{\partial \varphi}{\partial t} - \mathrm{div}\,((1 - \varphi)\mathbf{v_f}) = 0, \tag{2.2}$$

and from (1.79) at discontinuities:

$$[\varphi\,(\mathbf{v_s} \cdot \mathbf{e_I})] = \sigma[\varphi], \quad [(1 - \varphi)\,(\mathbf{v_f} \cdot \mathbf{e_I})] = \sigma[1 - \varphi]. \tag{2.3}$$

Subtraction of the two equations in (2.2) gives the conservation equation for the mixture,

$$\mathrm{div}\,\mathbf{q}(\mathbf{r}, t) = 0, \tag{2.4}$$

and subtraction of the equations in (2.3) the corresponding jump condition,

$$[\mathbf{q} \cdot \mathbf{e_I}] = 0, \tag{2.5}$$

where $\mathbf{q}(\mathbf{r}, t) = \varphi \mathbf{v_s} + (1 - \varphi)\mathbf{v_f}$. As we have seen in (1.83), $\mathbf{q}$ is the volume average velocity of the suspension. In terms of the relative solid-fluid velocity $\mathbf{u}$, the volume average velocity $\mathbf{q}$ can be expressed as

$$\mathbf{q} = \mathbf{v_s} - (1 - \varphi)\mathbf{u}. \tag{2.6}$$

Multiplying by φ yields

$$\varphi\mathbf{q} = \varphi\mathbf{v_s} - \varphi(1-\varphi)\mathbf{u}. \tag{2.7}$$

Let

$$\mathbf{f}(\varphi, t) = \varphi\mathbf{v_s} \tag{2.8}$$

denote the total flux density of the solid part and let

$$\mathbf{f_b}(\varphi) = \varphi(1-\varphi)\mathbf{u} \tag{2.9}$$

denote the so-called drift (or relative) flux density of the solid (Kluwick 1977, Wallis 1966). Equation (2.7) leads to

$$\mathbf{f}(\varphi, t) = \varphi\mathbf{q}(\mathbf{r}, t) + \mathbf{f_b}(\varphi), \tag{2.10}$$

that is, the total flux density of the solid consists of a linear part $\varphi\mathbf{q}(\mathbf{r}, t)$ corresponding to the global motion of the mixture plus a nonlinear part $f_b(\varphi)\mathbf{k}$ describing the local solid-fluid relative motion. With equation (2.10), the continuity equation for the solid component takes the form

$$\frac{\partial\varphi}{\partial t} + \operatorname{div}\mathbf{f}(\varphi, t) = 0. \tag{2.11}$$

At discontinuities, $[\mathbf{f} \cdot \mathbf{e_I}] = \sigma[\varphi]$.

2.2.1 Kynch theory of batch sedimentation

The first theory for batch sedimentation for incompressible materials was developed by Kynch under the assumptions:

(1) *The suspension is ideal.*

(2) *The concentration of particles is constant at any horizontal cross section in the column.*

(3) *The suspension has a homogeneous initial concentration ϕ_0.*

(4) *At the bottom of the container there is a continuous but extremely rapid increase of concentration from ϕ_0 to the final concentration φ_∞.*

From assumption (2) we deduce that the motion is one-dimensional and the concentration $\varphi = \varphi(z, t)$ is a function of z and t where z is the coordinate of the upwards pointing vertical direction and t is time. Since for batch sedimentation, the column is closed at the bottom, the average velocity q is zero at $z = 0$, and from equation (2.4) we obtain that q is a function of time only, hence $q(t) = 0$. From equations (2.10) and (2.11) with $q = 0$, we get

$$\frac{\partial\varphi}{\partial t} + \frac{\partial f_b(\varphi)}{\partial z} = 0 \quad \text{for } 0 < z < L, \ t > 0, \tag{2.12}$$

where L is the initial height of the mixture. Equation (2.12) is a first order quasilinear hyperbolic equation. From assumptions (3) and (4), Kynch obtained the initial and boundary conditions:

$$\varphi(z,0) = \phi_0 \quad \text{for } 0 < z < L, \tag{2.13a}$$

$$\varphi(0,t) = \varphi_\infty \quad \text{for } t > 0, \tag{2.13b}$$

$$\varphi(L,t) = \varphi_L = 0 \quad \text{for } t > 0. \tag{2.13c}$$

The concentrations φ_L, ϕ_0 and φ_∞ are constant and satisfy

$$\varphi_L = 0 < \phi_0 < \varphi_\infty < 1.$$

All the flux density functions studied by Kynch (1952) have the following properties:

$$f_b(\varphi) < 0 \text{ for } \varphi \in (0, \varphi_\infty), \quad f_b(0) = f_b(\varphi_\infty) = 0, \quad f_b'(0) < 0, \quad f_b'(\varphi_\infty) > 0. \tag{2.14}$$

A typical example is the flux density function determined by Shannon and Tory et al. (1963, 1964) for glass beads, see Figure 7.26 in Chapter 7.

The solution of the quasilinear hyperbolic equation (2.12) is $\varphi = constant$ along the characteristics with slope $dz/dt = f_b'(\varphi)$ in the z-t-plane (see Chapter 4). Since $f_b'(\varphi)$ is constant along such curves, the characteristics are straight lines. Consequently, we may extend the characteristics that originate in the t-axis with slope $f_b'(\varphi_\infty) > 0$ and those that originate in the line $z = L$ with slope $f_b'(0) < 0$ until they cut the z-axis. Then the initial-boundary value problem (2.12) and (2.13) can be reformulated as an initial value problem (Bustos and Concha 1988a, Concha and Bustos 1991). The sedimentation of an ideal suspension in a settling column is described by the field variables $\varphi(z,t)$ and $f_b(\varphi(z,t))$. These field variables constitute a kinematical batch sedimentation process if they obey the following field equation for regions where they are smooth:

$$\frac{\partial \varphi}{\partial t} + \frac{\partial f_b(\varphi)}{\partial z} = 0 \quad \text{for } z \in \mathbb{R}, \ t > 0, \tag{2.15}$$

and the constitutive equation $f_b = f_b(\varphi)$ with initial data

$$\varphi(z,0) = \varphi_I(z) \quad \text{for} \quad z \in \mathbb{R}, \tag{2.16}$$

where $\varphi_I(z)$ is given by

$$\varphi_I(z) = \begin{cases} \varphi_L = 0 & \text{for } L \leq z, \\ \phi_0 & \text{for } 0 \leq z \leq L, \\ \varphi_\infty & \text{for } z < 0. \end{cases} \tag{2.17}$$

To avoid confusion in what follows, we refer to $\varphi_I(z)$ as *initial data* and to ϕ_0 as *initial concentration*. Solutions of this single conservation law are in general discontinuous which in turn are not unique, so we consider entropy solutions, which is the subject of Chapters 4 and 5. At discontinuities, equation (2.15) is replaced by the mass jump balance $[f_b] = \sigma[\varphi]$.

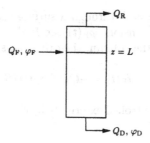

Figure 2.1: Ideal continuous thickener.

2.2.2 Kynch theory of continuous sedimentation

Consider an ideal suspension in a thickener in continuous operation, satisfying equations (2.4), (2.10) and (2.11):

$$\frac{\partial \varphi}{\partial t} + \operatorname{div} \mathbf{f}(\varphi, t) = 0, \quad \operatorname{div} \mathbf{q}(\mathbf{r}, t) = 0, \quad \mathbf{f}(\varphi, t) = \varphi \mathbf{q}(\mathbf{r}, t) + \mathbf{f_b}(\varphi). \tag{2.18}$$

Ideal continuous thickener

To model this problem, we make some idealizations in the operation of the thickener. An *ideal continuous thickener* (ICT) (Bustos *et al.* 1990, Concha and Bustos 1992, Shannon and Tory 1966) is a cylindrical column assumed to have the following properties (see Figure 2.1):

(1) The column has two sections, a clear liquid at the top and a settling section beneath. The suspension used is ideal.

(2) The concentration of particles is constant at any horizontal cross section of the column. Therefore, the flow is spatially one dimensional and the concentration $\varphi = \varphi(z, t)$ is a function of two independent variables: the height z and the time t. Therefore, from $\operatorname{div} \mathbf{q}(\mathbf{r}, t) = 0$ we get

$$\frac{\partial q}{\partial z} = 0, \tag{2.19}$$

that is, the volume average velocity $q = q(t)$ is a function of t only and is always negative. From the first equation of (2.18) and (2.19) we obtain

$$\frac{\partial \varphi}{\partial t} + \frac{\partial f(\varphi, t)}{\partial z} = 0, \tag{2.20}$$

$$f(\varphi, t) = \varphi q(t) + f_b(\varphi). \tag{2.21}$$

In sedimentation, the flux density function $f(\varphi, t)$ is given by a *constitutive equation*. For the vast majority of sedimentation experiments, this constitutive equation is a function with one (Richardson and Zaki 1954) or two (Shannon *et al.* 1963, 1964) inflection points.

(3) The column is fed at $z = L$ through a surface source. If $Q_F(t)$ is the feed volume flux of the suspension, $\varphi_F(t)$ the feed concentration and S is the cross sectional area of the column, then the feed solid flux density is

$$f_F(t) = -Q_F(t)\varphi_F(t)/S. \qquad (2.22)$$

This quantity can be controlled externally by changing $Q_F(t)$ or $\varphi_F(t)$, which are *given quantities*.

(4) At $z = 0$, a surface sink discharges the suspension of the column at a volumetric flow rate $Q_D(t)$. Let the unknown discharge concentration be $\varphi_D(t) = \varphi(0,t)$ and let $f_D(t) = f(\varphi(0,t),t)$ denote the solid flux density in the discharge stream, then

$$Q_D(t) = f_D(t)S/\varphi_D(t). \qquad (2.23)$$

From equation (2.21) at $z = 0$ we have that

$$f_D(t) = q(t)\varphi_D(t) + f_b(\varphi_D(t)). \qquad (2.24)$$

(5) Integrating equation (2.20) with respect to z we get the following relationship between $f_F(t)$ and $f_D(t)$:

$$\frac{d}{dt}\int_0^L \varphi(z,t)\,dz = f_D(t) - f_F(t) \qquad (2.25)$$

(6) The fluid crossing the feed level to the clear liquid section leaves the thickener at a volumetric rate $Q_R(t)$ so that the relationship

$$Q_F(t)\varphi_F(t) = Q_D(t)\varphi_D(t) + Q_R(t)\varphi_R(t) \qquad (2.26)$$

holds with $Q_R(t) \geq 0$.

Normal operation of an ideal continuous thickener

An ICT is operating *normally* if:

(1) The solid component is restricted to the settling section. In the case that the solid particles cross the feed level to the clear liquid section, we say that the thickener *overflows*.

(2) The discharge concentration $\varphi_D(t)$ is restricted to the values larger than the feed concentration $\varphi_F(t)$. Then, if $\varphi_D(t) \leq \varphi_F(t)$, we say that the thickener *empties*.

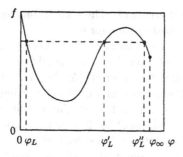

Figure 2.2: Concentration φ_L for a given flux density curve and solid flux density $f_F(t)$.

Concentration at $z = L$

The concentration $\varphi_L(t) = \varphi(L, t)$ at $z = L$ is determined by the flux density function $f(\varphi, t)$. To determine $\varphi_L(t)$, consider a mass balance at $z = L$. The solid flux entering the thickener operating normally per unit time at $z = L$ must be equal to the solid settling at $z = L$: $f_F(t) = f(\varphi_L, t)S$. From this and equation (2.24) we get

$$f_F(t) = q(t)\varphi_L(t) + f_b(\varphi_L(t)). \tag{2.27}$$

In general, equation (2.27) has more than one solution as shown in Figure 2.2. The appropriate value of $\varphi(L, t)$ can be selected by physical arguments. The suspension is instantaneously diluted when entering the thickener, hence the concentration at $z = L$ takes the following values:

$$\varphi(L, t) = \begin{cases} \varphi_L(t) & \text{if } \varphi_L(t) \leq \varphi_F(t) < \varphi'_L(t), \\ \varphi'_L(t) & \text{if } \varphi'_L(t) \leq \varphi_F(t) < \varphi''_L(t), \\ \varphi''_L(t) & \text{if } \varphi''_L(t) \leq \varphi_F(t) \leq \varphi_\infty. \end{cases} \tag{2.28}$$

Initial-boundary value problem

Collecting the previous results, we can say that the sedimentation of an ideal suspension in an ICT is described by the field variables $\varphi(z, t)$ and $f(\varphi(z, t), t)$. These field variables constitute a kinematical continuous sedimentation process if they obey the following field and constitutive equations in regions where the field variables are smooth:

$$\frac{\partial \varphi}{\partial t} + \frac{\partial f(\varphi, t)}{\partial z} = 0 \text{ for } 0 < z < L, \ t > 0; \ f(\varphi, t) = q\varphi + f_b(\varphi),$$

with initial data $\varphi(z, 0) = \varphi_I(z)$ for $0 < z < L$ and boundary conditions $\varphi(0, t) = \varphi_\infty(t)$ and $\varphi(L, t) = \varphi_L(t)$ for $t > 0$. At discontinuities, the jump balances

$[f(\varphi, t)] = \sigma[\varphi]$ and $[q] = 0$ are valid. These equations form an initial-boundary value problem which is studied in Chapter 6.

It should be mentioned that there are still other ways to generalize Kynch's theory to continuous sedimentation. Here, we model feed and discharge by boundary conditions, as Petty did in his paper (1975). In a recent series of papers (Diehl 1995, 1996, 1997, 1999), Diehl analyses a model of continuous sedimentation in a so-called settler-clarifier, in which the feed, discharge and overflow mechanisms are merely expressed by discontinuities of the flux-density function at the bottom and top of the vessel and at the (intermediate) feed level, and an additional singular source term at the feed level. Conservation of mass yields jump conditions valid across these discontinuities. The apparent advantages of this elegant model consist in the completeness of the treatment (since the overflow zone above the feed level, $z = L$, is not considered here) and in the avoidance of boundary conditions, such that only initial conditions need to be considered. Although exact solutions can be constructed for all practically relevant cases, a global existence and uniqueness result is not yet available, so we do not yet wish to elaborate on this sedimentation model here but recommend studying Diehl's original papers.

Chapter 3

Sedimentation with compression

3.1 Introduction

Kynch's theory describes quite well the sedimentation of incompressible particles, such as glass beads and the settling of non-flocculated mineral particles. It predicts that within the suspension, the lines of equal concentration in the z-t-plane are straight lines (Kynch 1952, Bustos and Concha 1988a, Concha and Bustos 1991). Unfortunately, the published experimental evidence on the distribution of concentration during batch sedimentation of flocculated suspensions (Scott 1968, Been and Sills 1981, Tiller *et al.* 1991) does not satisfy this model. Actually, the experimental data show that the lines of equal concentration, for the higher concentration range, are curves with decreasing slopes that become horizontal as time increases. In general, the sedimentation of flocculated suspensions, such as those encountered in the mineral industries, cannot be described by Kynch's theory since the consolidation of thick slurries involves the presence of forces not taken into account by the kinematical model. Careful observations of batch settling of flocculated suspensions have shown that two distinguishable zones appear in the settling column separated by a discontinuity. The upper zone, lasting only up to the critical time, is in hindered settling and the lower zone is in consolidation from the beginning till the end of the process. In a conventional continuous thickener, the upper part of the pulp is in hindered settling while the lower part is in consolidation, so that in this case both processes always coexist. The transition from hindered settling to consolidation is governed by the suspension concentration, with the critical concentration or critical concentration range (Pane and Schiffman 1985) as the turning point. Sedimentation models taking these phenomena into account have been proposed by several authors, among them Shirato *et al.* (1970), Adorján (1975, 1977), Concha and Bascur (1977), Kos (1977a,b, 1980, 1985), D'Avila (1976, 1978, 1982), D'Avila and Sampaio (1978), Sampaio and

D'Avila (1976, 1979), Concha and Barrientos (1980), Hill *et al.* (1980), Been and
Sills (1981), Tiller (1981), Tiller and Khatib (1984), Buscall and White (1987),
Auzerais *et al.* (1988), Font (1988), Bustos and Concha (1988b) and Concha *et
al.* (1996). The work of these and several other authors has set the foundation
for a quantitative and sound theory of thickening. It is therefore possible today
to set up a rational and unified theory describing the sedimentation of flocculated
or non-flocculated suspensions. It is the purpose of this chapter to show how an
axiomatic theory of sedimentation can be deduced, following the procedures of the
theory of mixtures of classical continuum mechanics (Drew and Lahey 1979, Drew
1983, Bedford and Drumheller 1983, Truesdell 1984, Ahmadi 1982, 1987) and, as
an example, to show how it can be applied to batch and continuous sedimentation
of flocculated suspensions.

In this chapter we develop a rational and rigorous theory of gravity thickening,
valid for all types of suspensions that can be treated as a continuum. In Section 3.2,
several restrictions have to be made: that the solid particles are small and incompressible having a unique density and being completely flocculated before settling,
that no mass transfer occurs between the solid and the fluid and that no body
forces exist other than gravity as noted in Chapter 1. The application of the two
axioms of mechanics: *the conservation of mass and linear momentum*, produces
a set of four integral equations known as the *macroscopic balance of mass and
momentum* for the solid and the fluid. These are the most general equations that
all suspensions must obey, no matter how different their particular properties are.
Since settling suspensions develop discontinuities, the macroscopic balances lead
to a set of four differential equations called *local balances*, valid in all points within
the suspension, except at discontinuities where, instead, the same macroscopic
balances yield four *jump balances*. We must emphasize that the jump balances or
jump conditions, as they are also called, follow rigorously from the macroscopic
balances for bodies having discontinuities. The derivations of the local balances
and jump conditions from the macroscopic balances were performed in Chapter 1.

In Section 3.3, we restrict the model to a particular container or equipment and
to a particular set of suspensions having similar properties. Constitutive equations
complete the set of differential equations necessary to solve a particular problem.
In our case, the idealization of the flow of a suspension in a sedimenting vessel
leads to the concept of an *ideal thickener*. The consequence of this idealization is
to restrict the possible motions that the suspension can perform to a *simple compressive motion*. This is known as a kinematical restriction. In turn, this result
yields through the gradient of deformation a very simple description of the type of
deformation that the suspension can undergo. A further implication of this result
is that the local field equations and jump conditions become one-dimensional in
the vertical direction. We characterize the particular behavior of a suspension
by its dynamical variables and by the kinematical response to dynamical stimu-
lations. The dynamical variables involved, besides the gravitational force, are the
contact and interaction forces; therefore, restrictions are imposed on the solid and
fluid stresses and on the solid-fluid interaction force. The fluid and the solid are
modeled as *elastic isotropic media* in the sense that they are compressible media

with no preferred spatial direction and have no memory of previous motions. This assumption could be considered strange for the solid component in thickening, since it implies that relieving the compressive force would return the suspension to its original concentration. As a matter of fact, this reasoning is partially correct because this phenomenon of stress relief does not occur during thickening. What indeed happens is that during a transient process in a continuous thickener, the concentration at a given level of the thickener can increase or diminish, and in that case the elastic assumption applies perfectly well because to each concentration we associate a corresponding solid stress.

Since the solid particles forming the suspensions are small, the sedimentation velocity is always small and the force exerted by the fluid on the solid is well represented by a linear function of the relative solid-fluid velocity and the concentration gradient. Some research workers have pointed out that flocculated suspensions are never at equilibrium. Perhaps this is correct if we consider an unlimited time interval, but for the thickening process the *end of sedimentation* is reached when the field variables do not change significantly with time. This end of sedimentation certainly can be considered *equilibrium*.

The introduction of all these restrictions into the field equations yields the one-dimensional local field equations which are discussed in Section 3.4. These equations still leave some freedom to represent different suspensions by admitting different functional forms of the solid pressure and of the flow-resistance coefficient. The local one-dimensional field equations can be further simplified by taking into account that some terms in the equations are much smaller than the rest. To perform this order-of-magnitude study, it is essential to define a characteristic time in which the process is evaluated. We define this time as the one necessary for a single floc to settle from the top to the bottom of the suspension in an unbounded medium. In this way the acceleration and the convective terms from both the solid and fluid momentum equations can be neglected.

In Section 3.5 we proceed to consider that for processes involving porous beds, such as consolidation in thickening, it is convenient to employ the concepts of *excess pore pressure* and *effective solid stress* instead of the defined fluid and solid pressures. The advantage is that the new variables can be measured experimentally.

Collecting the results, we define a *Dynamical Sedimentation Process (DSP)*. A DSP is a set of six field variables, describing the thickening process, which obey four field equations and two constitutive equations. At discontinuities they satisfy two jump conditions. It is interesting to note that the DSP can be considered as an extension of the Kynch Sedimentation Process (KSP) defined in Chapter 7 (Bustos and Concha 1988a, Concha and Bustos 1992), by including a diffusion term depending on the solid effective stress in the solid flux density function. If, for any reason, this term vanishes, we fall back to the KSP. This formulation of thickening represents a unified treatment of ideal and flocculated suspensions. Two parameters have to be determined experimentally: the resistance coefficient $\alpha(\varphi)$ and the solid effective stress $\sigma_e(\varphi)$. The functional form of the resistance coefficient depends on the range of concentration. For concentrations less than the critical, it has to be determined by hindered settling experiments and for concentrations greater than the critical it should be determined from percolation experiments in a porous bed. In our case, we use the same resistance coefficient obtained from hindered settling experiments in the whole range of concentrations. In a recent study (Bürger *et al.* 1999c), both constitutive functions, the resistance coefficient and the solid effective stress, have been determined from published sedimentation-consolidation experiments.

In Section 3.6 we apply the DSP to batch and continuous thickening of flocculated suspensions. The result is an initial-boundary value problem for one degenerate parabolic partial differential equation, which is analyzed in Chapter 9. Numerical solutions of this initial-boundary value problem are presented in Chapter 10.

The derivation of the mathematical model given by a DSP is based on a multidimensional formulation of the model equations, which are then reduced to one space dimension. Of course, this gives rise to the question whether the same balance equations and constitutive assumptions will also produce sound multidimensional sedimentation model equations. At present, however, no obvious simple generalization of a DSP to several space dimensions which provides a complete set of equations for the determination of all independent flow variables seems to be avilable. This problem is discussed briefly in Section 3.7.

3.2   Macroscopic balance

Consider the sedimentation under gravity of a suspension of small and flocculated particles under the following assumptions:

(1) *The solid particles are small (with respect to the container) and of the same density.*

(2) *The solid and the fluid component of the suspension are incompressible in their pure state.*

(3) *The suspension is completely flocculated before settling.*

(4) *There is no mass transfer between the solid and the fluid during sedimentation.*

(5) *Gravity is the only body force.*

These assumptions are usually satisfied by the feed of continuous thickeners in the mineral processing industries. The first assumption ensures that settling is slow and that no preconcentration occurs. The second assumption permits the definition of characteristic constant material densities. The third one is essential since it ensures that the solid particles, independent of their size and shape, settle as flocs having one characteristic size. Assumption (4) avoids the existence of volume sources or sinks in the field equations and the last assumption restricts the treatment to gravity thickening. A mixture obeying such assumptions may be regarded as a superimposed continuous medium with two incompressible components. It can be represented by the following field variables: the volume fraction of solids $\varphi(\mathbf{r}, t)$, the solid and fluid component velocities $\mathbf{v}_s(\mathbf{r}, t)$ and $\mathbf{v}_f(\mathbf{r}, t)$, the solid and fluid stresses $\mathbf{T}_s(\mathbf{r}, t)$ and $\mathbf{T}_f(\mathbf{r}, t)$ and the solid-fluid interaction force $\mathbf{m}(\mathbf{r}, t)$. In a fixed and bounded region R, these variables must obey the *macroscopic balance* of mass and linear momentum for the solid and the fluid:

$$\frac{d}{dt} \int_R \rho_s \varphi dV + \int_{\partial R} \rho_s \varphi \mathbf{v}_s \cdot \mathbf{n} dS = 0, \tag{3.1}$$

$$\frac{d}{dt} \int_R \rho_f (1 - \varphi) dV + \int_{\partial R} \rho_f (1 - \varphi) \mathbf{v}_f \cdot \mathbf{n} dS = 0, \tag{3.2}$$

$$\frac{d}{dt} \int_R \rho_s \varphi \mathbf{v}_s dV + \int_{\partial R} \rho_s \varphi \mathbf{v}_s \mathbf{v}_s \cdot \mathbf{n} dS = \int_{\partial R} \mathbf{T}_s \cdot \mathbf{n} dS + \int_R (\rho_s \varphi \mathbf{g} + \mathbf{m}) dV, \tag{3.3}$$

$$\frac{d}{dt} \int_R \rho_f (1 - \varphi) \mathbf{v}_f dV + \int_{\partial R} \rho_f (1 - \varphi) \mathbf{v}_f \mathbf{v}_f \cdot \mathbf{n} dS$$
$$= \int_{\partial R} \mathbf{T}_f \cdot \mathbf{n} dS + \int_R (\rho_f (1 - \varphi) \mathbf{g} - \mathbf{m}) dV, \tag{3.4}$$

where $\mathbf{r}$ is the position vector, ρ_s and ρ_f are the material densities of the solid and the fluid respectively, $\mathbf{g}$ is the acceleration of gravity, ∂R is the boundary of the fixed region R and $\mathbf{n}$ is the unit normal vector to the surface ∂R. In equations (3.1) and (3.2) the first terms represent the rate of change of mass in the fixed region R and the second terms, the net flux of mass into the region R through the boundary ∂R. In equations (3.3) and (3.4) the terms, from left to right, represent the rate of change of linear momentum in the fixed region R, the net flux of linear momentum through the boundary ∂R by convection, the net flux of linear momentum through the boundary ∂R due to contact forces and the rate of increase of linear momentum in R due to the body force and to the force excerted by the fluid on the solid by unit volume.

3.3 Local balances

Using the procedures of Chapter 1, the following *local conservation equations* are obtained:

$$\frac{\partial \varphi}{\partial t} + \mathrm{div}\,(\varphi \mathbf{v}_s) = 0 \ \text{ or } \ \varphi \det \mathbf{F}_s = \varphi_\chi, \tag{3.5}$$

$$\frac{\partial}{\partial t}(1 - \varphi) + \mathrm{div}\,((1 - \varphi)\mathbf{v}_f) = 0, \tag{3.6}$$

$$\rho_s \frac{\partial}{\partial t}(\varphi \mathbf{v}_s) + \rho_s \mathrm{div}\,(\varphi \mathbf{v}_s \mathbf{v}_s) = \mathrm{div}\,\mathbf{T}_s + \rho_s \varphi \mathbf{g} + \mathbf{m}, \tag{3.7}$$

$$\rho_f \frac{\partial}{\partial t}((1 - \varphi)\mathbf{v}_f) + \rho_f \mathrm{div}\,((1 - \varphi)\mathbf{v}_f \mathbf{v}_f) = \mathrm{div}\,\mathbf{T}_f + \rho_f (1 - \varphi)\mathbf{g} - \mathbf{m}, \tag{3.8}$$

where $\mathbf{F}_s$ is the rate of deformation tensor of the solid component and φ_χ is the volume fraction of solid at a chosen reference configuration. Equations (3.5) and

(3.6) represent the continuity equation for the solid and the fluid respectively while equations (3.7) and (3.8) are the linear momentum balance for the same components. Adding the first equation of (3.5) and (3.6), we obtain the continuity equation for the mixture,

$$\text{div} \, (\varphi \mathbf{v_s} + (1 - \varphi) \mathbf{v_f}) = 0. \tag{3.9}$$

Defining the *volume average velocity* $\mathbf{q} = \varphi \mathbf{v_s} + (1 - \varphi) \mathbf{v_f}$, the continuity equation for the mixture becomes

$$\text{div} \, \mathbf{q} = 0. \tag{3.10}$$

Equation (3.10) can replace one of the continuity equations, the first of (3.5) or (3.6).

It is well known that settling suspensions develop surfaces of discontinuities; therefore we consider a region R having a finite number of discontinuities. In this case equations (3.5)–(3.10) are valid only in those regions where the field variables are continuous. At discontinuities, the *jump balances* or *jump conditions* are valid:

$$[\varphi(\mathbf{v_s} \cdot \mathbf{e_I})] = \sigma[\varphi], \tag{3.11}$$

$$[(1 - \varphi)(\mathbf{v_f} \cdot \mathbf{e_I})] = \sigma[1 - \varphi], \tag{3.12}$$

$$[\varphi \mathbf{v_s}(\mathbf{v_s} \cdot \mathbf{e_I})] = \sigma[\varphi(\mathbf{v_s} \cdot \mathbf{e_I})] + \frac{1}{\rho_s} [\mathbf{T_s} \cdot \mathbf{e_I}], \tag{3.13}$$

$$[(1 - \varphi)\mathbf{v_f}(\mathbf{v_f} \cdot \mathbf{e_I})] = \sigma[(1 - \varphi)(\mathbf{v_f} \cdot \mathbf{e_I})] + \frac{1}{\rho_f} [\mathbf{T_f} \cdot \mathbf{e_I}], \tag{3.14}$$

where $[c]$ is the jump of a field variable $c(\mathbf{r}, t)$ across the surface of discontinuity S_s, $\sigma = \mathbf{v_I} \cdot \mathbf{e_I}$ is the displacement velocity of the discontinuity, $\mathbf{v_I}$ is the convective velocity of S_s and $\mathbf{e_I}$ is the unit normal vector to S_s. Equations (3.11) and (3.12) are the *mass jump balance* for the solid and the fluid component respectively, while equations (3.13) and (3.14) represent the *linear momentum jump balances*. This set of equations must be satisfied by the field variables at discontinuities. The sum of equations (3.11) and (3.12) leads to

$$[\mathbf{q} \cdot \mathbf{e_I}] = 0, \tag{3.15}$$

hence the volume average velocity suffers no jump across a discontinuity. In what follows, instead of using the continuity equations and jump conditions for the solid and the fluid, we use the continuity equations (3.5) for the solid, the continuity equation (3.10) for the mixture and the jump conditions (3.11) for the solid and (3.15) for the mixture.

3.4   Constitutive assumptions

The laws of conservation of mass and linear momentum for the components of a solid-fluid mixture, represented by the field equations (3.5), (3.7), (3.8) and (3.10) and the jump conditions (3.11), (3.13), (3.14) and (3.15) are common to all suspensions obeying assumptions (1)–(5). These equations are not sufficient to fully characterize the behavior of specific suspensions since there are more variables than equations. No distinction is made between different materials, different flocculation reagents and different sedimentation procedures or equipment. This variety in behavior is introduced by *constitutive assumptions*, which represent constraints on the material properties or on the motion (Gurtin 1981, Hinch and Leal 1975), which have to be added to the general assumptions stated in Section 3.2.

It is convenient to distinguish three types of constitutive assumptions representing kinematical constraints, that is, restrictions on the deformation or motion that the suspension may undergo; constraints on the form of the dynamical variables; and constraints on the functional relationship between the dynamical and kinematical variables, known as constitutive equations.

3.4.1 Kinematical constraints

The assumptions formulated at the beginning of this chapter already contain a kinematical restriction: assumption (2) allows only those suspensions formed by components which are incompressible in their pure state. This obviously constrains the types of motions that the suspension may suffer. An additional restriction on the motion is imposed by the shape and operating conditions of the sedimenting vessel. To model the effect of the container in sedimentation, the concept of *ideal thickener* (Shannon and Tory 1966, Bustos and Concha 1988a, Bustos *et al.* 1990, Concha and Bustos 1991, Concha and Bustos 1992) has been proposed and is defined as a cylindrical vessel fed through a surface source and discharging through surface sinks which does not present wall effects. These restrictions imply that the concentration is constant in each horizontal cross section of the container. As a consequence, the velocity field variables depend on only one space variable and on time. This definition of an ideal thickener restricts the type of motion of the suspension and, therefore, is a kinematical restriction that has to be stated as a constitutive assumption. The motion just described is called a *simple compressive motion* and is characterized by a *gradient of deformation tensor* $\mathbf{F_s}$ for the solid component of a suspension of the form

$$\mathbf{F_s} = \mathbf{B}^T \mathrm{diag}(1, 1, 1 + \delta)\mathbf{B}, \tag{3.16}$$

where $\mathbf{B}$ is the base of unit vectors, $\mathbf{B}^T = [\mathbf{i}, \mathbf{j}, \mathbf{k}]$, $\mathbf{i}$, $\mathbf{j}$ and $\mathbf{k}$ are the Cartesian unit vectors with $\mathbf{k}$ pointing in the vertical upward direction and $0 > \delta > -1$ is a parameter that measures the degree of compression in the medium. In general, this type of motion is called *simple extensional motion* (Truesdell and Noll 1965). When $\delta > 0$ it corresponds to an extension and when $\delta < 0$ it corresponds to a compression. These arguments can be stated as a new assumption:

(6) *The suspension can perform simple compressive motion only.*

From equation (3.16) and from the local conservation of mass, second equation of (3.5), it follows that $1 + \delta = \varphi_\chi/\varphi$. Finally, the gradient of deformation tensor

for the solid component of a suspension sedimenting in an ideal thickener can be
written in the form

$$\mathbf{F_s} = \mathbf{B}^T \text{diag}(1, 1, \varphi_\chi/\varphi)\mathbf{B} = \mathbf{ii} + \mathbf{jj} + (\varphi_\chi/\varphi)\mathbf{kk}. \tag{3.17}$$

3.4.2   Dynamical constraints

The dynamical variables involved in the process of sedimentation are the solid and
the fluid stresses, the body force and the solid-fluid interaction force. Assumption
(5) restricts body forces to the gravitational force, represented by the constant acceleration of gravity vector $\mathbf{g}$; therefore we now have to establish the form assumed
by $\mathbf{T_s}$, $\mathbf{T_f}$ and $\mathbf{m}$.

Fluid component stress tensor

As in the case of all fluids, the stress tensor $\mathbf{T_f}$ can be separated into a pressure
and a viscous or extra stress,

$$\mathbf{T_f} = -p_f \mathbf{I} + \mathbf{T}_f^E, \tag{3.18}$$

where p_f is the fluid phase pressure, $\mathbf{T}_f^E$ is the extra stress tensor that depends
on the motion and therefore vanishes at equilibrium, and $\mathbf{I}$ is the second-order
unit tensor. In suspensions there are two dynamical variables related to friction.
One of them is the viscous stress tensor $\mathbf{T}_f^E$, that is, the friction between the fluid
particles and the second is the solid-fluid interaction force $\mathbf{m}$, which includes a
term representing the friction exerted by the fluid on the solid particles. The
friction associated with the stress tensor $\mathbf{T}_f^E$ is much smaller than that associated
with the interaction force $\mathbf{m}$ and can be neglected (Marle 1967, Whitaker 1969,
1986). Therefore, equation (3.18) can be simplified to $\mathbf{T_f} = -p_f \mathbf{I}$. This restriction
can be stated as a new assumption:

(7) *The fluid component behaves as an elastic fluid* (Gurtin 1981).

Solid component stress tensor

Remember that we are modeling a collection of small solid particles as a continuum,
and therefore the discrete nature of the solid disappears. We assume that the solid
continuum does not present any anisotropy, therefore we state assumption (8) by:

(8) *The solid component is isotropic.*

A constitutive equation conforming with this assumption is given in the next
section.

Source forces

For each component, the interaction of other components constitutes a source
force. In this case, the source force is the solid-fluid interaction force $\mathbf{m}$.

3.4.3 Constitutive equations

We have seen that the six field variables φ, $\mathbf{v_s}$, $\mathbf{q}$, $\mathbf{T_s}$, $\mathbf{T_f}$ and $\mathbf{m}$ must obey the four local conservation equations (3.5), (3.7), (3.8) and (3.10) or the jump conditions (3.11), (3.13), (3.14) and (3.15). Since there are six field variables and only four equations, two new equations must be postulated for closing the system: one equation relating the dynamic variable $\mathbf{T_s}$ to $\mathbf{F_s}$ and a second equation relating the solid-fluid interaction force $\mathbf{m}$ to the motion.

Constitutive equation for the solid component stress tensor

By assumption (8), we know that the solid component is isotropic, therefore the constitutive equation for $\mathbf{T_s}$ is of the form $\mathbf{T_s} = \mathbf{T_s}(\mathbf{B_s})$ (Gurtin 1981), where

$$\mathbf{B_s} = \mathbf{F_s} \cdot \mathbf{F_s^T} \qquad (3.19)$$

is the left Cauchy-Green deformation tensor (to avoid an increase in notation, we write here and in analogous situations "$\mathbf{T_s} = \mathbf{T_s}(\mathbf{B_s})$" for "$\mathbf{T_s}$ is a function of $\mathbf{B_s}$" etc.). From equation (3.16) we conclude that for an ideal thickener $\mathbf{B_s}$ is given by

$$\mathbf{B_s} = \mathbf{B^T} \mathrm{diag}(1, 1, (\varphi_\chi/\varphi)^2)\mathbf{B} = \mathbf{ii} + \mathbf{jj} + (\varphi_\chi/\varphi)^2\mathbf{kk}. \qquad (3.20)$$

To keep the relationship between the solid stress $\mathbf{T_s}$ and the left Cauchy-Green deformation tensor as simple as possible, we assume an *isotropic linear* constitutive equation. This is assumption (9):

(9) *The solid stress tensor is an isotropic linear function of the left Cauchy-Green deformation tensor.*

Then

$$\mathbf{T_s} = -\beta\mathbf{I} - p_s(\varphi_\chi, \varphi)\mathbf{kk}, \qquad (3.21)$$

where β is a constant and p_s is called *solid pressure* and depends on the volume fraction of solids in the reference and in the present configuration.

We have stated in the introduction of this chapter that two distinct processes can be observed during the settling of a flocculated suspension: *hindered settling* and *consolidation*. During hindered settling, there is no contact between the solid particles and all momentum transfer between particles is done through the fluid. On the other hand, during consolidation, permanent contact is established between the solid particles acting through the solid pressure.

It is convenient to define as the critical concentration, φ_c, that value of φ at which particles first become in permanent contact. Then, for the hindered settling motion, $p_s = 0$. For the consolidation process, it is convenient to define as the reference a configuration where the concentration is constant and equal to $\varphi_\chi = \varphi_c$. With these definitions we can express the solid stress tensor in the form

$$\mathbf{T_s} = \begin{cases} -\beta\mathbf{I} & \text{if } \varphi \leq \varphi_c, \\ -\beta\mathbf{I} - p_s(\varphi, \varphi_c)\mathbf{kk} & \text{if } \varphi > \varphi_c, \end{cases} \qquad (3.22)$$

where β is a constant.

Constitutive equation for the solid-fluid interaction force

The solid-fluid interaction force corresponds to the force that the fluid component exerts on the flocs. This force is influenced by the solid and fluid velocity, by the solid concentration and by the concentration gradient:

$$\mathbf{m} = \mathbf{m}(\mathbf{v_s}, \mathbf{q}, \varphi, \operatorname{grad}\varphi).$$

Constitutive equations must be frame indifferent (Truesdell 1984, Hinch and Leal 1975); therefore the hydrodynamic interaction force should relate to a frame-indifferent quantity describing the motion. While the velocities $\mathbf{v_s}$ and $\mathbf{q}$ are not frame indifferent, their difference is; therefore it is convenient to write $\mathbf{m} = \mathbf{m}(\mathbf{v_s}-\mathbf{q}, \varphi, \operatorname{grad}\varphi)$. From $\mathbf{q} = \varphi\mathbf{v_s}+(1-\varphi)\mathbf{v_f}$ we can deduce that $\mathbf{v_s}-\mathbf{q} = (1-\varphi)\mathbf{u}$, where $\mathbf{u} = \mathbf{v_s}-\mathbf{v_f}$ is the solid-fluid relative velocity. Then the constitutive equation can be expressed in the form $\mathbf{m} = \mathbf{m}(\mathbf{u}, \varphi, \operatorname{grad}\varphi)$. Since the flocs constituting the suspension are small, their sedimentation velocity will also be small and a linear relationship may be assumed between the hydrodynamic force and the solid-fluid relative velocity $\mathbf{u}$ and the concentration gradient (Wang 1970, Smith 1971, Concha and Bustos 1985):

$$\mathbf{m} = -\alpha(\varphi)\mathbf{u} + \beta(\varphi)\operatorname{grad}\varphi. \tag{3.23}$$

The parameter $\alpha(\varphi)$ is a function of the *resistance coefficient* between the solid and fluid component in hindered settling and a function of the *permeability* of the sediment in consolidation. In both cases α depends on the viscosity of the fluid and on the volume fraction of solids. While for hindered settling, equation (3.23) is a special case of *Stokes' law*, for consolidation it is derived from *Darcy's law*.

3.5   Dimensional analysis

Collecting the results of the last section, we can write: $\mathbf{v_s} = v_s\mathbf{k}$, $\mathbf{v_f} = v_f\mathbf{k}$, $\mathbf{u} = u\mathbf{k}$, $\mathbf{m} = m\mathbf{k}$, $\mathbf{g} = -g\mathbf{k}$, $\mathbf{T_f} = -p_f\mathbf{I}$, $\mathbf{T_s} = -\beta\mathbf{I}-p_s(\varphi, \varphi_c)\mathbf{kk}$. Introducing these results into the field equations (3.5), (3.7), (3.8), (3.10) and jump conditions (3.11), (3.13), (3.14) and (3.15) yields the following set of equations:

$$\frac{\partial\varphi}{\partial t} + \frac{\partial}{\partial z}(\varphi v_s) = 0, \tag{3.24}$$

$$\frac{\partial q}{\partial z} = 0, \tag{3.25}$$

$$\rho_s\frac{\partial}{\partial t}(\varphi v_s) + \rho_s\frac{\partial}{\partial z}\left(\varphi v_s^2\right) = -\frac{\partial p_s}{\partial z} - \rho_s\varphi g + \beta(\varphi)\frac{\partial\varphi}{\partial z} - \alpha(\varphi)u, \tag{3.26}$$

$$\rho_f\frac{\partial}{\partial t}((1-\varphi)v_f) + \rho_f\frac{\partial}{\partial z}\left((1-\varphi)v_f^2\right) = -\frac{\partial p_f}{\partial z} - \rho_f(1-\varphi)g - \beta(\varphi)\frac{\partial\varphi}{\partial z} + \alpha(\varphi)u. \tag{3.27}$$

The corresponding jump conditions are $\sigma[\varphi] = [\varphi v_s]$, $[q] = 0$, $\sigma[\varphi v_s] = [\varphi v_s^2] + [p_s/\rho_s]$ and $\sigma[(1-\varphi)v_f] = [(1-\varphi)v_f^2]$. It is convenient to examine this set of field equations to detect if there are differences in the order of magnitude of their terms. Consider the following characteristic parameters: the height L of the suspension, the settling velocity u_∞ of an individual floc in an unbounded medium and a characteristic time t_0 to be defined later. With these parameters, we can define the following dimensionless variables and parameters:

$$z^* = z/L, \quad t^* = t/t_0, \quad \varphi^* = \varphi, \quad v_s^* = v_s/u_\infty, \quad v_f^* = v_f/u_\infty, \quad q^* = q/u_\infty, \quad (3.28)$$

$$u^* = u/u_\infty, \quad p_s^* = \frac{p_s}{\rho_s g L}, \quad p_f^* = \frac{p_f}{\rho_f g L}, \quad \frac{\partial}{\partial z^*} = L\frac{\partial}{\partial z}, \quad \frac{\partial}{\partial t^*} = t_0\frac{\partial}{\partial t}, \quad (3.29)$$

where g is the local acceleration of gravity. Substituting these new variables into equations (3.24)–(3.27) and simplifying yields

$$\frac{\partial \varphi^*}{\partial t^*} + \frac{t_0 u_\infty}{L} \cdot \frac{\partial}{\partial z^*}(\varphi^* v_s^*) = 0, \quad (3.30)$$

$$\frac{\partial q^*}{\partial z^*} = 0, \quad (3.31)$$

$$\frac{u_\infty}{g t_0} \cdot \frac{\partial}{\partial t^*}(\varphi^* v_s^*) + \frac{u_\infty^2}{Lg} \cdot \frac{\partial}{\partial z^*}\left(\varphi^*(v_s^*)^2\right) = -\frac{\partial p_s^*}{\partial z^*} - \varphi^* + \frac{\beta}{\rho_s g L} \cdot \frac{\partial \varphi^*}{\partial z^*} - \frac{\alpha u_\infty}{\rho_s g}u^*, \quad (3.32)$$

$$\frac{u_\infty}{g t_0} \cdot \frac{\partial}{\partial t^*}((1-\varphi^*)v_f^*) + \frac{u_\infty^2}{Lg} \cdot \frac{\partial}{\partial z^*}\left((1-\varphi^*)(v_f^*)^2\right) \quad (3.33)$$

$$= -\frac{\partial p_f^*}{\partial z^*} - (1-\varphi^*) - \frac{\beta}{\rho_f g L} \cdot \frac{\partial \varphi^*}{\partial z^*} + \frac{\alpha u_\infty}{\rho_f g}u^*.$$

To study the order of magnitude of each of the terms in these equations, we choose the characteristic parameters as follows: $L = 1\,[\text{m}]$, $u_\infty = 10^{-4}\,[\text{m/s}]$, and $g = 10\,[\text{m/s}^2]$. With these values, the parameters in the above equations become $u_\infty/L = 10^{-4}$, $u_\infty/g = 10^{-5}$, $u_\infty^2/(Lg) = 10^{-9}$. The choice of the characteristic time t_0 is essential to obtain a reasonable result. We consider only flows where the continuity equation is always satisfied; therefore, we choose $t_0 = L/u_\infty$ so that both terms of equation (3.30) are of the same order of magnitude (if $t_0 \ll L/u_\infty$, the second term in the continuity equation (3.30) is several orders of magnitude smaller that the first term). This becomes our tenth assumption:

(10) *The characteristic time is $t_0 = L/u_\infty$.*

The characteristic time so chosen is the time necessary for a single floc to settle from the top to the bottom of the suspension in an unbounded medium. With this selection of $t_0 = L/u_\infty$, equations (3.30)–(3.33) become:

$$\frac{\partial \varphi^*}{\partial t^*} + \frac{\partial}{\partial z^*}(\varphi^* v_s^*) = 0, \quad (3.34)$$

$$\frac{\partial q^*}{\partial z^*} = 0, \quad (3.35)$$

$$\text{Fr}\left(\frac{\partial}{\partial t^*}(\varphi^* v_s^*) + \frac{\partial}{\partial z^*}\left(\varphi^*(v_s^*)^2\right)\right) = -\frac{\partial p_s^*}{\partial z^*} - \varphi^* + \frac{\beta}{\rho_s g L} \cdot \frac{\partial \varphi^*}{\partial z^*} - \frac{\alpha u_\infty}{\rho_s g}u^*, \quad (3.36)$$

$$\text{Fr}\left(\frac{\partial}{\partial t^*}((1 - \varphi^*)v_{\text{f}}^*) + \frac{\partial}{\partial z^*}\left((1 - \varphi^*)(v_{\text{f}}^*)^2 \right) \right) \tag{3.37}$$

$$= -\frac{\partial p_{\text{f}}^*}{\partial z^*} - (1 - \varphi^*) - \frac{\beta}{\rho_{\text{f}} g L} \cdot \frac{\partial \varphi^*}{\partial z^*} + \frac{\alpha u_\infty}{\rho_{\text{f}} g} u^*.$$

The dimensionless parameter of the first terms of equations (3.36) and (3.37) is known as the *Froude number* defined by

$$\text{Fr} = u_\infty^2 / gL. \tag{3.38}$$

Since $\text{Fr} \approx \mathcal{O}(10^{-9})$, these terms can be neglected. With these considerations, the local momentum balance equations in dimensional form become

$$\frac{\partial p_{\text{s}}}{\partial z} = -\rho_{\text{s}} \varphi g + \beta \frac{\partial \varphi}{\partial z} - \alpha u, \tag{3.39}$$

$$\frac{\partial p_{\text{f}}}{\partial z} = -\rho_{\text{f}}(1 - \varphi)g - \beta \frac{\partial \varphi}{\partial z} + \alpha u. \tag{3.40}$$

3.6   Dynamical variables

Adding equations (3.39) and (3.40) we obtain

$$\frac{\partial p_{\text{t}}}{\partial z} = -\rho g \tag{3.41}$$

where

$$p_{\text{t}} = p_{\text{s}} + p_{\text{f}} \tag{3.42}$$

is the total vertical stress and $\rho = \rho_{\text{s}} \varphi + \rho_{\text{f}}(1 - \varphi)$ is the density of the suspension.

3.6.1   Solid effective stress and pore pressure

When a load is applied to a porous bed of particles, the total induced stress p_{t} is supported partially by the solid skeleton and partially by the fluid filling the interstices or pores. The stress in the liquid filling the pores is called *pore pressure* and is denoted by p and the stress supported by the solid skeleton is the *solid effective stress* σ_{e} or just *effective stress*.

Drawing from experience in soil mechanics (Schiffman *et al.* 1984), we consider the solid effective stress σ_{e} to be a nonnegative function of φ. The concept of solid effective stress was developed for cases in which the solid particles are in permament contact with each other. During hindered settling, solid stresses can also occur due to particle-particle interactions, however, this collision solid stress is assumed to be constant. Since φ_{c} is the critical concentration, the solid effective stress should then be given by a constitutive equation satisfying

$$\sigma_{\text{e}}'(\varphi) = 0 \text{ for } \varphi \le \varphi_{\text{c}}, \quad \sigma_{\text{e}}'(\varphi) > 0 \text{ for } \varphi > \varphi_{\text{c}}. \tag{3.43}$$

We have mentioned that layers of sediment having a concentration greater than the critical concentration accumulate at the bottom of the thickener during sedimentation. For this sediment the total stress p_t can be written in terms of the effective solid stress and the pore pressure in the form $p_t = \sigma_e + p$. Since the pore pressure p is defined *within the fluid in the voids* of the porous bed and the partial fluid stress p_f is defined *within the fluid component occupying the whole volume* of the mixture, the following balance of forces must be obeyed at any cross section of the porous bed, where the elements of area are related by $dS_f = \epsilon dS$:

$$\int_S p_f \, dS = \int_{S_f} p \, dS_f = \int_S p(\epsilon dS) \tag{3.44}$$

and where S_f is the cross-sectional area of the porous bed within the fluid only, S is the area including solid and fluid and ϵ is the superficial porosity. If we assume that the superficial and volume porosity are equal (Telles 1977), we can write $\epsilon = 1 - \varphi$, and the previous equation becomes

$$\int_S p_f \, dS = \int_S p(1 - \varphi) \, dS. \tag{3.45}$$

From this equation the relationship between the fluid component pressure and the pore pressure follows:

$$p_f = (1 - \varphi)p. \tag{3.46}$$

This last expresion is interesting since it establishes the relationship between the theoretical variable p_f and the experimental variable p. The pore pressure is continuous across semi-permeable membranes (Liu 1980); therefore, p can be measured with a manometer. Introducing equation (3.46) into $p_t = p_s + p_f$ and this equation into $p_t = \sigma_e + p$, we can express the solid component stress in terms of the solid effective stress and the pore pressure by

$$p_s = \sigma_e + \varphi p. \tag{3.47}$$

Here again the theoretical variable $p_s(\varphi, \varphi_c)$ is related to the experimental variables $\sigma_e(\varphi, \varphi_c)$ and p. Substituting equation (3.46) for the fluid component stress and equation (3.47) for the solid component stress into the linear momentum balances represented by equations (3.39) and (3.40) yields

$$\frac{\partial}{\partial z}(\varphi p) + \frac{\partial}{\partial z}(\sigma_e) = -\rho_s \varphi g + \beta \frac{\partial \varphi}{\partial z} - \alpha u, \tag{3.48}$$

$$\frac{\partial}{\partial z}((1 - \varphi)p) = -\rho_f(1 - \varphi)g - \beta \frac{\partial \varphi}{\partial z} + \alpha u. \tag{3.49}$$

Equations (3.48) and (3.49) represent the linear momentum equations for the solid and the fluid components in terms of the solid effective stress and the pore pressure.

3.6.2   Interaction force at equilibrium

To obtain the functional form of $\beta(\varphi)$, we use the special case of equilibrium, that is, at the end of the batch sedimentation. In this case, all velocities are zero and the equation of the fluid momentum (3.49) reduces to

$$(1 - \varphi)\frac{dp}{dz} - p\frac{d\varphi}{dz} = -\rho_f(1 - \varphi)g - \beta\frac{\partial\varphi}{\partial z}\Big|_{t\to\infty}. \tag{3.50}$$

It is well known that the gradient of the pore pressure at equilibrium is given by $dp/dz = -\rho_f g$; then from equation (3.50):

$$\beta\frac{\partial\varphi}{\partial z}\Big|_{t\to\infty} = p\frac{\partial\varphi}{\partial z}\Big|_{t\to\infty}. \tag{3.51}$$

This result at equilibrium suggests that, at any time, $\beta(\varphi)$ can be chosen as $p(\varphi)$,

$$\beta(\varphi) = p(\varphi). \tag{3.52}$$

The same functional form has been deduced by other authors (Drew and Segel 1971). Introducing equation (3.52) into equations (3.48) and (3.49) we obtain

$$\frac{\partial\sigma_e}{\partial z} = -\Delta\rho\varphi g - \frac{\alpha u}{1 - \varphi}, \tag{3.53}$$

$$\frac{\partial p}{\partial z} = -\rho_f g + \frac{\alpha u}{1 - \varphi}. \tag{3.54}$$

Here,

$$\Delta\rho = \rho_s - \rho_f \tag{3.55}$$

is the solid-fluid density difference, which is assumed to be positive.

3.6.3   Excess pore pressure

Processes which are based on the flow through porous beds usually depend on the fluid pressure in excess of the hydrostatic pressure $\rho_f g(L - z)$ and not on its absolute value. Therefore, it is convenient to introduce the concept of *excess pore pressure* into the balance equations. According to this definition, the excess pore pressure p_e is given by

$$p_e = p - \rho_f g(L - z), \tag{3.56}$$

where L is the height of the surface of the suspension at the beginning of the sedimentation. The excess pore pressure may be measured directly with a manometer in the same way as the pore pressure. From equation (3.56), the gradient of the excess pore pressure is

$$\frac{\partial p_e}{\partial z} = \frac{\partial p}{\partial z} + \rho_f g. \tag{3.57}$$

Substituting equation (3.57) into the momentum equation (3.54) for the fluid we get

$$\frac{\partial p_e}{\partial z} = \frac{\alpha u}{1 - \varphi}. \tag{3.58}$$

3.6.4 Solid flux density function

The volume average velocity q may be written in terms of the relative solid-fluid velocity u by $q = v_s - (1 - \varphi)u$. Then, the solid flux density $f = \varphi v_s$ is

$$f = q\varphi + \varphi(1 - \varphi)u \tag{3.59}$$

Calculating u from equation (3.53) and substituting into equation (3.59) we get

$$f = q\varphi - \frac{\varphi^2(1 - \varphi)^2 \Delta \rho g}{\alpha(\varphi)} \left(1 + \frac{\sigma_e'(\varphi)}{\Delta \rho \varphi g} \frac{\partial \varphi}{\partial z} \right) \tag{3.60}$$

where $\sigma_e' = d\sigma_e/d\varphi$. Defining the Kynch batch flux density function

$$f_{bk}(\varphi) = -\Delta \rho \varphi^2 (1 - \varphi)^2 g / \alpha(\varphi), \tag{3.61}$$

we can write f in the form

$$f = f\left(\varphi, \frac{\partial \varphi}{\partial z} \right) = q\varphi + f_{bk}(\varphi) \left(1 + \frac{\sigma_e'(\varphi)}{\Delta \rho \varphi g} \frac{\partial \varphi}{\partial z} \right). \tag{3.62}$$

Equation (3.62) represents the solid flux-density function for a flocculated compressible suspension. However, to make the difference from the governing equation (2.20) of the theory of sedimentation of ideal suspensions outlined in Chapter 2 apparent, we rewrite equation (3.24) in terms of the diffusion coefficient

$$a(\varphi) = -\frac{f_{bk}(\varphi)\sigma_e'(\varphi)}{\Delta \rho \varphi g} \tag{3.63}$$

and of the Kynch flux density function

$$f_k(\varphi, t) = q(t)\varphi + f_{bk}(\varphi) \tag{3.64}$$

as

$$\frac{\partial \varphi}{\partial t} + \frac{\partial}{\partial z} f_k(\varphi, t) = \frac{\partial}{\partial z} \left(a(\varphi) \frac{\partial \varphi}{\partial z} \right). \tag{3.65}$$

Note that equation (3.65) is of *parabolic* type for those values of φ for which $a(\varphi) > 0$; otherwise, it reduces to the *hyperbolic* equation (2.20) for ideal suspensions. In view of condition (3.43), $a(\varphi) = 0$ for $\varphi \leq \varphi_c$, hence equation (3.65) is hyperbolic for $\varphi \leq \varphi_c$. The equation for the excess pore pressure is most conveniently rewritten as

$$\frac{\partial p_e}{\partial z} = -\Delta \rho \varphi g - \sigma_e'(\varphi) \frac{\partial \varphi}{\partial z}. \tag{3.66}$$

3.6.5 Dynamical sedimentation processes

Collecting the previous results we can say that the sedimentation of a suspension of solid particles obeying assumptions (1)–(9) can be described by the following field variables: the volume fraction of solids $\varphi(z,t)$, the solid flux density function $f(z,t)$, the volume average velocity $q(z,t)$, the solid effective stress $\sigma_e(\varphi, \varphi_c)$, the excess pore pressure $p_e(z,t)$ and the drag force represented by the parameter $\alpha(\varphi)$. These six field variables constitute a *Dynamical Sedimentation Process* if, in the domain $\Omega = \{(z,t)|0 \le z \le L,\ 0 \le t\}$, they obey the field equations

$$\frac{\partial \varphi}{\partial t} + \frac{\partial}{\partial z} f_k(\varphi, t) = \frac{\partial}{\partial z}\left(a(\varphi)\frac{\partial \varphi}{\partial z}\right), \quad \frac{\partial q}{\partial z} = 0, \quad \frac{\partial p_e}{\partial z} = -\Delta\rho\varphi g - \sigma_e'(\varphi)\frac{\partial \varphi}{\partial z} \quad (3.67)$$

for times of the order of t_0, where

$$f_k(\varphi, t) = q\varphi + f_{bk}(\varphi), \quad a(\varphi) = -\frac{f_{bk}(\varphi)\sigma_e'(\varphi)}{\Delta\rho\varphi g},$$

together with constitutive relationships $\sigma_e = \sigma_e(\varphi; \varphi_c)$ and $\alpha = \alpha(\varphi; \varphi_c)$ or $f_{bk} = f_{bk}(\varphi; \varphi_c)$ and the jump conditions $\sigma[\varphi] = [f]$ and $[q] = 0$.

3.7 Extension to several space dimensions

The constitutive assumptions stated in Section 3.4 remain valid if the one-dimensionality of the motion is no longer imposed, that is, if a truly multidimensional framework is considered. It is then straightforward to perform the dimensional analysis, define the solid effective stress and the excess pore pressure and consider the interaction force at equilibrium to produce the following field equations, which replace (3.67):

$$\frac{\partial \varphi}{\partial t} + \text{div}\,(\varphi\mathbf{q} + f_{bk}(\varphi)\mathbf{k}) = \text{div}\,(a(\varphi)\text{grad}\,\varphi), \tag{3.68}$$

$$\text{div}\,\mathbf{q} = 0, \tag{3.69}$$

$$\text{grad}\,p_e = -\Delta\rho g\varphi\mathbf{k} - \sigma_e'(\varphi)\text{grad}\,\varphi. \tag{3.70}$$

In two or three space dimensions, the unknown flow variables are the concentration φ, the volume average flow velocity field $\mathbf{q}$ and the excess pore pressure p_e. The latter can be calculated a posteriori from the concentration distribution. In one space dimension, however, $q = q(t)$ is determined by a boundary condition, and only solving the scalar equation (3.65) for φ requires computational effort.

Schneider (1982, 1985) was the first to observe a remarkable property of equations (3.68)–(3.70): taking the curl of equation (3.70) reveals that φ depends only on the vertical space coordinate and on time wherever it is continuous. The same will then be true for the vertical component of $\mathbf{q}$. Under specific assumptions on the geometry of the vessel considered, the velocity field $\mathbf{q}$ can be determined from suitable boundary conditions. However, since equation (3.70) does not depend on q, equations (3.68)–(3.69) are in most circumstances not sufficent to determine that quantity.

The independence of equation (3.70) from q is, of course, a result of the simplifications performed based on the dimensional analysis. Bürger *et al.* (1998c) present a theory which is more general than the treatment of this chapter, in the sense that not only several space dimensions but also viscous stresses are considered, and that fewer terms are neglected. They obtain a set of equations similar to (3.68)–(3.70) but in which the analogue of equation (3.70) contains additional viscous and advective acceleration terms. That equation, together with (3.69), represents a nonlinear version of the well-known Navier-Stokes equations for an incompressible fluid, to which (in the proper sense) they reduce in the case of a pure fluid ($\varphi \equiv 0$). These model equations are sufficent for the computation of φ, q and p_e. However, some new source terms describing the interaction between the evolution of the concentration distribution, or kinematic waves, and the average flow field appear. In one space dimension, these terms affect only the excess pore pressure distribution, so that the interaction they describe becomes effective only in a truely two- or three-dimensional setup. Schneider (1985) pointed out that this interaction makes the two- or three-dimensional treatment qualitatively different from what is known in one space dimension.

The significance of these terms and their actual magnitude has not yet been analysed. Moreover, if the viscous stress tensors are not neglected, additional difficulties arise from the necessity to relate the solid and fluid phase viscosities, which are theoretical variables, to the effective viscosity of the mixture, which can be measured experimentally. Obviously additional steps of progress in the theoretical understanding of the phenomenological framework have to be made before a definite multidimensional model for sedimentation with compression can be advocated.

Chapter 4

The initial value problem for a scalar conservation law

4.1 Weak solutions for a scalar conservation law

Many physical applications involve one or several conservation laws. One typical example is the equation of continuity for Kynch's theory of sedimentation. Since the pioneering works of Lax (1957) and Oleinik (1957) on conservation laws, the field has been greatly developed as we show in this and the following chapters. We begin with some general properties of conservation laws. Let $f : [a_1, a_2] \to \mathbb{R}$, $f \in C^3$ be a nonlinear function of φ (in particular, f is Lipschitz continuous then), and let $\Omega = \{(z, t) | z \in \mathbb{R}, \ t > 0\}$. We consider the quasilinear hyperbolic equation

$$\frac{\partial \varphi}{\partial t} + \frac{\partial f(\varphi)}{\partial z} = 0 \ \text{in} \ \Omega \tag{4.1a}$$

and the initial condition

$$\varphi(z, 0) = \varphi_I(z) \ \text{for} \ z \in \mathbb{R}. \tag{4.1b}$$

Here z is the coordinate of the upwards pointing vertical axis. Problem (4.1) is called a *Cauchy problem*. Note that batch sedimentation of a slurry, as expressed by equations (2.15) and (2.16), is a Cauchy problem. For its physical applications, equation (4.1a) is also called a *conservation law*. In general, solutions of problem (4.1) develop discontinuities (Lax 1957) after a finite time even for smooth initial data, and cannot be continued as regular solutions. However, they can be continued as solutions in a generalized sense:

Definition 4.1 *A* weak solution *of problem (4.1) is a real-valued bounded and measurable function $\varphi(z, t)$ which satisfies the initial condition in a generalized*

52

*sense. This means that the following equality holds for any continuously differen-
tiable function $w(z,t)$ with compact support in $\mathbb{R} \times \mathbb{R}^+$:*

$$\int_0^\infty \int_{-\infty}^\infty \left(\varphi \frac{\partial w}{\partial t} + f(\varphi) \frac{\partial w}{\partial z} \right) dzdt + \int_{-\infty}^\infty w(z,0)\varphi_{\mathrm{I}}(z)dz = 0. \qquad (4.2)$$

A weak solution $\varphi(z,t)$ (Oleinik 1957) satisfies equation (4.1a) in all points of
continuity. On the lines of discontinuity $z = z(t)$, the function $\varphi(z,t)$ satisfies the
Rankine-Hugoniot condition (also called the jump condition)

$$\frac{dz(t)}{dt} = \sigma(\varphi^+, \varphi^-) = \frac{f(\varphi^+) - f(\varphi^-)}{\varphi^+ - \varphi^-}, \qquad (4.3)$$

where σ is the displacement velocity of the discontinuity $z(t)$ and φ^+ and φ^- are
the limiting values of the function $\varphi(z,t)$ in the point $(z(t),t)$ from above and
below, respectively; that is, $\varphi^\pm = \varphi(z(t) \pm 0, t)$ in the (z,t)-plane. In a graph of f
vs φ, the velocity $\sigma(\varphi^+, \varphi^-)$ represents the slope of the chord that joins the points
$(\varphi^+, f(\varphi^+))$ and $(\varphi^-, f(\varphi^-))$.

4.2   Method of characteristics

To solve problem (4.1), we use the method of characteristics. Characteristics are
curves $(z(t), t)$ in the (z,t)-plane along which the solution of (4.1) is constant.
The characteristic along which the initial value $\varphi_{\mathrm{I}}(z_0)$ propagates is given as the
solution $z(t)$ of the initial value problem

$$z'(t) = f'(\varphi(z(t), t)) \text{ for } t > 0, \quad z(0) = z_0 \qquad (4.4)$$

since then we have

$$\frac{d}{dt}\varphi(z(t), t) = \frac{\partial}{\partial t}\varphi(z(t), t) + \frac{\partial}{\partial z}\varphi(z(t), t)z'(t)$$

$$= \frac{\partial}{\partial t}\varphi(z(t), t) + f'(\varphi(z(t), t))\frac{\partial}{\partial z}\varphi(z(t), t)$$

$$= \frac{\partial}{\partial t}\varphi(z(t), t) + \frac{\partial}{\partial z}f(\varphi(z(t), t)) = 0.$$

Obviously, characteristics of equation (4.1a) are straight lines. For example, use
the method of characteristics to find the solution of the following linear initial
value problem:

$$\frac{\partial \varphi}{\partial t} + \frac{\partial \varphi}{\partial z} = 0 \text{ in } \Omega, \quad \varphi(z,0) = \begin{cases} 1 & \text{for } 0 \leq z, \\ 0 & \text{for } z < 0. \end{cases}$$

In this example, $f(\varphi)=\varphi$; therefore all the characteristics are parallel and have a
positive slope equal to one. The line of discontinuity $z(t) = t$ separates the half

plane $t > 0$ into two regions. To construct the solution, draw straight lines with slope 1 for $t > 0$, starting at $t = 0$. For $z \geq z(t) = t$, the constant value of φ along the characteristics is equal to one, satisfying the initial condition for $z \geq 0$. For $z < z(t) = t$, the constant value of φ along the characteristics is zero and satisfies the initial condition for $z < 0$. The solution is then

$$\varphi(z,0) = \begin{cases} 1 & \text{for } z(t) = t \leq z, \\ 0 & \text{for } z < z(t) = t. \end{cases}$$

Across the line of discontinuity $z(t) = t$ the Rankine-Hugoniot condition is satisfied since

$$\frac{f(1) - f(0)}{1 - 0} = \frac{1 - 0}{1 - 0} = 1,$$

which is precisely the value of the slope of the discontinuity $z(t) = t$. The solution is shown in Figure 4.1.

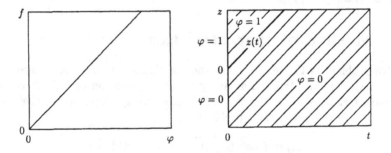

Figure 4.1: Graph of f vs φ and representation of the solution $\varphi(z,t)$.

4.3   Uniqueness of the solution

In the class of weak solutions, uniqueness of the solution of problem (4.1) is lost. This is illustrated by the following example. Consider the initial value problem

$$\frac{\partial \varphi}{\partial t} + \frac{\partial \varphi^3}{\partial z} = 0 \text{ in } \Omega, \quad \varphi(z,0) = \begin{cases} 1 & \text{for } 0 \leq z, \\ 0 & \text{for } z < 0. \end{cases}$$

Let $c > 1$ be a constant and consider the function

$$\varphi(z,t) = \begin{cases} 1 & \text{for} & z_1(t) \leq z, \\ c & \text{for} & z_2(t) \leq z < z_1(t), \\ 0 & \text{for} & z < z_2(t), \end{cases} \tag{4.5}$$

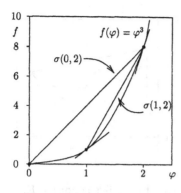 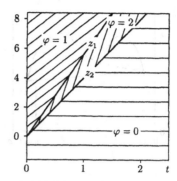

Figure 4.2: Graph of f vs φ and representation of the solution $\varphi(z,t)$.

where $z_1(t)$ and $z_2(t)$ are lines of discontinuity given by

$$z_1(t) = \frac{c^3 - 1}{c - 1}t, \quad z_2(t) = \frac{c^3 - 0}{c - 0}t = c^2 t.$$

This function $\varphi(z,t)$ is a solution of the given problem since it satisfies the initial condition and the differential equation at the points of continuity. At the discontinuities $z_1(t)$ and $z_2(t)$, it satisfies the Rankine-Hugoniot condition (4.3). The solution $\varphi(z,t)$ is not unique; in fact, there is an infinite number of solutions, since c can take any real value greater than one. In Figure 4.2, we show the graph of $f(\varphi) = \varphi^3$ vs φ and the representation of the solution $\varphi(z,t)$ in the (z,t)-plane for $c = 2$. To obtain the solution graphically, we draw first the discontinuities $z_1(t) = 7t$ and $z_2(t) = 4t$. Then starting at $t = 0$, $z > 0$, we draw the characteristics, which are parallel straight lines with slope $f'(1) = 3$, till they cut the discontinuity $z_1(t)$. Then, for $t = 0$ and $z < z_2(t)$, we draw the characteristics of slope $f'(0) = 0$. Finally, the gap between $z_1(t)$ and $z_2(t)$ is filled with characteristics of slope $f'(2) = 12$. A number of criteria motivated by mathematical or physical considerations have been proposed in order to single out an admissible weak solution. This condition is generally known as an *entropy condition*. The name comes from the equations of gas dynamics, where the requirement that the entropy should increase across discontinuities rules out nonadmissible solutions. Now some of the most important of those criteria are reviewed.

In the sequel, we will use the following notation for the interval between two numbers $a, b \in \mathbb{R}$:

$$I(a, b) = [\min\{a, b\}, \max\{a, b\}]. \tag{4.6}$$

4.3.1 Oleinik's condition E (Oleinik 1957)

Definition 4.2 *An entropy solution of problem* (4.1) *is a function which satisfies the initial condition* (4.1b), *the differential equation* (4.1a) *at the points of*

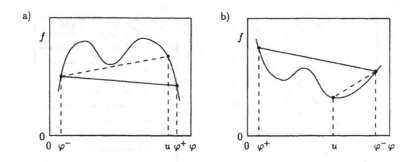

Figure 4.3: Geometrical interpretation of Oleinik's condition E.

continuity and the Rankine-Hugoniot condition (4.2) and Oleinik's condition E

$$\frac{f(u) - f(\varphi^-)}{u - \varphi^-} \geq \sigma(\varphi^+, \varphi^-) \geq \frac{f(u) - f(\varphi^+)}{u - \varphi^+} \quad \text{for all } u \in I(\varphi^-, \varphi^+) \qquad (4.7)$$

at discontinuities.

Geometrical interpretation of condition E

Lemma 4.1 *Condition E is satisfied if and only if the graph of f is on one side of the chord that joins the points $(\varphi^-, f(\varphi^-))$ and $(\varphi^+, f(\varphi^+))$ on a f versus φ plane.*

Proof. Let $\varphi^+ < \varphi^-$ and take the second inequality in (4.7),

$$\sigma(\varphi^+, \varphi^-) = \frac{f(\varphi^+) - f(\varphi^-)}{\varphi^+ - \varphi^-} \geq \frac{f(u) - f(\varphi^+)}{u - \varphi^+}, \qquad (4.8)$$

which also can be written as

$$f(u) \leq f(\varphi^+) + (u - \varphi^+)\frac{f(\varphi^+) - f(\varphi^-)}{\varphi^+ - \varphi^-} \qquad (4.9)$$

since $u - \varphi^+ > 0$. Choose $\alpha = (u - \varphi^-)/(\varphi^+ - \varphi^-)$ with $0 < \alpha < 1$. Then we get $u = \alpha\varphi^+ + (1 - \alpha)\varphi^-$. Replace u in equation (4.9) to get

$$f(\alpha\varphi^+ + (1 - \alpha)\varphi^-) \leq \alpha f(\varphi^+) + (1 - \alpha)f(\varphi^-). \qquad (4.10)$$

Then (4.8) is satisfied if and only if (4.10) is satisfied. Now consider the first inequality in (4.7),

$$\frac{f(u) - f(\varphi^-)}{u - \varphi^-} \geq \frac{f(\varphi^+) - f(\varphi^-)}{\varphi^+ - \varphi^-}. \qquad (4.11)$$

Now we have that $u - \varphi^- < 0$; therefore

$$f(u) \le f(\varphi^-) + (u - \varphi^-) \frac{f(\varphi^+) - f(\varphi^-)}{\varphi^+ - \varphi^-}.$$

Choosing α as before, we conclude that (4.11) is satisfied if and only if inequality (4.10) is satisfied. Inequality (4.10) shows that the graph of f over $[\varphi^+, \varphi^-]$ lies under the chord that joins the points $(\varphi^+, f(\varphi^+))$ and $(\varphi^-, f(\varphi^-))$ in a f versus φ plot. See Figure 4.3 (b). For $\varphi^- < \varphi^+$ it can be proved (in an analogous way) that (4.7) is satisfied if and only if

$$f(\alpha\varphi^+ + (1 - \alpha)\varphi^-) \ge \alpha f(\varphi^+) + (1 - \alpha)f(\varphi^-).$$

In this case, the graph of f lies above the chord as shown in Figure 4.3 (a). This proves the lemma. ■

4.3.2 Lax's shock admissibility criterion (Lax 1957, 1971, 1973)

Consider the initial value problem (4.1). A discontinuity $z(t)$ that propagates with speed $dz(t)/dt = \sigma(\varphi^-, \varphi^+)$ is said to satisfy *Lax's shock admissibility criterion* if

$$f'(\varphi^-) \ge \sigma(\varphi^+, \varphi^-) \ge f'(\varphi^+). \tag{4.12}$$

Definition 4.3 *A discontinuity that satisfies inequalities (4.7) and (4.12) is called a* shock.

Definition 4.4 *A shock that satisfies inequality (4.12) with one or both equalities holding, is called a* contact discontinuity.

The Lax shock admissibility criterion is a necessary but not a sufficient condition for the admissibility of a discontinuity. In this sense, Oleinik's condition E is more general, as we prove in the following lemma.

Lemma 4.2 *Let f be a continuous function of φ, then Oleinik's condition E implies Lax's shock admissibility criterion.*

Proof. We prove the lemma for $\varphi^- > \varphi^+$; the other case is proved analogously. Taking the limit of the first inequality (4.7) when $u \to \varphi^-$, we get

$$f'(\varphi^-) \ge \sigma(\varphi^+, \varphi^-). \tag{4.13}$$

Inequality (4.7) also can be written for $\varphi^- > \varphi^+$ as

$$\frac{f(u) - f(\varphi^+)}{u - \varphi^+} \le \sigma(\varphi^+, \varphi^-).$$

Taking the limit when $u \to \varphi^+$ of this last relation we obtain $f'(\varphi^+) \le \sigma(\varphi^+, \varphi^-)$, which together with (4.13) proves the lemma. ■

Remark Any discontinuity that satisfies condition E is therefore a shock. Moreover, for a concave function f, Oleinik's condition E and Lax's shock admissibility criterion are equivalent.

4.3.3   Entropy admissibility criterion (Lax 1971, 1973)

We seek conditions under which Lipschitz continuous solutions of (4.1) satisfy automatically an additional conservation law,

$$\frac{\partial \alpha(\varphi)}{\partial t} + \frac{\partial \beta(\varphi)}{\partial z} = 0, \tag{4.14}$$

where α and β are smooth real valued functions and $\alpha(\varphi)$ is convex. The function α is called an *entropy* and β an *entropy flux* for equation (4.1a); α and β are called an *entropy pair*.

Lemma 4.3 *Let*

$$\frac{\partial \varphi}{\partial t} + \frac{\partial f(\varphi)}{\partial z} = 0$$

be a conservation law. The function α satisfies the additional conservation law (4.14) if and only if

$$\beta'(\varphi) = \alpha'(\varphi) f'(\varphi). \tag{4.15}$$

Proof. Rewrite equation (4.14) as

$$\alpha'(\varphi)\frac{\partial \varphi}{\partial t} + \beta'(\varphi)\frac{\partial \varphi}{\partial z} = 0. \tag{4.16}$$

Multiplying equation (4.1a) by $\alpha'(\varphi)$, we obtain

$$\alpha'(\varphi)\frac{\partial \varphi}{\partial t} + f'(\varphi)\alpha'(\varphi)\frac{\partial \varphi}{\partial z} = 0.$$

Then α satisfies the conservation law (4.14) if and only if $\beta'(\varphi) = \alpha'(\varphi) f'(\varphi)$. ∎

Assuming that α and β satisfy the additional conservation law (4.14), the solution $\varphi(z, t)$ of problem (4.1) satisfies the *entropy admissibility criterion*

$$-\int_0^\infty \int_{-\infty}^\infty \left(\alpha(\varphi)\frac{\partial w}{\partial t} + \beta(\varphi)\frac{\partial w}{\partial z} \right) dz dt \leq 0 \tag{4.17}$$

for any continuously differentiable non-negative test function $w(z, t)$ with compact support in $\mathbb{R}^2$. If $\varphi(z, t)$ is piecewise smooth, then

$$\frac{\partial \alpha(\varphi)}{\partial t} + \frac{\partial \beta(\varphi)}{\partial z} \leq 0 \tag{4.18}$$

holds pointwise in the regions where φ is smooth, while across a discontinuity the inequality

$$\sigma(\varphi^+, \varphi^-)[\alpha(\varphi^+) - \alpha(\varphi^-)] - [\beta(\varphi^+) - \beta(\varphi^-)] \geq 0 \tag{4.19}$$

is satisfied for all convex entropy functions $\alpha(\varphi)$ and corresponding entropy fluxes $\beta(\varphi)$ satisfying (4.15).

Lemma 4.4 (Lax 1971) *For a single conservation law, (4.19) implies Oleinik's condition E.*

Proof. Suppose $\varphi^- > \varphi^+$, let u be any point between φ^+ and φ^-, and let

$$\alpha(\varphi) = \begin{cases} 0 & \text{for } \varphi < u, \\ \varphi - u & \text{for } u \leq \varphi. \end{cases} \tag{4.20}$$

From condition (4.15) we have that

$$\beta'(\varphi) = \begin{cases} 0 & \text{for } \varphi < u, \\ f'(\varphi) & \text{for } u \leq \varphi. \end{cases}$$

Therefore

$$\beta(\varphi) = \begin{cases} 0 & \text{for } \varphi < u, \\ f(\varphi) - f(u) & \text{for } u \leq \varphi. \end{cases} \tag{4.21}$$

Replacing α and β, as given by equations (4.20) and (4.21), in equation (4.19), we get

$$\sigma(\varphi^+, \varphi^-)[u - \varphi^-] - [f(u) - f(\varphi^-)] \geq 0. \tag{4.22}$$

Due to the fact that $u - \varphi^- < 0$, isolating σ from (4.22), we obtain

$$\sigma(\varphi^+, \varphi^-) \leq \frac{f(u) - f(\varphi^-)}{u - \varphi^-}. \tag{4.23}$$

Since u is any point between φ^+ and φ^-, (4.23) is Oleinik's condition E. The proof for $\varphi^- < \varphi^+$ is analogous. This completes the proof of Lemma 4.4. ∎

Remark The reverse is also true, that is, Oleinik's condition E implies (4.19). The proof involves the use of Kružkov entropy functions and fluxes which are introduced in § 4.3.5, see Holden and Risebro (1997) for details.

4.3.4   Viscosity admissibility criterion (Hopf 1969, Lax 1971)

Given equation (4.1a), we consider the following parabolic initial value problem:

$$\frac{\partial}{\partial t} + \frac{\partial f(\varphi)}{\partial z} = \mu \frac{\partial^2 \varphi}{\partial z^2} \quad \text{in } \Omega, \tag{4.24a}$$

$$\varphi(z, 0) = \varphi_I(z). \tag{4.24b}$$

with a viscosity parameter $\mu > 0$. We say that the solution $\varphi(z, t)$ of (4.1) satisfies the *viscosity admissibility criterion*, relative to (4.24a), if there is a set of solutions $\varphi_\mu(z, t)$ of problem (4.24) such that, when $\mu \to 0$, $\varphi_\mu \to \varphi(z, t)$ boundedly almost everywhere in Ω.

Lemma 4.5 *Every weak solution of (4.1) which satisfies the viscosity admissibility criterion also satisfies the entropy criterion (4.18).*

Proof. Rewrite equation (4.24a) as

$$\frac{\partial}{\partial t} + f'(\varphi)\frac{\partial}{\partial z} = \mu\frac{\partial^2 \varphi}{\partial z^2}$$

and multiply it by $\alpha'(\varphi)$ to obtain

$$\frac{\partial \alpha}{\partial t} + \frac{\partial \beta}{\partial z} = \mu\frac{\partial}{\partial z}\left(\frac{\partial \alpha}{\partial z}\right) - \mu\left(\frac{\partial}{\partial z}\right)^2 \alpha''.$$

Since $\alpha'' > 0$, we get

$$\frac{\partial \alpha}{\partial t} + \frac{\partial \beta}{\partial z} \le \mu\frac{\partial^2 \alpha}{\partial z^2}. \tag{4.25}$$

As $\mu \to 0$, the second member tends to zero in the topology of distributions. This completes the proof. ∎

4.3.5   Kružkov's formulation (Kružkov 1970)

Let $\Omega_T = \{(z, t)| z \in \mathbb{R}, t \in [0, T]\}$ with $T > 0$.

Definition 4.5 *A bounded measurable function $\varphi(z, t)$ is called a generalized solution of problem (4.1) in Ω_T if, for any constant $k \in \mathbb{R}$ and any smooth function $w(z, t) \ge 0$ which is finite in Ω_T (the support of w is strictly contained in Ω_T), the following inequality holds:*

$$\iint_{\Omega_T} \left\{ |\varphi(z, t) - k|\frac{\partial w}{\partial t} + \text{sgn}(\varphi(z, t) - k)(f(\varphi) - f(k))\frac{\partial w}{\partial z} \right\} \ge 0, \tag{4.26}$$

and there exists a set A of measure zero on $[0, T]$ such that for $t \in [0, T] \setminus A$, the function φ is defined almost everywhere in $\mathbb{R}$ and, for any interval $K_r = \{|z| \le r\} \subset \mathbb{R}$, we have

$$\lim_{t\to 0,\, t\in[0,T]\setminus A} \int_{K_r} |\varphi(z, t) - \varphi_I(z)| dz = 0. \tag{4.27}$$

Here sgn denotes the sign function,

$$\text{sgn}(y) = \begin{cases} 1 & \text{for} \quad 0 < y, \\ 0 & \text{for} \quad y = 0, \\ -1 & \text{for} \quad y < 0. \end{cases} \tag{4.28}$$

Remark This formulation contains not only the weak formulation (4.2) but also Oleinik's condition E. It should be pointed out that inequality (4.26) is produced by selecting the one-parameter family of entropy pairs

$$\alpha(\varphi) = |\varphi - k|, \quad \beta(\varphi) = \text{sgn}(\varphi - k)(f(\varphi) - f(k)), \quad k \in \mathbb{R} \tag{4.29}$$

in the weak formulation (4.17) of the entropy admissibility criterion. It is not entirely obvious that a non-smooth function such as $|\varphi - k|$ may be employed as an entropy, and that the whole set of entropy functions may be replaced by a one-parameter family. However, as stated e.g. by Godlewski and Raviart (1991), it is possible first to take smoothed convex approximations of the entropies $\alpha(\varphi)$ given by (4.29) and of the corresponding entropy fluxes and then to let the smoothing parameter tend to zero in order to obtain inequality (4.26). This shows that Kružkov's formulation follows from the entropy admissibility criterion expressed by inequality (4.17). Moreover, since any convex function belongs to the convex hull of the set of all affine functions and all functions of the form $\varphi \mapsto |\varphi - k|$, any weak solution of the initial value problem (4.1) satisfying inequality (4.26) for all $k \in \mathbb{R}$ and (4.27) is also an entropy weak solution in the sense of inequality (4.17). In other words, Kružkov's formulation and the entropy admissibility criterion introduced in § 4.3.3 are equivalent.

4.3.6 Uniqueness of the solution

This exhaustive review of existing conditions for the uniqueness of the solution shows that the uniqueness of the solution of the Cauchy problem (4.1) which satisfies an entropy condition has been proved by many authors. See for instance Keyfitz (1971), Kružkov (1970), Lax (1957, 1973) and Oleinik (1963a, 1963b). Here we give a theorem by Lax (1973) that proves the uniqueness of the solution of problem (4.1).

Theorem 4.1 (Lax 1973) *Let φ and ψ be two weak solutions of problem (4.1) and suppose that f is continuous and that φ and ψ satisfy Oleinik's condition E, then $\|\varphi(\cdot, t) - \psi(\cdot, t)\|_{L_1}$ is a decreasing function of t.*

Proof. Let

$$\|\varphi(\cdot, t) - \psi(\cdot, t)\|_{L_1} = \sum_n (-1)^n \int_{y_n}^{y_{n+1}} (\varphi(z, t) - \psi(z, t)) dz, \qquad (4.30)$$

where the points y_n and y_{n+1} are so chosen that

$$\operatorname{sgn}(\varphi(z, t) - \psi(z, t)) = (-1)^n \quad \text{for } y_n < z < y_{n+1}.$$

The limits y_n and y_{n+1} are functions of t. Now we take the derivative of equation (4.30) with respect to t to obtain

$$\frac{d}{dt} \|\varphi - \psi\|_{L_1} = \sum_n (-1)^n \left\{ \int_{y_n}^{y_{n+1}} \frac{\partial(\varphi - \psi)}{\partial t} dz \right.$$
$$\left. + (\varphi - \psi)\big|_{y_{n+1}} \frac{dy_{n+1}}{dt} - (\varphi - \psi)\big|_{y_n} \frac{dy_n}{dt} \right\}. \qquad (4.31)$$

Since

$$\frac{\partial}{\partial t}(\varphi - \psi) = -\frac{\partial}{\partial z}(f(\varphi) - f(\psi)),$$

is valid in the sense of distributions between y_n and y_{n+1} (Lax 1973), we can integrate the first term of the second member of equation (4.31) to obtain

$$\frac{d}{dt}\|\varphi - \psi\|_{L_1} = \sum_n (-1)^n \Big\{ f(\psi(y,t)) - f(\varphi(y,t))$$

$$+ (\varphi(y,t) - \psi(y,t))\frac{dy}{dt}\Big\}\Big|_{y_n}^{y_{n+1}}. \qquad (4.32)$$

Now we analyse the following cases:

(1) y_k is a point of continuity of both φ and ψ, then $\varphi(y_k,t) = \psi(y_k,t)$ and all the terms of the sum on the second menber of equation (4.32) are zero.

(2) y_{k+1} is a point of discontinuity of φ but not of ψ. Then, if this is the sole discontinuity, only two terms remain from the sum in equation (4.32):

$$\frac{d}{dt}\|\varphi - \psi\|_{L_1} = (-1)^k \Big\{ f(\psi(y_{k+1},t)) - f(\varphi(y_{k+1}^-,t))$$

$$+ (\varphi(y_{k+1}^-,t) - \psi(y_{k+1},t))\frac{dy_{k+1}}{dt}\Big\}$$

$$- (-1)^{(k+1)} \Big\{ f(\psi(y_{k+1},t)) - f(\varphi(y_{k+1}^+,t))$$

$$+ (\varphi(y_{k+1}^+,t) - \psi(y_{k+1},t))\frac{dy_{k+1}}{dt}\Big\}$$

Since φ has a discontinuity at $y = y_{k+1}$, ψ is between φ^- and φ^+; for example $\varphi^+ > \psi(y_{k+1}) > \varphi^-$, then $\varphi - \psi$ is negative in (y_k, y_{k+1}) and therefore k is an odd number:

$$\frac{d}{dt}\|\varphi - \psi\|_{L_1} = -\Big\{ f(\psi) - f(\varphi^-) + (\varphi^- - \psi)\frac{dy_{k+1}}{dt}\Big\}$$

$$- \Big\{ f(\psi) - f(\varphi^+) + (\varphi^+ - \psi)\frac{dy_{k+1}}{dt}\Big\} \qquad (4.33)$$

Since $y_{k+1}(t)$ is a discontinuity, the speed dy_{k+1}/dt of the discontinuity is given by the Rankine-Hugoniot condition (4.3):

$$\frac{dy_{k+1}}{dt} = \sigma(\varphi^+, \varphi^-).$$

Replacing this term in equation (4.33) and rearranging terms we get

$$\frac{d}{dt}\|\varphi - \psi\|_{L_1} = -(\psi - \varphi^-)\Big\{ \frac{f(\psi) - f(\varphi^-)}{\psi - \varphi^-} - \sigma(\varphi^+, \varphi^-)\Big\}$$

$$+ (\psi - \varphi^+)\Big\{ \sigma(\varphi^+, \varphi^-) - \frac{f(\psi) - f(\varphi^+)}{\psi - \varphi^+}\Big\}. \qquad (4.34)$$

Now we apply condition E, given in (4.7), to obtain

$$\frac{d}{dt}\|\varphi - \psi\|_{L_1} \leq 0.$$

(3) The case when y_k is a point of discontinuity of both φ and ψ is proved similarly. This completes the proof. ∎

Theorem 4.2 *Let φ, ψ and f be as in Theorem 4.1. If $\varphi = \psi$ at $t = 0$, then $\varphi(z,t) = \psi(z,t)$ for all t.*

Proof. The assertion is deduced immediately from Theorem 4.1. ∎

4.4   Existence of the global weak solution

The existence of the global weak solution for scalar conservation laws has been proved by many authors under different assumptions. See for example Ballou (1970), Conway and Smoller (1966), Kružkov (1969a, 1969b, 1970), Lax (1973) and Oleinik (1957, 1963a, 1963b, 1964) among others.

We present (Bustos 1984) an existence proof that consists of an adaptation of part of an existence proof given by Conway and Smoller (1966) for two space variables, together with four theorems due to Glimm (1965), Harten, Hyman and Lax (1976), Harten and Lax (1981) and Harten, Lax and van Leer (1983) for systems of conservation laws.

We consider the Cauchy problem (4.1) and we require that $\varphi_I(z)$ be bounded and have bounded total variation. Let the half plane $t \geq 0$ be covered by a uniform rectangular grid defined by the lines

$$t = nk, \ n = 0, 1, 2, \ldots; \quad z = jh, \ j \in \mathbb{Z}. \tag{4.35}$$

Here h and k are the space and time steps respectively. In the region $t > 0$ we consider the Lax-Friedrichs (L-F) finite difference scheme to approximate (4.1a) where we denote by v_j^n the mesh function that approximates the function $\varphi(z,t)$ in the point (nk, jh) of the mesh:

$$\frac{1}{k}\left[v_j^{n+1} - \frac{1}{2}(v_{j+1}^n + v_{j-1}^n)\right] + \frac{1}{2h}\left[f_{j+1}^n - f_{j-1}^n\right] = 0, \ j \in \mathbb{Z}, \ n = 1, 2, \ldots \tag{4.36}$$

$$v_j^0 = \varphi_I(jh). \tag{4.37}$$

Here we are using the standard notation $f_j^n = f(v_j^n)$.

4.4.1   Properties of the Lax-Friedrichs scheme

The L-F scheme can be written in conservation form

Definition 4.6 *A finite difference scheme is said to be in* conservation form *(Lax 1954) if it can be written in the form*

$$v_j^{n+1} = v_j^n - \lambda\left(g\left(z + h/2\right) - g\left(z - h/2\right)\right), \tag{4.38}$$

where

$$\lambda = k/h \tag{4.39}$$

is the mesh size ratio and

$$g\left(z + h/2\right) = g\left(v_{-\ell+1}^n, v_{-\ell+2}^n, ..., v_\ell^n\right), \quad g\left(z - h/2\right) = g\left(v_{-\ell}^n, v_{-\ell+1}^n, ..., v_{\ell-1}^n\right).$$

Here ℓ is a fixed index in the scheme. In order for (4.38) to be consistent with the equation (4.1a), g must be related to f as follows:

$$g(u, u, ..., u) = f(u) \tag{4.40}$$

The function g is called a numerical flux.

For the L-F scheme we have

$$v_j^{n+1} = v_j^n - \lambda(g(v_{j+1}^n, v_j^n) - g(v_j^n, v_{j-1}^n)), \tag{4.41}$$

where for all j and n

$$g(v_{j+1}^n, v_j^n) = \frac{1}{2}(f_{j+1}^n + f_j^n) - \frac{1}{2\lambda}(v_{j+1}^n - v_j^n). \tag{4.42}$$

The solutions v_j^n are bounded for all j and n

We prove this property following a lemma by Conway and Smoller (1966).

Lemma 4.6 *Let $M_0 \le v_j^0 \le M_1$ for all j and let*

$$A = \max_{M_0 \le v \le M_1} |f'(v)|.$$

If the stability requirement $\lambda A \le 1$ is fulfilled, then $M_0 \le v_j^n \le M_1$ for all values of j and n.

Proof. Using the mean value theorem, we write (4.36) in the form

$$v_j^{n+1} = \frac{1}{2}\left(v_{j+1}^n + v_{j-1}^n\right) - \frac{\lambda}{2}f'\left(\theta_j^n\right)\left(v_{j+1}^n - v_{j-1}^n\right), \tag{4.43}$$

where θ_j^n is some appropriate value between v_{j+1}^n and v_{j-1}^n. We write this equation as

$$v_j^{n+1} = v_{j+1}^n\left(\frac{1}{2} - \frac{\lambda}{2}f'\left(\theta_j^n\right)\right) + v_{j-1}^n\left(\frac{1}{2} + \frac{\lambda}{2}f'\left(\theta_j^n\right)\right). \tag{4.44}$$

Suppose inductively that $M_0 \le v_j^n \le M_1$ for all j, then, since the coefficients of v_{j+1}^n and v_{j-1}^n are non-negative and add up to one, we obtain $M_0 \le v_j^{n+1} \le M_1$ for all j. And since by hypothesis $M_0 \le v_j^0 \le M_1$, the assertion of the lemma is proved. ∎

In what follows, we assume that the stability condition $\lambda A \le 1$ is satisfied.

The L-F scheme is total variation stable

The *total variation* (TV) of any mesh function v_j^n is

$$TV(v^n) = \sum_j |v_{j+1}^n - v_j^n|. \qquad (4.45)$$

A numerical scheme is said to be *total variation stable* if the total variation of v_j^n is uniformly bounded. This means that there exists a constant B, not depending on n and j such that

$$TV(v^n) \le B \cdot TV(v^0) \qquad (4.46)$$

for all n. A numerical scheme is said to be *total variation diminishing* (TVD) if

$$TV(v^{n+1}) \le TV(v^n) \qquad (4.47)$$

for all n. To prove that a scheme is total variation stable it is sufficient to prove that it is total variation diminishing. We review the following lemma (Conway and Smoller 1966):

Lemma 4.7 *If k and h satisfy $h \le \delta k$ for some fixed $\delta > 0$, then for arbitrary X*

$$\sum_X |v_{j+1}^n - v_j^n| \le \sum_{X+\delta nk} |v_{j+1}^0 - v_j^0| \qquad (4.48)$$

where $\sum_L$ means summation over all j satisfying $h|j| \le L$.

Proof. Let $w_j^n = v_{j+1}^n - v_j^n$ for all j and n, then using (4.44), we get

$$w_j^{n+1} = \frac{1}{2}\left(v_{j+2}^n - v_{j+1}^n\right)\left(1 - \lambda f'\left(\theta_{j+1}^n\right)\right) + \frac{1}{2}\left(v_j^n - v_{j-1}^n\right)\left(1 + \lambda f'\left(\theta_{j-1}^n\right)\right)$$

where $v_i^n < \theta_i^n < v_{i+1}^n$ for $i = j + 1$ and $i = j - 1$. We rewrite this last equation as

$$w_j^{n+1} = \frac{1}{2}w_{j+1}^n\left(1 - \lambda f'\left(\theta_{j+1}^n\right)\right) + \frac{1}{2}w_{j-1}^n\left(1 + \lambda f'\left(\theta_{j-1}^n\right)\right). \qquad (4.49)$$

Due to the stability requirement, the coefficients of w_{j+1}^n and w_{j-1}^n in equation (4.49) are non-negative and have a sum equal to one, then

$$\sum_{h|j| \le X} |w_j^{n+1}| \le \sum_{h|j| \le X+h} |w_j^n|$$

and therefore the scheme is total variation stable. This completes the proof. ∎

A property of the finite difference approximation function

Lemma 4.8 *For the L-F scheme we have*

$$|v(z, t_{n+1}) - v(z, t_n)| = \mathcal{O}\left(\left|v_{j+1}^n - v_j^n\right| + \left|v_j^n - v_{j-1}^n\right|\right), \tag{4.50}$$

where $v(z, t)$ is the finite difference approximation function *defined as a real piecewise constant extension of the mesh function v_j^n by*

$$v(z, t) = v_j^n \quad \text{for} \quad (z, t) \in ((j - 1/2)\, h, (j + 1/2)\, h) \times [nk, (n + 1)k]. \tag{4.51}$$

Proof. By definition $v(z, t) = v_j^n$ for some j, then

$$|v(z, t_{n+1}) - v(z, t_n)| = |v_j^{n+1} - v_j^n|.$$

Equation (4.36) can be written as

$$v_j^{n+1} = \frac{1}{2}\left(v_{j+1}^n + v_{j-1}^n\right) - \frac{\lambda}{2}\left(f_{j+1}^n - f_{j-1}^n\right)$$

where λ is kept constant. From this equation and the fact that f is Lipschitz continuous, we find with aid of Lemma 4.6 that

$$|v_j^{n+1} - v_j^n| \le \frac{\lambda}{2}\left|f_{j+1}^n - f_j^n\right| + \frac{\lambda}{2}\left|f_j^n - f_{j-1}^n\right|$$
$$+ \frac{1}{2}\left|v_{j+1}^n - v_j^n\right| + \frac{1}{2}\left|v_j^n - v_{j-1}^n\right| = \mathcal{O}\left(\left|v_{j+1}^n - v_j^n\right| + \left|v_j^n - v_{j-1}^n\right|\right)$$

This completes the proof. ∎

The L-F scheme is strictly monotone

A finite difference scheme

$$v_j^{n+1} = H\left(v_{j-k}^n, v_{j-k+1}^n, ..., v_{j+k}^n\right) \tag{4.52}$$

is said to be *strictly monotone* if H is a strictly monotone increasing function of each of its arguments. We prove this property in the following lemma:

Lemma 4.9 *The L-F scheme is strictly monotone if $\lambda|f'(v_j^n)| \le 1$ for all j and n.*

Proof. The L-F scheme can be written as $v_j^{n+1} = H\left(v_{j-1}^n, v_{j+1}^n\right)$ where

$$H\left(v_{j-1}^n, v_{j+1}^n\right) = \frac{1}{2}\left(v_{j+1}^n + v_{j-1}^n\right) - \frac{\lambda}{2}\left(f_{j+1}^n - f_{j-1}^n\right).$$

Consider a positive increment ϵ in the first argument of H. Then the Taylor formula yields

$$H\left(v_{j-1}^n + \epsilon, v_{j+1}^n\right) = \frac{1}{2}\left(v_{j+1}^n + v_{j-1}^n + \epsilon\right) - \frac{\lambda}{2}\left(f_{j+1}^n - f\left(v_{j-1}^n + \epsilon\right)\right)$$

$$= H\left(v_{j-1}^n, v_{j+1}^n\right) + \frac{\epsilon}{2}\left(1 + \lambda f'\left(\gamma_{j-1}^n\right)\right) \tag{4.53}$$

with $v_{j-1}^n < \gamma_{j-1}^n < v_{j-1}^n + \epsilon$. Similarly,

$$H\left(v_{j-1}^n, v_{j+1}^n + \epsilon\right) = H\left(v_{j-1}^n, v_{j+1}^n\right) + \frac{\epsilon}{2}\left(1 - \lambda f'\left(\gamma_{j+1}^n\right)\right) \tag{4.54}$$

with $v_{j+1}^n < \gamma_{j+1}^n < v_{j+1}^n + \epsilon$. Since the second terms of the right-hand side of (4.53) and (4.54) are positive due to the stability condition, the scheme is monotone. This completes the proof. ∎

The L-F scheme is consistent with the entropy condition

Definition 4.7 *A finite difference scheme is said to be* consistent *with the entropy condition* (4.18) *if an inequality of the following form is satisfied:*

$$\alpha_j^{n+1} < \alpha_j^n - \lambda F\left(v_{-\ell+1}^n, ..., v_\ell^n\right) + \lambda F\left(v_{-\ell}^n, ..., v_{\ell-1}^n\right) \tag{4.55}$$

where α is a convex entropy function and F is a numerical entropy flux, consistent with the entropy flux:

$$F(u, u, ..., u) = \beta(u). \tag{4.56}$$

We review the following theorem by Harten, Hyman and Lax (1976):

Theorem 4.3 *Let*

$$v_j^{n+1} = H\left(v_{j-k}^n, v_{j-k+1}^n, \cdots, v_{j+k}^n\right), \tag{4.57}$$

$$v_j^{n+1} = v_j^n - \lambda\left(g\left(v_{j-k+1}^n, \cdots, v_{j+k}^n\right) - g\left(v_{j-k}^n, \cdots, v_{j+k-1}^n\right)\right) \tag{4.58}$$

be a finite difference approximation to (4.1) *in conservation form with*

$$g(v, v, \dots, v) = f(v) \tag{4.59}$$

which is monotone, that is

$$\frac{\partial H}{\partial w_i^n}\left(w_{-k}^n, \cdots, w_k^n\right) \geq 0 \tag{4.60}$$

for all i such that $-k \leq i \leq k$, then the difference scheme is consistent with the entropy condition (4.18).

Proof. Define α and β as in (4.20) and (4.21) where u is an arbitrarily chosen number. Let F be defined by

$$F\left(v_{-k+1}^n, \cdots, v_k^n\right) = \sum_{j=-k+1}^{k} \dot\alpha(v_j^n)(v_j^n - u) \int_0^1 \frac{\partial}{\partial v_j(\theta)} g\left(v_{-k+1}(\theta), \dots, v_k(\theta)\right) d\theta. \tag{4.61}$$

Here, $v_j(\theta) = u + \theta(v_j^n - u)$ denotes the straight line connecting u with v_j^n and

$$\dot{\alpha}(v_j^n) = \frac{d}{dv_j^n}\alpha(v_j^n) = \begin{cases} 0 & \text{for } v_j^n < u, \\ 1 & \text{for } u \leq v_j^n. \end{cases}$$

The function F, as defined by (4.61), is continuous and consistent with β. In fact,

$$F(v_j^n, \ldots, v_j^n) = \dot{\alpha}(v_j^n) \int_0^1 \sum_{\ell=-k+1}^k \frac{\partial}{\partial v_\ell(\theta)} g(v_j(\theta), \ldots, v_j(\theta))(v_j^n - u)\, d\theta$$

$$= \dot{\alpha}(v_j^n) \int_0^1 \frac{d}{d\theta} g(v_j(\theta), \ldots, v_j(\theta))\, d\theta$$

$$= \dot{\alpha}(v_j^n)(g(v_j^n, \ldots, v_j^n) - g(u, \ldots, u))$$

$$= \dot{\alpha}(v_j^n)(f(v_j^n) - f(u)) = \beta(v_j^n).$$

Define

$$v(\theta) = H(v_{j-k}(\theta), \ldots, v_{j+k}(\theta)). \qquad (4.62)$$

Then $v(0) = u$ and $v(1) = v_j^{n+1}$. With these definitions, α_j^n and α_j^{n+1} can be written as

$$\alpha_j^n = \alpha_j^n - \alpha(u) = \int_0^1 \dot{\alpha}(v_j(\theta))(v_j^n - u)\, d\theta = \int_0^1 \frac{d}{d\theta}\alpha(v_j(\theta))\, d\theta, \qquad (4.63)$$

$$\alpha_j^{n+1} = \alpha_j^{n+1} - \alpha(u) = \int_0^1 \frac{d}{d\theta}\alpha(v(\theta))$$

$$= \int_0^1 \dot{\alpha}(v(\theta)) \sum_{i=-k}^k \frac{\partial}{\partial v_{j+i}(\theta)} H(v_{j-k}(\theta), \ldots, v_{j+k}(\theta))(v_{j+i} - u)\, d\theta. \qquad (4.64)$$

Now we prove that

$$A_j = \alpha_j^{n+1} - \alpha_j^n + \lambda\left(F(v_{j-k+1}^n, \ldots, v_{j+k}^n) - F(v_{j-k}^n, \ldots, v_{j+k-1}^n)\right) \leq 0. \qquad (4.65)$$

We insert (4.61), (4.63) and (4.64) into (4.65), to obtain

$$A_j = \sum_{i=-k}^k (v_{j+i}^n - u) \int_0^1 \frac{\partial}{\partial v_{j+i}(\theta)} H(v_{j-k}(\theta), \ldots, v_{j+k}(\theta)) \times$$

$$\times\, (\dot{\alpha}(v(\theta)) - \dot{\alpha}(v_{j+i}(\theta)))\, d\theta. \qquad (4.66)$$

By assumption,

$$\frac{\partial}{\partial v_{j+i}(\theta)} H(v_{j-k}(\theta), \ldots, v_{j+k}(\theta)) \geq 0 \quad \text{for} \quad -k \leq i \leq k.$$

Furthermore,

$$(\dot{\alpha}(v(\theta)) - \dot{\alpha}(v_j^n(\theta)))(v_j^n - u) \leq 0,$$

since if $v_j^n - u < 0$, then $\dot{\alpha}(v_j^n(\theta)) = 0$ and $\dot{\alpha}(v(\theta)) \geq 0$ for all $0 \leq \theta \leq 1$. If on the other hand, $v_j^n - u > 0$, then $\dot{\alpha}(v_j^n(\theta)) = 1$ and $\dot{\alpha}(v(\theta)) \leq 1$ for all $0 \leq \theta \leq 1$. This completes the proof. ∎

Since the L-F scheme can be written in conservation form and is monotone, then from Theorem 4.3 we conclude that the L-F scheme is consistent with the entropy condition.

4.4.2   Convergence of the Lax-Friedrichs scheme

Before proving the convergence of the L-F scheme, we need a lemma by Glimm (1965).

Lemma 4.10 *Since the L-F scheme is consistent with the conservation law (4.1a) and is total variation stable, then for all n and m:*

$$\|v(\cdot, t_m) - v(\cdot, t_n)\|_{L_1} = \mathcal{O}(|t_m - t_n|). \tag{4.67}$$

Proof. From Lemma 4.8 we have

$$\|v(\cdot, t_{n+1}) - v(\cdot, t_n)\|_{L_1} = \sum_j \int_{I_j} |v(z, t_{n+1}) - v(z, t_n)| \, dz = \mathcal{O}(h) \sum_j |v_{j+1}^n - v_j^n|,$$

where $I_j = [z_{j-1}, z_j]$. But the total variation of v^n at time t_n is uniformly bounded by hypothesis, then:

$$\|v(\cdot, t_{n+1}) - v(\cdot, t_n)\|_{L_1} = \mathcal{O}(h).$$

By the triangle inequality it follows that

$$\|v(\cdot, t_m) - v(\cdot, t_n)\|_{L_1} \leq |v(\cdot, t_m) - v(\cdot, t_{m-1})| + |v(\cdot, t_{m-1}) - v(\cdot, t_{m-2})|$$
$$+ \ldots + |v(\cdot, t_{n+1}) - v(\cdot, t_n)|,$$
$$\|v(\cdot, t_m) - v(\cdot, t_n)\|_{L_1} = (m - n)\mathcal{O}(h) = \mathcal{O}((m - n)h).$$

Now we use the fact that h and k are proportional to obtain

$$\|v(\cdot, t_m) - v(\cdot, t_n)\|_{L_1} = \mathcal{O}((m - n)k) = \mathcal{O}(|t_m - t_n|),$$

and this completes the proof. ∎

Now we prove that the L-F scheme is convergent. The following theorem due to Harten and Lax (1981) is based on a theorem by Glimm (1965).

Theorem 4.4 *Suppose that the finite difference scheme is consistent with the con-servation law* (4.1a) *and the corresponding entropy inequality* (4.18). *If the result-ing numerical approximation is* TV *stable and uniformly bounded, then the scheme is convergent and the limit of the finite difference approximation function is the unique weak solution that satisfies the entropy inequality.*

Proof. The finite difference approximation function $v(z, t)$ was defined in (4.51) as a piecewise constant function. For $h \to 0$, we have a corresponding family of approximating functions $\{v^{i_0}(z,t)\}_{i_0 \in \mathbb{N}}$, such that $i_0 \to \infty$ for $h \to 0$ and $k \to 0$. By hypothesis, the $v^{i_0}(z,t)$ are uniformly bounded and have bounded total variation, then by Helly's theorem (see pages 34 and 109 in Smirnov (1962)) we have that a subsequence $\{v^{i_1}(z,t)\}_{i_1 \in \mathbb{N}_1 \subset \mathbb{N}}$ converges on bounded intervals on any horizontal line $t = const.$, to a function $\varphi(z, t)$, that is,

$$\lim_{i_1 \to \infty} v^{i_1}(z,t) = \varphi(z,t) \text{ for almost all } z \in \mathbb{R}. \tag{4.68}$$

The functions $v^{i_1}(z,t)$ are measurable and bounded almost everywhere; then by Lebesgue's theorem we get:

$$\lim_{i_1 \to \infty} \int_{-\infty}^{\infty} v^{i_1}(z,t)dz = \int_{-\infty}^{\infty} \varphi(z,t)dz$$

and moreover

$$\lim_{i_1 \to \infty} \int_{-\infty}^{\infty} |v^{i_1}(z,t) - \varphi(z,t)|dz = 0, \tag{4.69}$$

that is, $v^{i_1}(z,t)$ converges in the L_1 norm to $\varphi(z,t)$ in the horizontal line.

By a diagonal process, the same result can be achieved for a countable number of horizontal lines located at say, a rational number $t_\ell \in \mathbb{Q}$. Let $t_1 \in \mathbb{Q}$; then the sequence $\{v^{i_1}(z,t_1)\}$ converges to the value $\varphi(z,t_1)$. Now let $t_2 > t_1$; then $\{v^{i_1}(z,t_2)\}$ is a bounded sequence and therefore it has a convergent subsequence $\{v^{i_2}(z,t_2)\}$ with $i_2 \in \mathbb{N}_2 \subset \mathbb{N}_1$, that converges to a value $\varphi(z,t_2)$. Let $t_3 > t_2$; then the sequence $\{v^{i_2}(z,t_3)\}$ is bounded and therefore has a convergent subsequence $\{v^{i_3}(z,t_3)\}_{i_3 \in \mathbb{N}_3 \subset \mathbb{N}_2}$ that converges to a value $\varphi(z,t)$ and so on. Now consider the diagonal sequence $\{u_i(z,t_i)\}$ formed by the functions $u_i(z,t_i) = i^{th}$ term of $\{v^{i_i}(z,t_i)\}$. By construction, $\{u_i(z,t_i)\}$ is obtained from $\{v^{i_1}(z,t_1)\}$. Also, except for the first term, it is obtained from $\{v^{i_2}(z,t_2)\}$. Except for the first two terms, it is obtained from $\{v^{i_3}(z,t_3)\}$ and so forth. Then $\{u_i(z,t_1)\}$ converges to $\varphi(z,t_1)$, $\{u_i(z,t_2)\}$ converges to $\varphi(z,t_2)$, $\{u_i(z,t_3)\}$ converges to $\varphi(z,t_3)$ and so on. That is,

$$\lim_{i \to \infty} u_i(z,t_\ell) = \varphi(z,t_\ell) \text{ for every } t_\ell \in \mathbb{Q}$$

and

$$\lim_{i \to \infty} \int_{-\infty}^{\infty} |u_i(z,t_\ell) - \varphi(z,t_\ell)| = 0 \quad \text{for} \quad t_\ell \in \mathbb{Q}. \tag{4.70}$$

We prove now that $\{u_i(z,t)\}$ converges in the L_1 norm for arbitrary t. Let $t_\ell \in \mathbb{Q}$ and t arbitrary; then:

$$\int_{-\infty}^{\infty} |u_i(z,t) - u_n(z,t)|\, dz \leq \int_{-\infty}^{\infty} |u_i(z,t) - u_i(z,t_\ell)|\, dz$$

$$+ \int_{-\infty}^{\infty} |u_i(z,t_\ell) - u_n(z,t_\ell)|\, dz + \int_{-\infty}^{\infty} |u_n(z,t_\ell) - u_n(z,t)|\, dz.$$

Now we apply Lemma 4.10 and obtain

$$\int_{-\infty}^{\infty} |u_i(z,t) - u_n(z,t)|\, dz \leq \int_{-\infty}^{\infty} |u_i(z,t_\ell) - u_n(z,t_\ell)|\, dz + C\,|t - t_\ell|$$

where C is a positive constant. Since $\{u_i(z,t_\ell)\}$ is a convergent sequence, then given $\epsilon > 0$, there exists N_1 such that for $i > N_1$ and $n > N_1$ we have

$$\int_{-\infty}^{\infty} |u_i(z,t_\ell) - u_n(z,t_\ell)|\, dz < \frac{\epsilon}{2}.$$

Also, there exists N_2 such that for $\ell > N_2$ we have $C|t - t_\ell| < \epsilon/2$. Then the following inequality holds:

$$\int_{-\infty}^{\infty} |u_i(z,t) - u_n(z,t)|dz < \epsilon$$

for $i, n > \max(N_1, N_2)$. We have thus proved that the subsequence $\{u_i(z,t)\}$ converges to $\varphi(z,t)$ in the L_1 norm. Moreover, since we assumed that the finite difference scheme is consistent with the conservation law (4.1a) and the entropy inequality (4.18), $\varphi(z,t)$ is the unique weak solution of problem (4.1). This completes the proof. ∎

Since the L-F scheme and the finite difference approximation function obtained therefrom satisfy the hypothesis of Theorem 4.4, we conclude that the L-F scheme is convergent and the limit of the finite difference approximation function is the unique weak solution of the Cauchy problem (4.1). We have thus proved the following existence theorem:

Theorem 4.5 Let $\varphi_I(z)$ be bounded and have bounded total variation; then there exists a unique weak solution $\varphi(z,t)$ of the Cauchy problem (4.1) for all t, $t \geq 0$. For each fixed t, the function $\varphi(z,t)$ is bounded and has bounded total variation.

Chapter 5

The Riemann problem for a scalar conservation law

5.1 Introduction

In this chapter we consider a special initial value problem in which the initial datum is piecewise constant. As in Chapter 4, let $f : [a_1, a_2] \to \mathbb{R}$, $f \in C^3$ be a nonlinear function of φ (f is then, in particular, Lipschitz continuous), and let $\Omega = \{(z, t) | z \in \mathbb{R}, \ t > 0\}$.

Definition 5.1 *The* Riemann problem *is the following initial value problem in which the initial datum consists of two constant states* φ_-, $\varphi_+ \in [a_1, a_2]$:

$$\frac{\partial \varphi}{\partial t} + \frac{\partial f(\varphi)}{\partial z} = 0 \quad \text{in } \Omega, \tag{5.1a}$$

$$\varphi(z, 0) = \begin{cases} \varphi_+ & \text{for } z > 0, \\ \varphi_- & \text{for } z < 0. \end{cases} \tag{5.1b}$$

The sedimentation problem has at least three initial states. However, it is instructive to consider the simpler problem first. Also, a piecewise application of solutions of the Riemann problem can be used to construct solutions of part of the more general problem. As we saw in Chapter 4, the solution of problem (5.1) is discontinuous; therefore we should consider the solution in a generalized sense. Where the solution is continuous, it satisfies the differential equation (5.1a). On the lines of discontinuity $z = z(t)$, the function φ satisfies the Rankine-Hugoniot condition

$$\frac{dz(t)}{dt} = \sigma(\varphi^+, \varphi^-) = \frac{f(\varphi^+) - f(\varphi^-)}{\varphi^+ - \varphi^-} \tag{5.2}$$

and Oleinik's condition E,

$$\frac{f(u) - f(\varphi^-)}{u - \varphi^-} \geq \sigma(\varphi^+, \varphi^-) \geq \frac{f(u) - f(\varphi^+)}{u - \varphi^+} \quad \text{for all } u \in I(\varphi^+, \varphi^-), \tag{5.3}$$

where $I(a, b)$ was defined in (4.6). Here $\varphi^\pm = \varphi(z(t) \pm 0, t)$ are the values of φ at each side of the discontinuity $z = z(t)$. These values should not be confused with φ_+ for $z > 0$ and φ_- for $z < 0$ used as constant states for the initial value (5.1b). We also saw in Chapter 4 that Oleinik's condition E implies Lax's shock admissibility criterion,

$$f'(\varphi^-) \geq \sigma(\varphi^+, \varphi^-) \geq f'(\varphi^+). \tag{5.4}$$

In what follows, we construct the global weak solution of problem (5.1)) for a convex flux density function f, a function f with one inflection point and finally a function f with two inflection points.

5.2   The Riemann problem for a convex flux density function

Consider a convex function f in (5.1a), i.e. $f''(\varphi) > 0$ for all $\varphi \in [a_1, a_2]$.

Theorem 5.1 *For a convex flux density function f, the Riemann problem (5.1) has an explicit global weak solution.*

Proof. The global weak solution is constructed in such a way that it satisfies the initial condition (5.1b), the differential equation (5.1a) at the points of continuity and the Rankine-Hugoniot condition (5.2) and Oleinik's condition E (5.3) at discontinuities. This can be checked by direct substitution. Two types of solution can be obtained depending on the relative values of φ_+ and φ_-.

(a) **The value of φ_+ is less than φ_-**

   Consider $\varphi_+ < \varphi_-$ as shown in the graph of f vs φ (see Figure 5.1 (a), where the tangents at $(\varphi_+, f(\varphi_+))$ and $(\varphi_-, f(\varphi_-))$ are also drawn). As we saw in Chapter 4, the solution φ of problem (5.1) is constant along the characteristics with slope $f'(\varphi)$. To construct the solution, draw the characteristics starting at the axis $t = 0$ in the (z, t) plane as follows: For $z > 0$, the characteristics satisfy the equation $z = x + f'(\varphi_+)t$ for $0 < x$. They are parallel to the tangent of f at $(\varphi_+, f(\varphi_+))$, which has a slope equal to $f'(\varphi_+)$. For $z < 0$, the characteristics satisfy the equation $z = x + f'(\varphi_-)t$ for $0 > x$. They are parallel to the tangent of f at $(\varphi_-, f(\varphi_-))$, which has a slope equal to $f'(\varphi_-)$. When we draw these characteristics, there is a region between the lines $z = f'(\varphi_+)t$ and $z = f'(\varphi_-)t$ in the (z, t) plane where the characteristics cross and the solution has two values, namely $\varphi = \varphi_+$ and $\varphi = \varphi_-$, as shown in Figure 5.1 (b). To overcome this difficulty, before drawing the characteristics, we trace a line of discontinuity $z_1(t)$ where the solution jumps from the value φ_+ to φ_-. From the Rankine-Hugoniot condition (5.2), we obtain the equation $z_1(t) = \sigma(\varphi_+, \varphi_-)t$ for the discontinuity, where $\sigma(\varphi_+, \varphi_-)$ is the slope of the chord that joins the points $(\varphi_-, f(\varphi_-))$

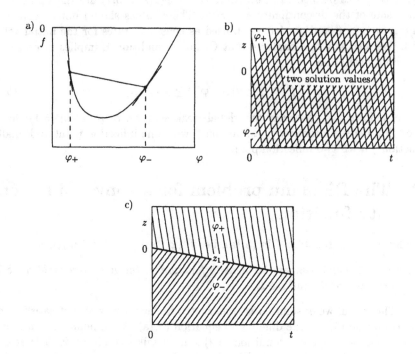

Figure 5.1: a) Graph of f vs φ, b) intersection of characteristics and c) representation of the global weak solution $\varphi(z,t)$ for $\varphi_+ < \varphi_-$.

and $(\varphi_+, f(\varphi_+))$ on the graph of f vs φ. The function f lies on one side of the chord, which implies that the discontinuity $z_1(t)$ satisfies Oleinik's condition E (5.3). Besides, $z_1(t)$ also satisfies Lax's criterion (5.4) and it is therefore a *shock*. Now we can draw the characteristics for $t \geq 0$, $z \geq z_1(t)$ with slope $f'(\varphi_+)$. Along these characteristics the solution, which has a constant value $\varphi = \varphi_+$, is continuous and satisfies the differential equation (5.1a) and the initial condition (5.1b) for $z > 0$. Then, for $t \geq 0$, $z \leq z_1(t)$, we draw the characteristics with slope $f'(\varphi_-)$. Along these characteristics the solution, which has a constant value $\varphi = \varphi_-$, is continuous and satisfies the differential equation (5.1a) and the initial condition (5.1b) for $z < 0$. The solution is a *shock wave*, two constant states separated by a shock:

$$\varphi(z,t) = \begin{cases} \varphi_+ & \text{for } z_1(t) \leq z, \\ \varphi_- & \text{for } z < z_1(t). \end{cases} \tag{5.5}$$

A representation of the solution is shown in Figure 5.1 (c).

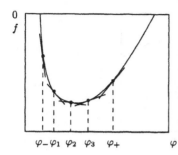 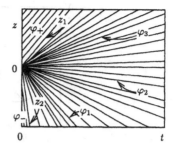

Figure 5.2: Graph of f vs φ and representation of the global weak solution $\varphi(z,t)$ for $\varphi_- < \varphi_+$. The values φ_1, φ_2 and φ_3 correspond to the marked rays in the rarefaction wave.

Remark In this construction we see the advantage of selecting the z axis pointing upward and the t axis pointing to the right. The characteristics and shocks are drawn easily by taking them parallel to the corresponding tangents and chords in the f vs φ graph.

(b) **The value of φ_- is less than φ_+**

To construct the solution, draw the characteristics (starting at the line $t = 0$ on the (z,t) plane), with equations $z = x + f'(\varphi_+)t$ for $x > 0$ and $z = x + f'(\varphi_-)t$ for $x < 0$. Denote by $z_1(t)$ and $z_2(t)$ the two characteristics that start at the origin and have slopes of $f'(\varphi_+)$ and $f'(\varphi_-)$ respectively, i.e. $z_1(t) = f'(\varphi_+)t$ and $z_2(t) = f'(\varphi_-)t$. Since the slope of $z_1(t)$ is greater than the slope of $z_2(t)$, a gap develops between the lines $z_1(t)$ and $z_2(t)$. This gap is filled with a fan of rays centered at the origin of the (z,t) plane with slopes equal to $f'(\varphi)$ for $\varphi_- \leq \varphi \leq \varphi_+$. Let h be the inverse of f'; then the solution of problem (5.1) is

$$\varphi(z,t) = \begin{cases} \varphi_+ & \text{for} \quad z_1(t) \leq z, \\ h(z/t) & \text{for} \quad z_2(t) \leq z < z_1(t), \\ \varphi_- & \text{for} \quad z \leq z_2(t), \end{cases} \qquad (5.6)$$

which can be checked by direct substitution. The solution is continuous and we say the φ_+ is connected to φ_- by a *rarefaction wave*. Both the function f and a representation of the global weak solution are shown in Figure 5.2. This completes the proof of the theorem. ∎

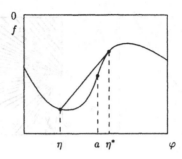

Figure 5.3: Graph of f vs φ showing η and η^*.

5.3   The Riemann problem for a flux density function with one inflection point

Let the function f have one inflection point a. Without loss of generality, we assume that $f''(\varphi) > 0$ for $\varphi < a$ and $f''(\varphi) < 0$ for $\varphi > a$.

5.3.1   Properties of the flux density function

In what follows, we state some properties of the flux density function (Ballou 1970, Cheng 1981) before solving the Riemann problem.

Definition 5.2 *Given $\eta < a$, define $\eta^* = \eta^*(\eta)$ by*

$$\eta^* = \sup\{u > \eta : \sigma(\eta, v) < \sigma(\eta, u) \quad \text{for all} \quad v \in (\eta, u)\}. \tag{5.7}$$

The location of η^* is shown in Figure 5.4. Given $\mu > a$, define $\mu_* = \mu_*(\mu)$ by

$$\mu_* = \inf\{u < \mu : \sigma(v, \mu) < \sigma(u, \mu) \text{ for all } v \in (u, \mu)\}. \tag{5.8}$$

See Figure 5.5.

The sets indicated above are not empty, since they contain a. Note that both $\eta^* = +\infty$ and $\mu_* = -\infty$ are possible. Here σ is as usual

$$\sigma(u, v) = \frac{f(u) - f(v)}{u - v}.$$

Definition 5.3 *Let $\beta > a$, then $\beta^{**} < a$ is the unique number satisfying $\beta = (\beta^{**})^*$.*

This is shown in Figure 5.5.

Lemma 5.1 *Let $\eta < a$, then*

$$f'(\eta) < \sigma(\eta, u) < f'(u) \text{ for all } u \in (\eta, \eta^*). \tag{5.9}$$

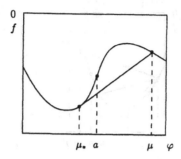

Figure 5.4: Graph of f vs φ showing μ and μ_*.

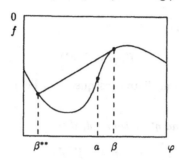

Figure 5.5: Graph of f vs φ showing β and β^{**}.

This inequality is illustrated in Figure 5.6.

Proof. To prove the first inequality, let $u \in (\eta, \eta^*)$; then by Definition 5.2, we have

$$\sigma(\eta, v) < \sigma(\eta, u) \text{ for all } v \in (\eta, u). \tag{5.10}$$

Taking the limit of inequality (5.10) when $v \to \eta$, we obtain $\sigma(\eta, v) \to f'(\eta)$, so that $f'(\eta) < \sigma(\eta, u)$. To prove the second inequality of (5.9), we rewrite inequality (5.10) as $\sigma(\eta, u) < \sigma(v, u)$. Taking the limit when $v \to u$, we obtain $\sigma(v, u) \to f'(u)$ and the lemma is proved. ■

Corollary 5.1 *For $\eta < a$,*

$$\frac{d\sigma(\eta, u)}{du} > 0 \text{ for all } u \in (\eta, \eta^*). \tag{5.11}$$

Proof. Taking the derivative of $\sigma(\eta, u)$ with respect to u, we get

$$\frac{d\sigma(\eta, u)}{du} = \frac{1}{u - \eta}(f'(u) - \sigma(\eta, u)).$$

Since $\eta < u$ and the expression in parenthesis is greater than zero, by Lemma 5.1, the corollary is proved. ■

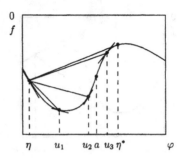

Figure 5.6: Graph of f vs φ showing properties stated in Lemma 5.1.

Corollary 5.2 *For $\eta < a$,*

$$f'(\eta) < f'(\eta^*). \tag{5.12}$$

Proof. This result is an immediate consequence of Lemma 5.1 and Corollary 5.1. ∎

Lemma 5.2 *Let $\eta < a$ and η^* be bounded, then*

$$\sigma(\eta, \eta^*) = f'(\eta^*). \tag{5.13}$$

Proof. Since σ and f' are continuous functions, Lemma 5.1 implies that $\sigma(\eta, \eta^*) \leq f'(\eta^*)$. If the inequality were not satisfied, then there would exist $\bar{\eta} > \eta^*$ for which $\sigma(\eta, \bar{\eta}) = f'(\bar{\eta})$. Therefore

$$\frac{d\sigma(\eta, u)}{du} > 0 \qquad \text{for all} \quad u \in (\eta, \bar{\eta})$$

and then $\sigma(\eta, u) < \sigma(\eta, \bar{\eta})$ for all $u \in (\eta, \bar{\eta})$, which implies that $\bar{\eta} \leq \eta^*$ by Definition 5.2. We have arrived at a contradiction that proves the assertion. ∎

Lemma 5.3 *Let η_1 and η_2 be given, with $\eta_1 < \eta_2 < a$. If $\eta_2^* > a$ is bounded, then $\eta_2^* < \eta_1^*$.*

Proof. Suppose η_1^* is bounded and assume that $\eta_1^* \leq \eta_2^*$; see Figure 5.7. Since $\eta_1^* \in (\eta_2, \eta_2^*)$, we obtain from Lemma 5.1 and 5.2 that

$$\sigma(\eta_2, \eta_1^*) \leq f'(\eta_1^*) = \sigma(\eta_1, \eta_1^*). \tag{5.14}$$

But, since the graph of f is completely under the chord $\sigma(\eta_1, \eta_1^*)$ then, for $\eta_2 > \eta_1$, we have that $\sigma(\eta_1, \eta_1^*) < \sigma(\eta_2, \eta_1^*)$. Combining this last inequality with (5.14) yields

$$\sigma(\eta_2, \eta_1^*) \leq \sigma(\eta_1, \eta_1^*) < \sigma(\eta_2, \eta_1^*),$$

a contradiction that proves the lemma. ∎

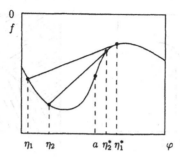

Figure 5.7: Graph of f vs φ showing properties stated in Lemma 5.3.

5.3.2 Construction of the global weak solution

Next we construct the global weak solution for the Riemann problem for different values of φ_+ and φ_-.

Theorem 5.2 *The Riemann problem* (5.1) *has an explicit global weak solution if the inverse functions* h_1 *and* h_2 *of* f', *restricted to the intervals* $(-\infty, a)$ *and* (a, ∞), *respectively, exist.*

Proof. The global weak solution is constructed in such a way that it satisfies the initial condition (5.1b), the differential equation (5.1a) at the points of continuity and the Rankine-Hugoniot condition (5.2) and Oleinik's condition E (5.3) at points of discontinuity. This can be checked by direct substitution.

(a) **The value of φ_+ is less than φ_-**

Three types of solutions can be obtained depending on the relative positions of φ_+ and φ_- with respect to the graph of f. We analyze each case separately.

Case I: $\varphi_+ < a$ and $\varphi_+ < \varphi_- < \varphi_+^*$

The global weak solution is

$$\varphi(z, t) = \begin{cases} \varphi_+ & \text{for } z_1(t) \leq z, \\ \varphi_- & \text{for } z < z_1(t), \end{cases} \qquad (5.15)$$

where $z_1(t) = \sigma(\varphi_+, \varphi_-)t$. The states φ_+ and φ_- are separated by the discontinuity $z_1(t)$. This discontinuity is a shock, since it satisfies Oleinik's condition E (5.3): the chord $\sigma(\varphi_+, \varphi_-)$ lies completely on one side of the graph of f and it satisfies Lax's criterion (5.4), both inequalities holding. This is shown in Figure 5.8.

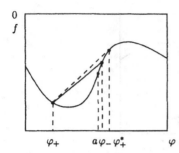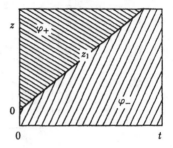

Figure 5.8: Graph of f vs φ and representation of the global weak solution $\varphi_+ < a$ and $\varphi_+ < \varphi_- < \varphi_+^*$.

Case II: $\varphi_+ < a < \varphi_+^* < \varphi_-$

The global weak solution is

$$\varphi(z,t) = \begin{cases} \varphi_+ & \text{for} \quad z_1(t) \leq z, \\ h_2(z/t) & \text{for} \quad z_2(t) \leq z < z_1(t), \\ \varphi_- & \text{for} \quad z < z_2(t), \end{cases} \qquad (5.16)$$

with

$$z_1(t) = f'(\varphi_+^*)t = \sigma(\varphi_+, \varphi_+^*)t$$

and $z_2(t) = f'(\varphi_-)t$. The states φ_+ and φ_- are separated by the discontinuity $z_1(t)$ and by a rarefaction wave as can be seen in Figure 5.9. The line $z_1(t)$ is a contact discontinuity since it satisfies Oleinik's condition E (5.3): the chord $\sigma(\varphi_+, \varphi_+^*)$ lies completely on one side of the graph of f. Besides, by Corollary 5.2 and Lemma 5.2 we have that

$$f'(\varphi_+) < \sigma(\varphi_+, \varphi_+^*) = f'(\varphi_+^*),$$

that is, Lax's criterion (5.4) is satisfied with one equality holding, which is the definition of a *contact discontinuity*. The line $z_2(t)$ is a line of continuity.

Case III: $a < \varphi_+ < \varphi_-$

The global weak solution is

$$\varphi(z,t) = \begin{cases} \varphi_+ & \text{for} \quad z_1(t) \leq z, \\ h_2(z/t) & \text{for} \quad z_2(t) \leq z < z_1(t), \\ \varphi_- & \text{for} \quad z < z_2(t). \end{cases} \qquad (5.17)$$

Here both $z_1(t) = f'(\varphi_+)t$ and $z_2(t) = f'(\varphi_-)t$ are lines of continuity. The solution is continuous and consists of the two states φ_+ and φ_- separated by a rarefaction wave. See Figure 5.10.

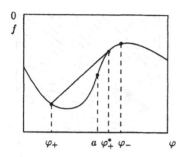
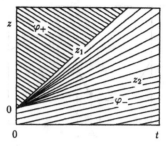

Figure 5.9: Graph of f vs φ and representation of the global weak solution $\varphi(z,t)$ for $\varphi_+ < a < \varphi_+^* < \varphi_-$.

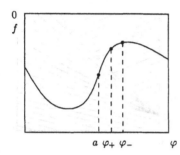
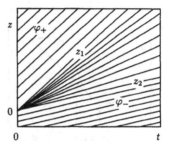

Figure 5.10: Graph of f vs φ and representation of the global weak solution $\varphi(z,t)$ for $a < \varphi_+ < \varphi_-$.

(b) The value of φ_+ is greater than φ_-

Again we obtain three types of solutions depending on the relative positions of φ_+ and φ_- with respect to the graph of f.

Case IV: $\varphi_- < \varphi_+ < a$

The global weak solution is

$$\varphi(z,t) = \begin{cases} \varphi_+ & \text{for} \quad z_1(t) \leq z, \\ h_1(z/t) & \text{for} \quad z_2(t) \leq z < z_1(t), \\ \varphi_- & \text{for} \quad z < z_2(t), \end{cases} \tag{5.18}$$

where $z_1(t) = f'(\varphi_+)t$ and $z_2(t) = f'(\varphi_-)t$ are lines of continuity. The solution is continuous and consists of two constant states φ_+ and φ_- separated by a rarefaction wave. This shown in Figure 5.11.

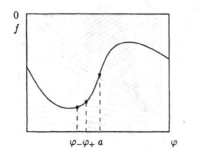 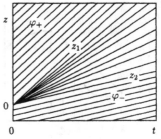

Figure 5.11: Graph of f vs φ and representation of the global weak solution $\varphi(z,t)$ for $\varphi_- < \varphi_+ < a$.

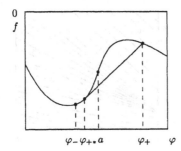 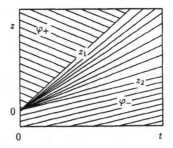

Figure 5.12: Graph of f vs φ and representation of the global weak solution $\varphi(z,t)$ for $\varphi_- < \varphi_{+*} < a < \varphi_+$.

Case V: $\varphi_- < \varphi_{+*} < a < \varphi_+$

The global weak solution is

$$\varphi(z,t) = \begin{cases} \varphi_+ & \text{for} \quad z_1(t) \le z, \\ h_1(z/t) & \text{for} \quad z_2(t) \le z < z_1(t), \\ \varphi_- & \text{for} \quad z < z_2(t) \end{cases} \qquad (5.19)$$

with

$$z_1(t) = f'(\varphi_{+*})t = \sigma(\varphi_{+*}, \varphi_+)t$$

and $z_2(t) = f'(\varphi_-)t$. This solution consists of two constant states φ_- and φ_+ separated by the contact discontinuity $z_1(t)$ and a rarefaction wave as can be seen in Figure 5.12. The proof that $z_1(t)$ is a contact discontinuity is similar to the one given in Case II. The line $z_2(t)$ is a line of continuity.

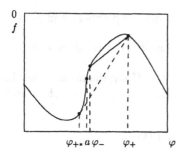
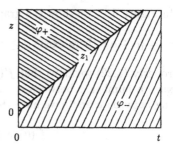

Figure 5.13: Graph of f vs φ and representation of the global weak solution $\varphi(z,t)$ for $a < \varphi_+$ and $\varphi_{+*} < \varphi_- < \varphi_+$.

Case VI: $a < \varphi_+$ and $\varphi_{+*} < \varphi_- < \varphi_+$

The global weak solution is

$$\varphi(z,t) = \begin{cases} \varphi_+ & \text{for} \quad z_1(t) \leq z, \\ \varphi_- & \text{for} \quad z < z_1(t). \end{cases} \tag{5.20}$$

Here $z_1(t) = \sigma(\varphi_-, \varphi_+)t$. The states φ_+ and φ_- are separated by the line $z_1(t)$, which is a shock since the graph of f lies on one side of the chord $\sigma(\varphi_-, \varphi_+)$ and Lax' criterion (5.4) is satisfied with inequalities holding. See Figure 5.13. This completes the proof of the theorem. ■

5.4 The Riemann problem for a flux density function with two inflection points (Cheng 1981, 1986)

Let the function f possess two inflection points $a < b$. We assume without loss of generality that

$$f''(\varphi) > 0 \quad \text{for } \varphi < a \text{ or } \varphi > b, \tag{5.21}$$

$$f''(\varphi) < 0 \quad \text{for } a < \varphi < b, \tag{5.22}$$

$$\lim_{\varphi \to -\infty} f(\varphi) = \lim_{\varphi \to +\infty} f(\varphi) = +\infty. \tag{5.23}$$

5.4.1 Geometrical properties of the flux density function

Lemma 5.4 *Let f be a function with two inflection points that satisfies (5.21)–(5.23), then there exist numbers c_1 and c_2 satisfying*

$$f'(c_1) = \sigma(c_1, c_2) = f'(c_2).$$

Proof. In Figure 5.14 we show the graph of f and the points c_1 and c_2. Let

$$S = \{c \le a : \text{ there exists } d \ge b \text{ such that } f'(c) > \sigma(c,d) > f'(d)\}.$$

From property (5.22) we know that $f'(a) > f'(b)$. Using the mean value theorem, there exists a number x between a and b such that $f'(x) = \sigma(a,b)$. Since f' is decreasing between a and b we have that

$$f'(a) > f'(x) > f'(b)$$

and hence

$$f'(a) > \sigma(a,b) > f'(b).$$

Therefore $a \in S$. Let $c_1 = \inf S$. Then, by property (5.23), c_1 is finite. Let $\{C_n\}_{n=1}^\infty$ be a sequence in S such that $\lim_{n \to \infty} C_n = c_1$. Let $\{d_n\}_{n=1}^\infty$, $d_n > b$ for all $n = 1, 2, \ldots$ be a sequence associated with $\{C_n\}_{n=1}^\infty$ by

$$f'(C_n) > \sigma(C_n, d_n) > f'(d_n).$$

By property (5.23), the sequence $\{d_n\}_{n=1}^\infty$ is bounded; then let $\lim_{n \to \infty} d_n = c_2$. Assume that $f'(c_1) > \sigma(c_1, c_2)$. Then by property (5.21), there exists $\epsilon > 0$ small enough such that

$$f'(c_1) > f'(c_1 - \epsilon) > \sigma(c_1 - \epsilon, c_2) > \sigma(c_1, c_2) \ge f'(c_2).$$

This contradicts the definition of $c_1 = \inf S$. Now assume that $\sigma(c_1, c_2) > f'(c_2)$. Then there exists $\epsilon > 0$ small enough such that

$$f'(c_1) \ge \sigma(c_1, c_2) > \sigma(c_1, c_2 + \epsilon) > f'(c_2 + \epsilon) > f'(c_2).$$

Then there exists $\delta > 0$ small enough such that

$$f'(c_1) > f'(c_1 - \delta) > \sigma(c_1 - \delta, c_2 + \epsilon) > \sigma(c_1, c_2 + \epsilon) > f'(c_2 + \epsilon).$$

This implies that $c_1 - \delta \in S$ which contradicts the definition of c_1. Therefore $\sigma(c_1, c_2) = f'(c_2)$, which completes the proof. ■

Lemma 5.5 *For $b < \alpha < c_2$ there exists a unique number $\tilde{\alpha} < b$ with*

$$f'(\tilde{\alpha}) \ge \sigma(\tilde{\alpha}, \alpha) = f'(\alpha).$$

Proof. The tangent line $y = f'(\alpha)\varphi + m$ at $(\alpha, f(\alpha))$ intersects the graph of f in two points, in $(d, f(d))$ and in $(\tilde{\alpha}, f(\tilde{\alpha}))$ with $d < c_1$ and $c_1 < \tilde{\alpha}$, see Figure 5.15. Since $\alpha \in (b, c_2)$ and $f'' > 0$ in this interval, the tangent to the graph at $(\alpha, f(\alpha))$ remains under the graph of f in the interval $(\tilde{\alpha}, \alpha)$, therefore $f(\varphi) \ge f'(\alpha)\varphi + m$ for $\tilde{\alpha} \le \varphi \le \alpha$.

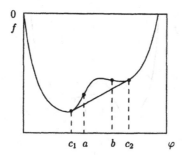

Figure 5.14: Graph of the function f with two inflection points, showing c_1 and c_2.

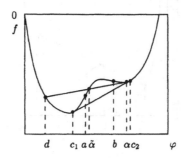

Figure 5.15: Graph of the function f showing $\tilde{\alpha}$, α and d.

(1) Suppose that $\tilde{\alpha} \in (a, b)$. In this interval, $f'' < 0$, hence the tangent line $y = f'(\tilde{\alpha})\varphi + n$ remains over the graph of f in the interval (a, b). Moreover, $f'(\tilde{\alpha})\varphi + n \geq f'(a)\varphi + m$ for φ between $\tilde{\alpha}$ and b. Since both straight lines intersect at the point $(\tilde{\alpha}, f(\tilde{\alpha}))$, the linear function

$$g(\varphi) = (f'(\tilde{\alpha}) - f'(a))\varphi + (n - m) \geq 0$$

is increasing and therefore $f'(\tilde{\alpha}) - f'(a) \geq 0$, which proves the assertion.

(2) Suppose $\tilde{\alpha} \in (c_1, a)$. By property (5.21) and Lemma 5.4 we have that

$$f'(\tilde{\alpha}) > f'(c_1) = f'(c_2) > f'(\alpha).$$

This completes the proof. ■

Definition 5.4 *Let $\delta < b$, then $\delta^{\ddagger} > b$ is the unique number satisfying $\delta = \widetilde{(\delta^{\ddagger})}$.*

See Figure 5.16.

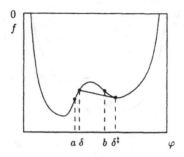

Figure 5.16: Graph of f vs φ showing δ and $\delta^{\ddagger}$.

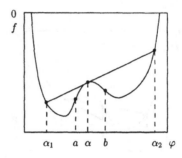

Figure 5.17: Graph of the function f showing α_1, α and α_2.

Lemma 5.6 *For $a < \alpha < b$ there exist unique numbers $\alpha_1 < a$ and $\alpha_2 > b$ such that*

$$f'(\alpha_1) < \sigma(\alpha_1, \alpha) = f'(\alpha) = \sigma(\alpha, \alpha_2) < f'(\alpha_2). \qquad (5.24)$$

Proof. Let $y = f'(\alpha)\varphi + m$ be the tangent line to the graph of f at the point $(\alpha, f(\alpha))$, see Figure 5.17. Since $f'' < 0$ between a and b, this tangent line lies above the graph of f between a and b and then cuts the graph in two points, in $(\alpha_1, f(\alpha_1))$ and in $(\alpha_2, f(\alpha_2))$ with $\alpha_1 < a$ and $\alpha_2 > b$. Then by construction $f'(\alpha) = \sigma(\alpha_1, \alpha) = \sigma(\alpha, \alpha_2)$.

(1) We prove that

$$f'(\alpha_1) < f'(\alpha) = \sigma(\alpha_1, \alpha). \qquad (5.25)$$

First we have that

$$f(\varphi) \le f'(\alpha)\varphi + m \quad \text{for } \alpha_1 \le \varphi \le \alpha. \qquad (5.26)$$

On the other hand, the tangent $y = f'(\alpha_1)\varphi + n$ at $(\alpha_1, f(\alpha_1))$ lies under the graph of f between α_1 and a since in this interval $f'' > 0$; then

$$f(\varphi) \ge f'(\alpha_1)\varphi + n \quad \text{for } \alpha_1 \le \varphi \le a. \qquad (5.27)$$

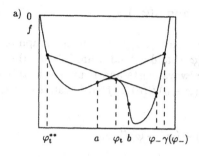 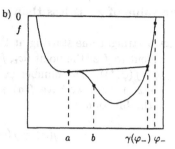

Figure 5.18: Illustration of the two possible cases of Definition 5.5.

From (5.26), (5.27) and the fact that both tangent lines intersect at $(\alpha_1, f(\alpha_1))$, the linear function

$$g(\varphi) = (f'(\alpha) - f'(\alpha_1))\varphi + m - n \geq 0 \quad \text{for } \alpha_1 \leq \varphi \leq a$$

is increasing and therefore $f'(\alpha) > f'(\alpha_1)$, which proves (5.25).

(2) The second inequality is proved similarly. This completes the proof. ∎

5.4.2 Construction of the global weak solution

To solve the different Riemann problems, we consider only the cases $\varphi_+ > b$ or $\varphi_- > b$; the solutions of the Riemann problems of the remaining cases, in which there is at most one of the inflection points between φ_- and φ_+, correspond to the solutions for a flux density function with one inflection point which were determined in the previous section. The following definition is relevant to the case $\varphi_- > b$.

Definition 5.5 *For $\varphi_+ < a < b < \varphi_-$ let*

$$\gamma(\varphi_-) = \min\{\varphi_-\} \cup \{\varphi > a : f(a) + (\varphi - a)f'(a) \leq f(\varphi)\}.$$

Obviously, the cases (i) $\gamma(\varphi_-) = \varphi_-$ and (ii) $\gamma(\varphi_-) < \varphi_-$ are possible; see Figure 5.18.

Theorem 5.3 *If the inverse functions h_3, h_4 and h_5 of f' restricted to the respective intervals $(-\infty, a)$, (a, b) and (b, ∞) exist, the Riemann problem (5.1) has an explicit global weak solution.*

Proof. The global weak solution is constructed in a way which is similar to those of Theorem 5.2. This solution satisfies the initial condition (5.1b) and the differential equation (5.1a) at points of continuity. At points of discontinuity, it satisfies the Rankine-Hugoniot condition (5.2) and Oleinik's condition E (5.3). Therefore, it is the unique weak solution of each Riemann problem considered.

(a) **The value of φ_+ is less than φ_- and $\gamma(\varphi_-) = \varphi_-$**

Draw a straight line starting at the point $(\varphi_-, f(\varphi_-))$ in the graph of f vs φ, tangent to f at the point $(\varphi_t, f(\varphi_t))$ and extend it until it cuts the curve at $(\varphi_t^{**}, f(\varphi_t^{**}))$. The number φ_t always exists and is unique in this case: from $\gamma(\varphi_-) = \varphi_-$ we have $f(a) + (\varphi - a)f'(a) > f(\varphi)$ for all $a < \varphi < \varphi_-$, hence the function

$$g(\varphi) = f(\varphi) + (\varphi_- - \varphi)f'(\varphi)$$

satisfies $g(a) \geq f(\varphi_-)$. Moreover,

$$g'(\varphi) = (\varphi_- - \varphi)f''(\varphi) < 0 \text{ for } \varphi \in (a, b),$$

and, since $f''(\varphi) > 0$ for $\varphi > b$,

$$g(b) = f(b) + (\varphi_- - b)f'(b) \leq f(\varphi_-).$$

This implies that there exists a unique number $\varphi_t \in [a, b]$ satisfying

$$g(\varphi_t) = f(\varphi_t) + (\varphi_- - \varphi_t)f'(\varphi_t) = f(\varphi_-);$$

hence

$$f'(\varphi_t) = \frac{f(\varphi_-) - f(\varphi_t)}{\varphi_- - \varphi_t} = \sigma(\varphi_t, \varphi_-).$$

By Lemma 5.6 we have

$$f'(\varphi_t^{**}) < \sigma(\varphi_t^{**}, \varphi_t) = f'(\varphi_t) = \sigma(\varphi_t, \varphi_-) < f'(\varphi_-). \qquad (5.28)$$

We obtain four types of solutions depending on the relative position of φ_+ and φ_- with respect to the graph of f.

Case I: $\varphi_+ < \varphi_t^{**} < a < b < \varphi_-$

The global weak solution is

$$\varphi(z, t) = \begin{cases} \varphi_+ & \text{for } z \geq z_1(t), \\ \varphi_- & \text{for } z < z_1(t), \end{cases} \qquad (5.29)$$

where $z_1(t) = \sigma(\varphi_+, \varphi_-)t$. The states φ_+ and φ_- are separated by the discontinuity z_1. This discontinuity is a shock since it satisfies condition E (5.3): the chord $\sigma(\varphi_+, \varphi_-)$ lies completely on one side of the graph of f and Lax's criterion (5.4) with both inequalities holding. This is shown in Figure 5.19.

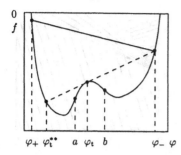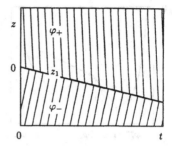

Figure 5.19: Graph of f vs φ and representation of the global weak solution $\varphi(z,t)$ for $\varphi_+ < \varphi_t^{**}$.

Case II: $\varphi_t^{**} < \varphi_+ < a < b < \varphi_-$

The global weak solution is

$$\varphi(z,t) = \begin{cases} \varphi_+ & \text{for} & z_1(t) \leq z, \\ h_4(z/t) & \text{for} & z_2(t) \leq z < z_1(t), \\ \varphi_- & \text{for} & z < z_2(t). \end{cases} \tag{5.30}$$

Here

$$z_1(t) = f'(\varphi_+^*)t = \sigma(\varphi_+, \varphi_+^*)t$$

and

$$z_2(t) = f'(\varphi_t)t = \sigma(\varphi_t, \varphi_-)t.$$

This solution consists of two constant states φ_- and φ_+ separated by two contact discontinuities $z_1(t)$ and $z_2(t)$ and by a rarefaction wave as can be seen in Figure 5.20. The line $z_1(t)$ is a contact discontinuity since it satisfies Oleinik's condition E (5.3): the chord $\sigma(\varphi_+, \varphi_+^*)$ lies completely on one side of the graph of f. Besides, by Corollary 5.2 and Lemma 5.2 we have that

$$f'(\varphi_+) < \sigma(\varphi_+, \varphi_+^*) = f'(\varphi_+^*).$$

That is, Lax's criterion (5.4) is satisfied with one equality holding, which is the definition of a contact discontinuity. The line $z_2(t)$ is also a contact discontinuity since it satisfies Oleinik's condition E: the chord $\sigma(\varphi_+^*, \varphi_-)$ lies completely on one side of the graph of f. Besides, by Lemma 5.6 we have that

$$f'(\varphi_t) = \sigma(\varphi_t, \varphi_-) < f'(\varphi_-),$$

that is, Lax's criterion is satisfied with one equality holding.

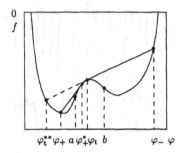
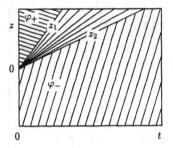

Figure 5.20: Graph of f vs φ and representation of the global weak solution $\varphi(z,t)$ for $\varphi_t^{**} < \varphi_+ < a$.

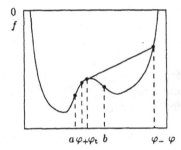
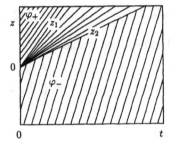

Figure 5.21: Graph of f vs φ and representation of the global weak solution $\varphi(z,t)$ for $a < \varphi_+ < \varphi_t$.

Case III: $a < \varphi_+ < \varphi_t < \varphi_-$

The global weak solution is

$$\varphi(z,t) = \begin{cases} \varphi_+ & \text{for} & z_1(t) \leq z, \\ h_4(z/t) & \text{for} & z_2(t) \leq z < z_1(t), \\ \varphi_- & \text{for} & z < z_2(t). \end{cases} \tag{5.31}$$

Here, $z_1(t) = f'(\varphi_+)t$ is a line of continuity and

$$z_2(t) = f'(\varphi_t)t = \sigma(\varphi_t, \varphi_-)t$$

is a discontinuity. This solution consists of two constant states φ_- and φ_+ separated by a rarefaction wave and by the contact discontinuity $z_2(t)$, which is the same one as in Case II. See Figure 5.21.

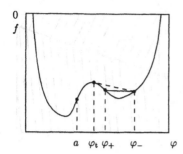 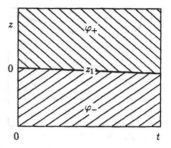

Figure 5.22: Graph of f vs φ and representation of the global weak solution $\varphi(z,t)$ for $\varphi_t < \varphi_+ < \varphi_-$.

Case IV: $\varphi_t < \varphi_+ < \varphi_-$

The global weak solution is

$$\varphi(z,t) = \begin{cases} \varphi_+ & \text{for} \quad z_1(t) \leq z, \\ \varphi_- & \text{for} \quad z < z_1(t), \end{cases} \tag{5.32}$$

where $z_1(t) = \sigma(\varphi_+, \varphi_-)t$. The states φ_+ and φ_- are separated by the discontinuity $z_1(t)$. This discontinuity is a shock since it satisfies condition E (5.3), the chord $\sigma(\varphi_+, \varphi_-)$ lies completely on one side of the graph of f and Lax's criterion (5.4) is satisfied with both inequalities holding. This is shown in Figure 5.22.

(b) **The value of φ_+ is less than φ_- and $\gamma(\varphi_-) < \varphi_-$**

It is easy to see that in this case the tangential point constructed in (a) does not exist. We obtain one type of solution.

Case V: $\varphi_+ < a < b < \gamma(\varphi_-) < \varphi_-$

The global weak solution is

$$\varphi(z,t) = \begin{cases} \varphi_+ & \text{for} \quad z \geq z_1(t), \\ \varphi_- & \text{for} \quad z < z_1(t), \end{cases} \tag{5.33}$$

where $z_1(t) = \sigma(\varphi_+, \varphi_-)t$ is a discontinuity separating the states φ_+ and φ_-. The discontinuity is a shock since it satisfies condition E (5.3), the chord $\sigma(\varphi_+, \varphi_-)$ lies completely on one side of the graph of f and Lax's criterion (5.4) is satisfied with both inequalities holding. This is shown in Figure 5.23.

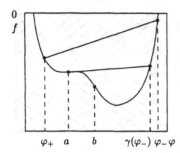
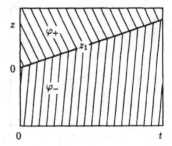

Figure 5.23: Graph of f vs φ and representation of the global weak solution $\varphi(z,t)$ for $\varphi_t < \varphi_+ < \varphi_-$.

(c) The value of φ_+ is greater than φ_-

Three types of solutions are obtained depending on the relative position of φ_+ and φ_- with respect to the graph of f.

Case VI: $\varphi_- < c_1 < c_2 < \varphi_+$

The global weak solution is

$$\varphi(z,t) = \begin{cases} \varphi_+ & \text{for} \quad z_1(t) \leq z, \\ h_5(z/t) & \text{for} \quad z_2(t) \leq z < z_1(t), \\ h_3(z/t) & \text{for} \quad z_3(t) \leq z < z_2(t) \\ \varphi_- & \text{for} \quad z < z_3(t). \end{cases} \tag{5.34}$$

Here $z_1(t) = f'(\varphi_+)t$ is a line of continuity,

$$z_2(t) = f'(c_2)t = \sigma(c_1, c_2)t = f'(c_1)t$$

and $z_3(t) = f'(\varphi_-)t$ which is also a line of continuity. This solution consists of two constant states φ_- and φ_+ separated by a rarefaction wave and by the discontinuity $z_2(t)$. See Figure 5.24. The discontinuity $z_2(t)$ is a *double contact discontinuity* because it satisfies Oleinik's condition E since the chord $\sigma(c_1, c_2)$ is on one side of the graph of f and furthermore, from Lemma 5.4, it satisfies Lax's criterion with both equalities holding.

Case VII: $c_1 < \varphi_- < b < c_2 < \varphi_+$

The global weak solution is

$$\varphi(z,t) = \begin{cases} \varphi_+ & \text{for} \quad z_1(t) \leq z, \\ h_5(z/t) & \text{for} \quad z_2(t) \leq z < z_1(t), \\ \varphi_- & \text{for} \quad z < z_2(t). \end{cases} \tag{5.35}$$

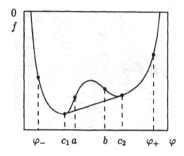
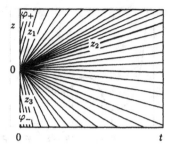

Figure 5.24: Graph of f vs φ and representation of the global weak solution $\varphi(z,t)$ for $\varphi_- < c_1 < \varphi_+$.

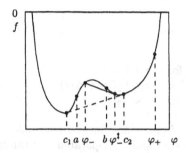
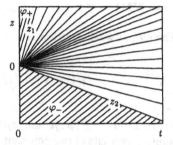

Figure 5.25: Graph of f vs φ and representation of the global weak solution $\varphi(z,t)$ for $c_1 < \varphi_- < b < \varphi_+$.

Here $z_1(t) = f'(\varphi_+)t$ is a line of continuity and

$$z_2(t) = f'(\varphi_-^\dagger)t = \sigma(\varphi_-, \varphi_-^\dagger)t.$$

This solution consists of two constant states φ_- and φ_+ separated by a rarefaction wave and by the discontinuity $z_2(t)$. This discontinuity is a contact discontinuity because, first, the graph of f stays on one side of the chord $\sigma(\varphi_-, \varphi_-^\dagger)$ and, second, from Lemma 5.5 we have that

$$f'(\varphi_-) \geq \sigma(\varphi_-, \varphi_-^\dagger) = f'(\varphi_-^\dagger)$$

and Lax's criterion is satisfied with one equality holding. See Figure 5.25.

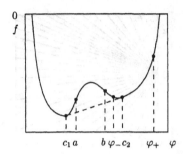 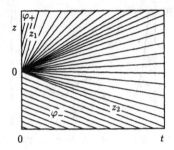

Figure 5.26: Graph of f vs φ and representation of the global weak solution $\varphi(z,t)$ for $a < b < \varphi_- < \varphi_+$.

Case VIII: $a < b < \varphi_- < \varphi_+$

The global weak solution is

$$
\varphi(z,t) = \begin{cases} \varphi_+ & \text{for} \quad z_1(t) \le z, \\ h_5(z/t) & \text{for} \quad z_2(t) \le z < z_1(t), \\ \varphi_- & \text{for} \quad z < z_2(t), \end{cases} \tag{5.36}
$$

where $z_1(t) = f'(\varphi_+)t$ and $z_2(t) = f'(\varphi_-)t$ are lines of continuity. This solution consists of two constant states φ_- and φ_+ separated by a rarefaction wave and the solution is continuous. This solution is shown in Figure 5.26. This completes the proof. ∎

Chapter 6

The initial-boundary value problem for a scalar conservation law

6.1 Formulation of the problem

We consider the initial-boundary value problem for a scalar conservation law. For nonlinear problems, it has turned out that classical initial boundary value problems are not well posed, requiring an extension of the concept of boundary conditions. Bardos, Le Roux and Nedelec (1979) proved existence and uniqueness of the solution for a scalar conservation law in several space variables with an initial condition and one generalized boundary condition. To prove existence they use the vanishing viscosity method and to prove uniqueness, they define an entropy condition also at the boundary. Le Roux (1977, 1979a, 1979b, 1981) also has proved existence and uniqueness of the initial-boundary value problem using the Godunov finite difference scheme for one first-order quasilinear hyperbolic equation in one dimension and the vanishing viscosity method for the first order in several dimensions.

Let $\Omega = (0, L)$, $\mathcal{T} = (0, T)$, $Q_T = \Omega \times \mathcal{T} \subset \mathbb{R}^2$ with $L, T > 0$. We consider the initial-boundary value problem for a one-dimensional conservation law with $f \in C^1(\mathbb{R} \times \mathcal{T})$: Find a function $\varphi \in BV(Q_T)$ which satisfies the following conservation law with initial and boundary conditions:

$$\frac{\partial \varphi}{\partial t} + \frac{\partial f(\varphi, t)}{\partial z} = 0 \ \text{ for } (z, t) \in Q_T, \tag{6.1a}$$

$$\varphi(z, 0) = \phi_0(z) \ \text{ for } z \in \Omega, \tag{6.1b}$$

$$\varphi(0, t) = \varphi_0(t) \ \text{ for } t \in \mathcal{T}, \tag{6.1c}$$

$$\varphi(L, t) = \varphi_L(t) \ \text{ for } t \in \mathcal{T} \tag{6.1d}$$

with $\phi_0 \in BV(\Omega)$ and φ_0 and $\varphi_L \in BV(\mathcal{T})$. By $BV(Q_T)$ we denote the space of functions defined on Q_T which are of bounded variation in the sense of Tonelli and Cesari.

This is the problem of continuous sedimentation as stated in Chapter 2 if we set $\varphi_0(t) = \varphi_\infty$, where φ_∞ is the constant maximum solid concentration. We shall always consider this particular constant boundary datum in Chapter 8 where the theory to be developed here is applied. However, we treat here the more general case of a time-dependent boundary function for $z = 0$. To make this distinction apparent, we choose the symbol $\varphi_0(t)$ instead of φ_∞. In particular, we do not require any interrelation between $\varphi_0(t)$ and the flux density function f, as implied by the choice $\varphi_0 = \varphi_\infty = const.$ in view of the relationships (2.14) and (2.21).

As we saw in Chapters 4 and 5, the solution of the nonlinear Cauchy problem is in general discontinuous, hence solutions in a generalized sense have to be considered. Weak solutions are not unique and we need an additional condition, called the *entropy condition*, to select the physically relevant one. With the initial-boundary value problem (6.1), the situation is quite similar. To obtain uniqueness and existence we need to generalize the concept of boundary conditions at $z = 0$ and $z = L$. Next, we give the definition of an *entropy solution* of the problem (6.1).

Definition 6.1 *A function $\varphi \in BV(Q_T)$ is an* entropy solution *of problem* (6.1) *if*

$$\varphi(z,0) = \phi_0(z) \quad \text{for almost all} \quad z \in \Omega \tag{6.2}$$

and if for all $k \in \mathbb{R}$ and all non-negative test functions $w \in C_0^2(\overline{\Omega} \times \mathcal{T})$

$$\int_0^T \int_0^L \left\{ |\varphi - k| \frac{\partial w}{\partial t} + \operatorname{sgn}(\varphi - k)[f(\varphi,t) - f(k,t)] \frac{\partial w}{\partial z} \right\} dz dt$$

$$\geq \int_0^T \{ \operatorname{sgn}(\varphi_L(t) - k)[f(\gamma\varphi(L,t),t) - f(k,t)]w(L,t)$$

$$- \operatorname{sgn}(\varphi_0(t) - k)[f(\gamma\varphi(0,t),t) - f(k,t)]w(0,t) \} dt. \tag{6.3}$$

By $C_0^2(\overline{\Omega} \times \mathcal{T})$ we denote the space of all twice differentiable functions in $\mathbb{R}^2$ with support in $\overline{\Omega} \times \mathcal{T}$. The traces $\gamma\varphi(0,t)$ and $\gamma\varphi(L,t)$ are defined as

$$\gamma\varphi(0,t) = \lim_{x \to 0,\, x > 0,\, x \notin E_0} \varphi(x,t), \quad \gamma\varphi(L,t) = \lim_{x \to L,\, x < L,\, x \notin E_L} \varphi(x,t), \tag{6.4}$$

where E_0 and E_L are sets of one-dimensional measure zero. These limits exist since φ is of bounded variation in Q_T in the sense of Tonelli and Cesari. The sign function is defined in (4.28).

6.2 Characterization of the entropy solution

Proposition 6.1 (Bustos, Paiva and Wendland 1996) *Let $\varphi \in BV(\overline{\Omega} \times \overline{\mathcal{T}})$. Then the following statements are equivalent, where $I(a,b)$ is defined in (4.6):*

(a) For every $k \in \mathbb{R}$, there holds

$$[\text{sgn}(\gamma\varphi(0,t) - k) + \text{sgn}(k - \varphi_0(t))][f(\gamma\varphi(0,t),t) - f(k,t)] \leq 0.$$

(b) For every $k \in I[\varphi_0(t), \gamma\varphi(0,t)]$, there holds

$$\text{sgn}(\gamma\varphi(0,t) - \varphi_0(t))[f(\gamma\varphi(0,t),t) - f(k,t)] \leq 0.$$

(c) For every $k \in I[\varphi_0(t), \gamma\varphi(0,t)]$, there holds

$$\text{sgn}(\gamma\varphi(0,t) - k)[f(\gamma\varphi(0,t),t) - f(k,t)] \leq 0.$$

Proof. Conditions (a) and (b) are generalizations of the conditions for the homogeneous problem (i.e. $\varphi_0(t) = \varphi_L(t) = 0$) as stated by Touré (1982). For the inhomogeneous problem (6.1), we define

$$I_0(t) = I(\varphi_0(t), \gamma\varphi(0,t)) \text{ for } t \in \mathcal{T}. \tag{6.5}$$

To establish the equivalence, we note:

1. For $k \in \mathbb{R}$ we have

$$\text{sgn}(\gamma\varphi(0,t) - k) + \text{sgn}(k - \varphi_0(t))$$

$$= \begin{cases} 0 & \text{if } k \notin I_0(t), \\ \text{sgn}(\gamma\varphi(0,t) - \varphi_0(t)) & \text{if } k \in \{\varphi_0(t), \gamma\varphi(0,t)\}, \\ 2\text{sgn}(\gamma\varphi(0,t) - \varphi_0(t)) & \text{if } k \in I_0(t)\backslash\{\varphi_0(t), \gamma\varphi(0,t)\}. \end{cases} \tag{6.6}$$

This relation already implies the equivalence of (a) and (b).

2. If $k \in I_0(t)\backslash\{\varphi_0(t), \gamma\varphi(0,t)\}$, then

$$\text{sgn}(\gamma\varphi(0,t) - k) = \text{sgn}(k - \varphi_0(t)) = \text{sgn}(\gamma\varphi(0,t) - \varphi_0(t)),$$

which shows the equivalence of (b) and (c) for these k. To establish the equivalence also at the endpoints of $k \in I_0(t)$, we proceed by continuity. This completes the proof. ∎

Proposition 6.2 *Let $\varphi \in BV(\overline{\Omega} \times \overline{\mathcal{T}})$. Then the following statements are equivalent:*

(a) For every $k \in \mathbb{R}$, there holds

$$[\text{sgn}(\gamma\varphi(L,t) - k) + \text{sgn}(k - \varphi_L(t))][f(\gamma\varphi(L,t),t) - f(k,t)] \geq 0.$$

(b) For every $k \in I(\varphi_L(t), \gamma\varphi(L,t))$, there holds

$$\text{sgn}(\gamma\varphi(L,t) - \varphi_L(t))[f(\gamma\varphi(L,t),t) - f(k,t)] \geq 0.$$

(c) For every $k \in I(\varphi_L(t), \gamma\varphi(L,t))$, there holds

$$\operatorname{sgn}(\gamma\varphi(L,t) - k)[f(\gamma\varphi(L,t),t) - f(k,t)] \geq 0.$$

Proof. Similar to the proof of Proposition 6.1 taking

$$I_L(t) = I(\gamma\varphi(L,t), \varphi_L(t)) \quad \text{for } t \in \mathcal{T} \tag{6.7}$$

and obtaining inter alia

$$\operatorname{sgn}(\gamma\varphi(L,t) - k) + \operatorname{sgn}(k - \varphi_L(t)) = 0 \quad \text{if } \ k \notin I_L(t). \tag{6.8}$$

This completes the proof. ∎

Lemma 6.1 (Wu and Wang 1982) *If $\varphi \in BV(Q_T)$, then for every $k \in \mathbb{R}$, $\gamma\operatorname{sgn}(\varphi - k)(y,t)$ exists f.a.a. $t \in \mathcal{T}$ and $\gamma\operatorname{sgn}(\varphi - k)(y,t) = \operatorname{sgn}(\gamma\varphi(y,t) - k)$ for $y = 0$ and $y = L$.*

Theorem 6.1 *A function $\varphi \in BV(Q_T)$ is an entropy solution of the initial-boundary value problem (6.1) if and only if*

$$\varphi(z,0) = \phi_0(z) \quad \text{for almost all } z \in \Omega, \tag{6.9}$$

and if for every $k \in \mathbb{R}$ and all non-negative test functions $w \in C_0^2(Q_T)$:

$$-\int_0^T \int_0^L \left\{ |\varphi - k| \frac{\partial w}{\partial t} + \operatorname{sgn}(\varphi - k)[f(\varphi,t) - f(k,t)]\frac{\partial w}{\partial z} \right\} dz\, dt \leq 0, \tag{6.10}$$

$$\sup_{k \in I_0(t)} (\operatorname{sgn}(\gamma\varphi(0,t) - k)[f(\gamma\varphi(0,t),t) - f(k,t)]) = 0, \tag{6.11}$$

$$\inf_{k \in I_L(t)} (\operatorname{sgn}(\gamma\varphi(L,t) - k)[f(\gamma\varphi(L,t),t) - f(k,t)]) = 0. \tag{6.12}$$

Inequality (6.10) is known as *Kružkov's inequality* (Kružkov, 1970). Equations (6.11) and (6.12) are the *generalized boundary conditions* at $z = 0$ and $z = L$, respectively.

Proof of Theorem 6.1. Define the function

$$J(\varphi,k,t) = \operatorname{sgn}(\varphi - k)(f(\varphi,t) - f(k,t)) \tag{6.13}$$

and consider the family of test functions (Kružkov, 1970) $\{\mu_\delta\}_{\delta>0}$, $\{\nu_\delta\}_{\delta>0}$ and $\{\rho_\delta\}_{\delta>0}$ with

$$\mu_\delta \in C^2(\overline{\Omega}; [0,1]), \quad \operatorname{supp} \mu_\delta \subset [0,\delta], \quad 0 \leq \mu_\delta \leq 1, \quad \mu_\delta(0) = 1, \quad |\mu_\delta'(\cdot)| \leq c/\delta \tag{6.14}$$

where c is a constant independent of δ,

$$\nu_\delta(z) = \mu_\delta(L - z) \quad \text{for } z \in \overline{\Omega}, \tag{6.15}$$

$$\rho_\delta(z) = \mu_\delta(z) + \nu_\delta(z). \tag{6.16}$$

Note that $\nu_\delta(L) = \rho_\delta(L) = \rho_\delta(0) = 1$ and that $\operatorname{supp} \rho_\delta \subset [0,\delta) \cup (L - \delta, L]$.

i) Suppose first that φ is an entropy solution. Since relation (6.9) is satisfied, we have to prove (6.10), (6.11) and (6.12). Let $w \in C_0^2(Q_T)$ with $w \geq 0$ and add

$$\int_0^T \{\operatorname{sgn}(\gamma\varphi(0,t) - k)[f(\gamma\varphi(0,t),t) - f(k,t)]w(0,t)$$

$$- \operatorname{sgn}(\varphi(L,t) - k)[f(\gamma\varphi(L,t),t) - f(k,t)]w(L,t)\} \, dt$$

to both sides of (6.3); then Propositions 6.1 and 6.2 together with (6.6) and (6.8) imply (6.10) for $k \in \mathbb{R}$. To obtain (6.11), choose the test function $w = v\mu_\delta$ with arbitrary nonnegative $v \in C_0^2(\mathcal{T})$. Replacing w in (6.3) yields the inequality

$$\int_0^T \int_0^\delta |\varphi - k|v'(t)\mu_\delta(z) \, dzdt + \int_0^T \int_0^\delta J(\varphi,k,t)v(t)\mu_\delta'(z) \, dzdt$$

$$\geq -\int_0^T \operatorname{sgn}(\varphi_0(t) - k)[f(\gamma\varphi(0,t),t) - f(k,t)]v(t) \, dt. \quad (6.17)$$

On the other hand, with Lebesgue's theorem we obtain

$$\lim_{\delta \to 0} \int_0^T \int_0^\delta J(\varphi,k,t)v(t)\mu_\delta'(z)dzdt$$

$$= \int_0^T \operatorname{sgn}(\gamma\varphi(0,t) - k)[f(\gamma\varphi(0,t),t) - f(k,t)]v(t) \, dt.$$

Taking the limit $\delta \to 0$ in (6.17), we get

$$\int_0^T (\operatorname{sgn}(\gamma\varphi(0,t) - k) + \operatorname{sgn}(k - \varphi_0(t))) \times$$

$$\times [f(\gamma\varphi(0,t),t) - f(k,t)]v(t) \, dt \leq 0.$$

Since v is arbitrary, we obtain (6.11) due to Proposition 6.1. In the same manner we obtain (6.12) by using the family $\{v_\delta\}_{\delta>0}$ and Proposition 6.2.

ii) Let $\varphi \in BV(Q_T)$ satisfy (6.9)–(6.12). Since relation (6.2) is satisfied, we have to show (6.3). Choose a non-negative test function $\tilde{w} = w(1-\rho_\delta) \in C_0^2(\overline{\Omega} \times \mathcal{T})$ satisfying $w \geq 0$. Inserting $\tilde{w}$ in (6.10) and integrating by parts leads to

$$\int_0^T \int_0^L \left(|\varphi - k|\frac{\partial w}{\partial t} + J(\varphi,k,t)\frac{\partial w}{\partial z}\right)(1 - \rho_\delta) \, dzdt$$

$$\geq \int_0^T J(\gamma\varphi(L,t),k,t)w(L,t)dt - \int_0^T J(\gamma\varphi(0,t),k,t)w(0,t)dt$$

$$- \int_0^T \int_0^L \frac{\partial}{\partial z}(\operatorname{sgn}(\varphi - k)[f(\varphi,t) - f(k,t)])w\rho_\delta \, dzdt.$$

For $\delta \to 0$, the last integral in this inequality tends to zero due to Lebesgue's theorem. Similarly, ρ_s vanishes from the first integral. Add

$$\int_0^T \text{sgn}(k - \varphi_L(t))[f(\gamma\varphi(L,t),t) - f(k,t)]w(L,t)\,dt$$

$$- \int_0^T \text{sgn}(k - \varphi_0(t))[f(\gamma\varphi(0,t),t) - f(k,t)]w(0,t)\,dt$$

to both sides of the resulting inequality. Then inequality (6.3) follows from Propositions 6.1 and 6.2. This completes the proof. ∎

6.3   Entropy conditions

Inequality (6.10) in Theorem 6.1 implies that the weak solution φ satisfies the differential equation (6.1a) in the distributional sense wherever it is continuous. If $z(t)$ is a smooth line of discontinuity with parametric representation, then the one-sided limits of a piecewise continuous solution satisfy the *Rankine-Hugoniot condition*

$$\sigma(\varphi^+, \varphi^-, t) = \frac{f(\varphi^+, t) - f(\varphi^-, t)}{\varphi^+ - \varphi^-}, \qquad (6.18)$$

where σ is the speed of propagation of the discontinuity and $\varphi^\pm = \varphi(z(t) \pm 0, t)$ as before in Chapter 4. Inequality (6.10) is equivalent to

$$\frac{\partial}{\partial t}|\varphi - k| + \frac{\partial}{\partial z}(f(\varphi,t) - f(k,t))\text{sgn}(\varphi - k) \leq 0 \qquad (6.19)$$

in the distributional sense. As we saw in Chapter 4, this last inequality is equivalent to Oleinik's condition E across the discontinuity, namely

$$\frac{f(\varphi,t) - f(\varphi^-,t)}{\varphi - \varphi^-} \leq \frac{f(\varphi^+,t) - f(\varphi^-,t)}{\varphi^+ - \varphi^-} \quad \text{for all } \varphi \in I(\varphi^+, \varphi^-), \qquad (6.20)$$

where $I(a,b)$ was defined in equation (4.6). Accordingly, the *generalized* boundary conditions (6.11) and (6.12) correspond to a *boundary entropy condition* and Propositions 6.1 and 6.2 are characterizations of these conditions at $z = 0$ and at $z = L$. Now we prove the above statement. A classical characterization of the physically relevant solution of problem (6.1) consists of taking the solution as the limit of solutions of a family of regularized parabolic equations, which is called the *vanishing viscosity method*. To this end, consider the following initial-boundary value problem: Find φ^ϵ satisfying

$$\frac{\partial \varphi^\epsilon}{\partial t} + \frac{\partial f(\varphi^\epsilon, t)}{\partial z} = \epsilon \frac{\partial^2 \varphi^\epsilon}{\partial z^2} \quad \text{for } (z,t) \in Q_T, \qquad (6.21a)$$

$$\varphi^\epsilon(z,0) = \phi_0(z), \quad z \in \Omega, \qquad (6.21b)$$

$$\varphi^\epsilon(0,t) = \varphi_0(t), \quad t \in \mathcal{T}, \qquad (6.21c)$$

$$\varphi^\epsilon(0,t) = \varphi_L(t), \quad t \in \mathcal{T}, \qquad (6.21d)$$

where $\epsilon > 0$ is a small viscosity parameter. From Dubois and Le Floch (1988), we have the following lemma:

Lemma 6.2 *Let* $G \in C_0^2(\mathcal{I})$ *and the family of solutions* $\{\varphi^\epsilon\}_\epsilon > 0$ *be bounded in* $W^{1,1}(Q_T)$ *uniformly with respect to* $\epsilon > 0$, *i.e.*

$$\|\varphi^\epsilon\|_{L^\infty(Q_T)} + \left\|\frac{\partial\varphi^\epsilon}{\partial z}\right\|_{L^1(Q_T)} + \left\|\frac{\partial\varphi^\epsilon}{\partial t}\right\|_{L^1(Q_T)} \le k_0 \tag{6.22}$$

where k_0 *is a constant independent of* ϵ. *Then we have*

$$\lim_{\epsilon\to0} \epsilon \int_0^T G(t)\frac{\partial\varphi^\epsilon}{\partial z}(0,t)\,dt = \int_0^T G(t)[f(\varphi_0(t),t) - f(\gamma\varphi(0,t),t)]\,dt, \tag{6.23}$$

$$\lim_{\epsilon\to0} \epsilon \int_0^T G(t)\frac{\partial\varphi^\epsilon}{\partial z}(L,t)\,dt = \int_0^T G(t)[f(\varphi_L(t),t) - f(\gamma\varphi(L,t),t)]\,dt. \tag{6.24}$$

Proof. Consider the following integral:

$$\mathcal{I}(\delta) = \epsilon \int_0^T G(t) \int_0^\delta \frac{\partial^2\varphi^\epsilon}{\partial z^2}\rho_\delta(z)\,dzdt.$$

Integrating $\mathcal{I}(\delta)$ by parts yields

$$\mathcal{I}(\delta) = -\epsilon \int_0^T G(t)\frac{\partial\varphi^\epsilon}{\partial z}(0,t)dt - \epsilon \int_0^T \int_0^\delta G(t)\frac{\partial\varphi^\epsilon}{\partial z}\mu_\delta'(z)\,dzdt. \tag{6.25}$$

Now take the limit of (6.25) first when $\epsilon \to 0$ and then when $\delta \to 0$. From the left-hand side of (6.25), using (6.21a), we get

$$\lim_{\epsilon\to0} \mathcal{I}(\delta) = \lim_{\epsilon\to0} \epsilon \int_0^T G(t) \int_0^\delta \frac{\partial^2\varphi^\epsilon}{\partial z^2}\mu_\delta(z)dzdt$$

$$= \lim_{\epsilon\to0} \int_0^T \int_0^\delta G(t)\left(\frac{\partial\varphi^\epsilon}{\partial t} + \frac{\partial f(\varphi^\epsilon,t)}{\partial z}\right)\mu_\delta(z)\,dzdt. \tag{6.26}$$

From the right-hand side of (6.25), we get

$$\lim_{\epsilon\to0} \mathcal{I}(\delta) = -\lim_{\epsilon\to0} \epsilon \int_0^T G(t)\frac{\partial\varphi^\epsilon}{\partial z}(0,t)\,dt, \tag{6.27}$$

since for each $\delta > 0$ we have

$$\left|\epsilon \int_0^T \int_0^\delta G(t)\frac{\partial\varphi^\epsilon}{\partial z}\mu_\delta'dzdt\right| \le \epsilon\,\text{const}\,\|\varphi^\epsilon\|_{W_{loc}^{1,1}} \to 0.$$

From equations (6.26) and (6.27), we get

$$\lim_{\epsilon\to0} \epsilon \int_0^T G(t)\frac{\partial\varphi^\epsilon}{\partial z}(0,t)\,dt = -\lim_{\epsilon\to0} \int_0^T \int_0^\delta G(t)\left(\frac{\partial\varphi^\epsilon}{\partial t} + \frac{\partial f(\varphi^\epsilon,t)}{\partial z}\right)\mu_\delta(z)\,dzdt. \tag{6.28}$$

Now we integrate by parts the second member of (6.28) to obtain

$$\lim_{\epsilon \to 0} \epsilon \int_0^T G(t) \frac{\partial \varphi^\epsilon}{\partial z}(0, t)\, dt$$

$$= \lim_{\epsilon \to 0} \left(\int_0^T \int_0^\delta G'(t)\varphi^\epsilon \mu_\delta\, dz dt + \int_0^T \int_0^\delta G(t)f(\varphi^\epsilon, t)\mu_\delta'\, dz dt \right.$$

$$\left. + \int_0^T G(t)f(\varphi_0(t), t)\, dt \right)$$

$$= \int_0^T \int_0^\delta G'(t)\varphi\mu_\delta\, dz dt + \int_0^T \int_0^\delta G(t)f(\varphi, t)\mu_\delta' dz dt + \int_0^T G(t)f(\varphi_0(t), t)\, dt.$$

Now we take the limit for $\delta \to 0$ and use for $t > 0$ that

$$\lim_{\delta \to 0} \int_0^\delta f(\varphi(z, t), t)\mu_\delta'(z)\, dz = -f(\gamma\varphi(0, t), t)$$

to obtain

$$\lim_{\epsilon \to 0} \epsilon \int_0^T G(t) \frac{\partial \varphi^\epsilon}{\partial z}(0, t)\, dt = \int_0^T G(t)[f(\varphi_0(t), t) - f(\gamma\varphi(0, t), t)]\, dt$$

which proves (6.23). The proof of equation (6.24) is analogous, choosing a family of functions $\{\nu_\delta\}_{\delta > 0}$ at the boundary $z = L$. This completes the proof. ∎

Lemma 6.3 *Let*

$$\frac{\partial \varphi}{\partial t} + \frac{\partial f(\varphi, t)}{\partial z} = 0 \text{ for } (z, t) \in Q_T \tag{6.29}$$

be a conservation law. The convex function $\alpha(\varphi)$ satisfies the additional conservation law .

$$\frac{\partial \alpha(\varphi)}{\partial t} + \frac{\partial \beta(\varphi, t)}{\partial z} = 0 \text{ for } (z, t) \in Q_T \tag{6.30}$$

if and only if

$$\frac{\partial \beta}{\partial \varphi} = \alpha'(\varphi) \frac{\partial f(\varphi, t)}{\partial \varphi}. \tag{6.31}$$

Proof. The proof is analogous to the proof of Lemma 4.3. ∎

Theorem 6.2 *For every $t > 0$, the traces $\gamma\varphi(0, t)$ and $\gamma\varphi(L, t)$ of a limit function φ of the sequence of viscosity solutions φ^ϵ of problem (6.21) are related to the boundary values $\varphi_0(t)$ and $\varphi_L(t)$ by means of the respective inequalities*

$$\beta(\gamma\varphi(0, t), t) - \beta(\varphi_0(t), t) - \alpha'(\varphi_0(t))[f(\gamma\varphi(0, t), t) - f(\varphi_0(t), t)] \leq 0, \tag{6.32}$$

$$\beta(\gamma\varphi(L, t), t) - \beta(\varphi_L(t), t) - \alpha'(\varphi_L(t))[f(\gamma\varphi(L, t), t) - f(\varphi_L(t), t)] \geq 0 \tag{6.33}$$

for every entropy pair (α, β) in the sense of Lax (1971, 1973).

Proof. Multiply equation (6.21a) by $\alpha'(\varphi^\epsilon(z,t))\psi(t)\mu_\delta(z)$ with an arbitrary function $\psi \in C_0^2(\mathcal{T})$ and integrate:

$$\int_0^T \int_0^\delta \left(\frac{\partial \varphi^\epsilon}{\partial t} + \frac{\partial f(\varphi^\epsilon, t)}{\partial z}\right) \alpha'(\varphi^\epsilon)\psi(t)\mu_\delta(z) \, dzdt$$

$$- \epsilon \int_0^T \int_0^\delta \frac{\partial^2 \varphi^\epsilon}{\partial z^2}\alpha'(\varphi^\epsilon)\psi(t)\mu_\delta(z) \, dzdt = 0.$$

Using equation (6.31), we obtain

$$\int_0^T \int_0^\delta \left(\frac{\partial \alpha^\epsilon}{\partial t} + \frac{\partial \beta(\varphi^\epsilon, t)}{\partial z}\right) \psi(t)\mu_\delta(z) \, dzdt$$

$$- \epsilon \int_0^T \int_0^\delta \frac{\partial^2 \varphi^\epsilon}{\partial z^2}\alpha'(\varphi^\epsilon)\psi(t)\mu_\delta(z) \, dzdt = 0.$$

Integrating the first term by parts and rewriting the second, we obtain $A + B = 0$ with

$$A = \int_0^T \int_0^\delta \alpha(\varphi^\epsilon)\psi'(t)\mu_\delta(z) \, dzdt + \int_0^T \int_0^\delta \beta(\varphi^\epsilon, t)\psi(t)\mu_\delta'(z) \, dzdt$$

$$+ \int_0^T \beta(\varphi_0(t), t)\psi(t) \, dt,$$

$$B = \epsilon \int_0^T \int_0^\delta \frac{\partial}{\partial z}\left(\frac{\partial \varphi^\epsilon}{\partial z}\alpha'(\varphi^\epsilon)\right) \psi(t)\mu_\delta(z)dzdt$$

$$- \epsilon \int_0^T \int_0^\delta \left(\frac{\partial \varphi^\epsilon}{\partial z}\right)^2 \alpha''(\varphi^\epsilon)\psi(t)\mu_\delta(z)dzdt.$$

The limit of A when $\epsilon \to 0$ is

$$A \xrightarrow{\epsilon \to 0} \int_0^T \int_0^\delta \alpha(\varphi)\psi'(t)\mu_\delta(z)dzdt$$

$$+ \int_0^T \int_0^\delta \beta(\varphi, t)\psi(t)\mu_\delta'(z)dzdt + \int_0^T \beta(\varphi_0(t), t)\psi(t)dt.$$

Now we take the limit when $\delta \to 0$ in the last expression to obtain

$$\lim_{\delta \to 0}\lim_{\epsilon \to 0} A = \int_0^T [\beta(\varphi_0(t), t) - \beta(\gamma\varphi(0, t), t)]\psi(t) \, dt. \qquad (6.34)$$

Note that B can be written as

$$B = -\epsilon\left\{\int_0^T \int_0^\delta \frac{\partial \varphi^\epsilon}{\partial z}\alpha'(\varphi^\epsilon)\mu_\delta'\psi(t)dzdt + \int_0^T \alpha'(\varphi_0(t))\frac{\partial \varphi^\epsilon}{\partial z}(0, t)\psi(t)dt\right.$$

$$+ \int_0^T \int_0^\delta \left(\frac{\partial \varphi^\epsilon}{\partial z} \right)^2 \alpha''(\varphi^\epsilon) \psi(t) \mu_\delta(z) dz dt \bigg\}. \quad (6.35)$$

For each $\delta > 0$,

$$\left| \epsilon \int_0^T \int_0^\delta \alpha'(\varphi^\epsilon) \frac{\partial \varphi^\epsilon}{\partial z} \mu'_\delta dz dt \right| \leq \epsilon \, \text{const} \, \|\varphi^\epsilon\|_{W^{1,1}_{\text{loc}}},$$

which tends to zero when $\epsilon \to 0$. For the second term of (6.35), we apply Lemma 6.2 to obtain

$$B \overset{\epsilon \to 0}{\to} - \int_0^T \alpha'(\varphi_0(t))[f(\varphi_0(t), t) - f(\gamma\varphi(0, t), t)]\psi(t) \, dt$$

$$- \lim_{\epsilon \to 0} \int_0^T \int_0^\delta \left(\frac{\partial \varphi^\epsilon}{\partial z} \right)^2 \alpha''(\varphi^\epsilon) \psi(t) \rho_\delta(z) \, dz dt. \quad (6.36)$$

Adding equations (6.34) and (6.36) and remarking that the second term on the right-hand side of (6.36) is negative since $\alpha'' > 0$ due to the convexity of α, we get

$$\int_0^T \{[\beta(\gamma\varphi(0, t), t) - \beta(\varphi_0(t), t)]$$

$$+ \alpha'(\varphi_0(t))[f(\gamma\varphi(0, t), t) - f(\varphi_0(t), t)]\}\psi(t) dt \leq 0,$$

which proves inequality (6.32). To prove inequality (6.33), we follow the same steps with μ_δ replaced by $\nu_\delta(z)$. This completes the proof. ■

Consider Kružkov's entropy pair (α, β) defined for all $k \in \mathbb{R}$ by

$$\alpha(\varphi) = |\varphi - k| \quad \text{and} \quad \beta(\varphi, t) = \text{sgn}(\varphi - k)(f(\varphi, t) - f(k, t)). \quad (6.37)$$

Then $\alpha'(\varphi) = \text{sgn}(\varphi - k)$. Introducing α' and β into (6.32) and (6.33), we get

$$\text{sgn}(\gamma\varphi(0, t) - k)[f(\gamma\varphi(0, t), t) - f(k, t)] - \text{sgn}(\varphi_0(t) - k) \times$$

$$\times [f(\varphi_0(t), t) - f(k, t)] - \text{sgn}(\varphi_0(t) - k)[f(\gamma\varphi(0, t), t) - f(\varphi_0(t), t)] \leq 0.$$

Rearranging the terms,

$$\{\text{sgn}(\gamma\varphi(0, t) - k) + \text{sgn}(k - \varphi_0(t))\}[f(\gamma\varphi(0, t), t) - f(k, t)] \leq 0.$$

From Proposition 6.1 we obtain

$$\text{sgn}(\gamma\varphi(0, t) - k)[f(\gamma\varphi(0, t), t) - f(k, t)] \leq 0,$$

which is the generalized boundary condition (6.10) at $z = 0$. Starting from inequality (6.33) and using Proposition 6.2 instead, we obtain the generalized boundary condition (6.11).

6.4   Existence of the entropy solution

A classical method used to find the physically relevant solution of problem (6.1) is the vanishing viscosity method. Consider the problem (6.21) for $\epsilon \geq 0$.

Theorem 6.3 (Bardos, Le Roux and Nedelec 1979) *The family $\{\varphi^\epsilon\}_{\epsilon>0}$ of solutions of problem (6.21) is sequentially compact in $L^1(Q_T)$ and each limit function φ for $\epsilon \to 0$ belongs to $BV(Q_T)$ and its trace for $t = 0$ satisfies (6.1c).*

While this theorem ensures that the limit function exists, we still have to show that this limit is actually an entropy solution of the initial-boundary value problem (6.1). In general the function φ does not satisfy the boundary conditions (6.1c) and (6.1d). This limit function satisfies the initial condition (6.9) and entropy boundary conditions (6.11) and (6.12).

Theorem 6.4 (Bustos and Paiva 1993) *There exists an entropy solution of the problem*

$$\frac{\partial \varphi}{\partial t} + \frac{\partial f(\varphi, t)}{\partial z} = 0 \qquad \text{for} \quad (z,t) \in Q_T, \tag{6.38a}$$

$$\varphi(z,0) = \phi_0(z), \qquad z \in \Omega, \tag{6.38b}$$

$$\sup_{k \in I(\gamma\varphi(0,t),\varphi_0(t))} [\operatorname{sgn}(\gamma\varphi(0,t) - k)(f(\gamma\varphi(0,t),t) - f(k,t))] = 0 \quad \text{a.e. in } \mathcal{T},$$
$$\tag{6.38c}$$

$$\inf_{k \in I(\gamma\varphi(L,t),\varphi_L(t))} [\operatorname{sgn}(\gamma\varphi(L,t) - k)(f(\gamma\varphi(L,t),t) - f(k,t))] = 0 \quad \text{a.e. in } \mathcal{T},$$
$$\tag{6.38d}$$

and φ is the limit function in $L^1(Q_T)$ of the family $\{\varphi^\epsilon\}_{\epsilon>0}$ of solutions of the regularized problem (6.21).

Proof. Let φ be the $L^1(Q_T)$ limit of a converging sequence $\{\varphi^{\epsilon_m}\}_{m\in\mathbb{N}}$ of solutions of problem (6.21). The proof of the theorem uses as a main argument that $\{\varphi^\epsilon\}_{\epsilon>0}$ is bounded in $W^{1,1}(Q_T)$ uniformly with respect to $\epsilon > 0$, i.e. that Condition (6.22) holds (see Bardos *et al.* 1979), which implies that $\{\varphi^\epsilon\}_{\epsilon>0}$ is sequentially compact in $L^1(Q_T)$. By Theorem 6.3, $\varphi \in BV(Q_T)$ satisfies the initial condition so we have to prove that φ satisfies (6.3) and then the proof of Theorem 6.4 follows from the characterization Theorem 6.1. Consider an approximation sgn_n for $n \in \mathbb{R}$, $n > 0$ of the sign function sgn:

$$\operatorname{sgn}_n(z) = \begin{cases} \operatorname{sgn}(z) & \text{if } |z| > n, \\ z/n & \text{if } |z| \leq n. \end{cases}$$

Consider a number $k \in \mathbb{R}$ and a test function $w \in C^2(Q_T)$ with compact support in $[0, L] \times \mathcal{T}$. Multiplying equation (6.21a) by $-\operatorname{sgn}_n(\varphi^\epsilon - k)w$ for $n > 0$ and $\epsilon \in \{\epsilon_m\}_{m\in\mathbb{N}}$ and integrating by parts yields

$$\int_0^T \int_0^L \left\{ \left(\int_k^{\varphi^\epsilon} \operatorname{sgn}_n(s-k)ds \right) \frac{\partial w}{\partial t} + \operatorname{sgn}_n(\varphi^\epsilon - k)[f(\varphi^\epsilon, t) - f(k,t)]\frac{\partial w}{\partial z} \right.$$

$$+\text{sgn}'_n(\varphi^\epsilon - k)[f(\varphi^\epsilon, t) - f(k, t)]w\frac{\partial\varphi^\epsilon}{\partial z}\Big\} \, dz dt$$

$$= \epsilon \int_0^T \int_0^L \text{sgn}'_n(\varphi^\epsilon - k)w\left(\frac{\partial\varphi^\epsilon}{\partial z}\right)^2 dz dt \int_0^T \int_0^L \text{sgn}_n(\varphi^\epsilon - k)\frac{\partial w}{\partial z}\frac{\partial\varphi^\epsilon}{\partial z} \, dz dt$$

$$+ \epsilon \int_0^T \text{sgn}_n(\varphi_0(t) - k)\frac{\partial\varphi^\epsilon}{\partial z}(0, t)w(0, t)dt \tag{6.39}$$

$$- \int_0^T \text{sgn}_n(\varphi_0(t) - k)[f(\varphi_0(t), t) - f(k, t)]w(0, t)dt$$

$$- \epsilon \int_0^T \text{sgn}_n(\varphi_L(t) - k)\frac{\partial\varphi^\epsilon}{\partial z}(L, t)w(L, t)dt$$

$$+ \int_0^T \text{sgn}_n(\varphi_L(t) - k)[f(\varphi_L(t), t) - f(k, t)]w(L, t) \, dt.$$

Now, the following step is the passage to the limit in equation (6.39) when n tends to zero: the first two terms on the left-hand side of equation (6.39) give the first two terms of inequality (6.3) with φ^ϵ instead of φ (the function $\text{sgn}_n(\varphi^\epsilon - k)$ converges in all points to $\text{sgn}(\varphi^\epsilon - k)$, and sgn_n is uniformly bounded by 1, then by Lebesgue's theorem, the convergence is in $L^1(Q_T)$. For the third term, the n-limit is zero; in fact,

$$\left|\int_0^L \int_0^T \text{sgn}'_n(\varphi^\epsilon - k)(f(\varphi^\epsilon, t) - f(k, t))w\frac{\partial\varphi^\epsilon}{\partial z}dz dt\right| \leq C \iint_A \left|\frac{\partial(\varphi^\epsilon - k)}{\partial z}\right| dz dt$$

and by a lemma by Saks (1937),

$$\lim_{n\to 0} \iint_A \left|\frac{\partial(\varphi^\epsilon - k)}{\partial z}\right| dz dt = 0,$$

where $A = \{(z, t) \in Q_T | \, |\varphi^\epsilon(z, t) - k| \leq n\}$ and C is a constant that depends only on the Lipschitz constant of the function f and on $\|w\|_{L^\infty(Q_T)}$. The first term on the right-hand side of equation (6.39) is non-negative. By Lebesgue's theorem the n-limit is easily obtained for each of the remaining terms. We get

$$\int_0^T \int_0^L \left\{|\varphi^\epsilon - k|\frac{\partial w}{\partial t} + \text{sgn}(\varphi^\epsilon - k)[f(\varphi^\epsilon, t) - f(k, t)]\frac{\partial w}{\partial z}\right\} dz dt$$

$$\geq \epsilon \int_0^T \int_0^L \text{sgn}(\varphi^\epsilon - k)\frac{\partial w}{\partial z}\frac{\partial\varphi^\epsilon}{\partial z}dz dt + \epsilon \int_0^T \text{sgn}(\varphi_0(t) - k)\frac{\partial\varphi^\epsilon}{\partial z}(0, t)w(0, t)dt$$

$$- \int_0^T \text{sgn}(\varphi_0(t) - k)[f(\varphi_0(t), t) - f(k, t)]w(0, t)dt \tag{6.40}$$

$$- \epsilon \int_0^T \text{sgn}(\varphi_L(t) - k)\frac{\partial\varphi^\epsilon}{\partial z}(L, t)w(L, t)dt$$

$$+ \int_0^T \text{sgn}(\varphi_L(t) - k)[f(\varphi_L(t), t) - f(k, t)]w(L, t)dt.$$

Consider the limit of (6.40) for ϵ tend to zero. For the first term on the right-hand side of (6.40), the ϵ-limit is zero. In fact, from (6.22),

$$\left| \epsilon \int_0^L \int_0^T \text{sgn}(\varphi - k) \frac{\partial w}{\partial z} \frac{\partial \varphi^\epsilon}{\partial z} dz dt \right| \leq \epsilon \tilde{k} \left\| \frac{\partial w}{\partial z} \right\|_{L^\infty(Q_T)}. \tag{6.41}$$

Letting $\epsilon \to 0$, we apply Lemma 6.2 to the rest of the terms of the right-hand side of (6.40) and obtain the right-hand side of inequality (6.3). As the sign function is discontinuous, we proceed by regularization, in a similar way to Le Roux (1979a) using Lemma 2 of his thesis. This proves the theorem. ∎

6.5 Uniqueness of the solution and admissible states at the boundaries

Theorem 6.5 *The entropy solution of problem* (6.38) *is unique.*

Proof. Let φ and $\hat{\varphi} \in BV(Q_T)$ be two entropy solutions. Due to Theorem 6.1 they satisfy inequality (6.10). Moreover, since $f \in C^1(\mathbb{R})$, these solutions also satisfy the inequality (Kružkov, 1970)

$$\int_0^T \int_0^L \left(|\varphi - \hat{\varphi}| \frac{\partial w}{\partial t} + J(\varphi, \hat{\varphi}, t) \frac{\partial w}{\partial z} \right) dz dt \geq 0 \tag{6.42}$$

for all non-negative test functions $w \in C_0^2(Q_T)$, where

$$J(\varphi, \hat{\varphi}, t) = \text{sgn}(\varphi - \hat{\varphi})[f(\varphi, t) - f(\hat{\varphi}, t)].$$

Choose a test function $w = (1 - \rho_\delta)u$ with arbitrary $u \in C_0^2(\mathcal{T})$ and $u \geq 0$, where $\rho_\delta \in C^2(\overline{\Omega})$ is defined in equation (4.14). For this test function, (6.42) yields the inequality

$$\int_0^T \int_0^L |\varphi - \hat{\varphi}| u'(t)(1 - \rho_\delta) dz dt - \int_0^T \int_0^L J(\varphi, \hat{\varphi}, t) u(t) \rho_\delta'(z) dz dt \geq 0. \tag{6.43}$$

Now by Lebesgue's theorem and Lemma 6.1, we obtain

$$\lim_{\delta \to 0} \int_0^T \int_0^L J(\varphi, \hat{\varphi}, t) u(t) \rho_\delta'(z) dz dt$$
$$= \int_0^T [J(\gamma\varphi, \gamma\hat{\varphi}, t)(L, t) - J(\gamma\varphi, \gamma\hat{\varphi}, t)(0, t)]u(t) dt.$$

Taking the limit when $\delta \to 0$ in (6.43), we get

$$\int_0^T \int_0^L |\varphi - \hat{\varphi}| u'(t) dz dt \geq \int_0^T [J(\gamma\varphi, \gamma\hat{\varphi}, t)(L, t) - J(\gamma\varphi, \gamma\hat{\varphi}, t)(0, t)]u(t) dt. \tag{6.44}$$

On the other hand, for $t \in \mathcal{T}$ we define

$$g(0,t) = \begin{cases} \gamma\varphi(0,t) & \text{if } \gamma\varphi(0,t) \in I(\gamma\hat{\varphi}(0,t),\varphi_0(t)), \\ \varphi_0(t) & \text{if } \varphi_0(t) \in I(\gamma\varphi(0,t),\gamma\hat{\varphi}(0,t)), \\ \gamma\hat{\varphi}(0,t) & \text{if } \gamma\hat{\varphi}(0,t) \in I(\gamma\varphi(0,t),\varphi_0(t)). \end{cases} \qquad (6.45)$$

Then we find

$$g(0,t) \in I(\varphi_0(t),\gamma\hat{\varphi}(0,t)) \cap I(\varphi_0(t),\gamma\varphi(0,t)) \quad \text{for all } t \in \mathcal{T}.$$

Since $J(u,v,t)$ can be expressed as

$$J(u,v,t) = \operatorname{sgn}(u-v)[f(u,t)-f(k,t)] + \operatorname{sgn}(v-u)[f(v,t)-f(k,t)]$$

for all $k \in \mathbb{R}$, we can rewrite $J(\gamma\varphi,\gamma\hat{\varphi},t)$ as

$$J(\gamma\varphi,\gamma\hat{\varphi},t)(0,t) = J(\gamma\varphi(0,t),g(0,t),t) + J(\gamma\hat{\varphi}(0,t),g(0,t),t).$$

Then, using the entropy condition (6.11) and Proposition 6.1, we get

$$\int_0^T J(\gamma\varphi,\gamma\hat{\varphi},t)(0,t)u(t)dt \leq 0. \qquad (6.46)$$

Similarly, we define $g(L,t)$ as

$$g(L,t) = \begin{cases} \gamma\varphi(L,t) & \text{if } \gamma\varphi(L,t) \in I(\gamma\hat{\varphi}(L,t),\varphi_L(t)), \\ \varphi_L(t) & \text{if } \varphi_L(t) \in I(\gamma\varphi(L,t),\gamma\hat{\varphi}(L,t)), \\ \gamma\hat{\varphi}(L,t) & \text{if } \gamma\hat{\varphi}(L,t) \in I(\gamma\varphi(L,t),\varphi_L(t)). \end{cases} \qquad (6.47)$$

Using the entropy condition (6.12) and Proposition 6.2, we obtain

$$\int_0^T J(\gamma\varphi,\gamma\hat{\varphi},t)(L,t)u(t)dt \geq 0. \qquad (6.48)$$

Substituting (6.46) and (6.48) into (6.44), we obtain

$$\int_0^T \int_0^L |\varphi - \hat{\varphi}|u'(t)dzdt \geq 0 \quad \text{for all non-negative } u \in C_0^2(\mathcal{T}).$$

This relation implies that $\varphi = \hat{\varphi}$ on Q_T. This completes the proof. ∎

6.5.1 Admissible states at the boundaries

We now proceed to specify the admissible states at the boundaries for all entropy pairs (α,β) in the sense of Dubois and Le Floch (1988).

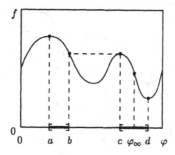

Figure 6.1: Graph of f vs φ showing the set $\epsilon_0(\varphi_0(t)) = [a,b] \cup [c,d]$.

Definition 6.2 *For any given states $\varphi_0(t)$ and $\varphi_L(t)$, the sets $\epsilon_0(\varphi_0(t))$ and $\epsilon_L(\varphi_L(t))$ of admissible states at the boundaries $z = 0$ and $z = L$ are defined by*

$$\epsilon_0(\varphi_0(t)) = \{\varphi \in \mathbb{R} \,|\, \beta(\varphi) - \beta(\varphi_0) - \alpha'(\varphi_0)[f(\varphi,t) - f(\varphi_0,t)] \leq 0\}, \qquad (6.49)$$

$$\epsilon_L(\varphi_L(t)) = \{\varphi \in \mathbb{R} \,|\, \beta(\varphi) - \beta(\varphi_L) - \alpha'(\varphi_L)[f(\varphi,t) - f(\varphi_L,t)] \geq 0\}, \qquad (6.50)$$

respectively, for every entropy pair (α, β).

Now we are in a position to reformulate the initial-boundary value problem (6.38) as follows: find a function $\varphi \in BV(Q_T)$ which satisfies

$$\frac{\partial \varphi}{\partial t} + \frac{\partial f(\varphi,t)}{\partial z} = 0 \quad \text{for a.e.} \quad (z,t) \in Q_T \quad \text{weakly}, \qquad (6.51a)$$

$$\varphi(z,0) = \phi_0(z) \quad \text{for a.e.} \quad z \in \Omega, \qquad (6.51b)$$

$$\gamma\varphi(0,t) \in \epsilon_0(\varphi_0(t)) \quad \text{for a.e.} \quad t \in \mathfrak{I}, \qquad (6.51c)$$

$$\gamma\varphi(L,t) \in \epsilon_L(\varphi_L(t)) \quad \text{for a.e.} \quad t \in \mathfrak{I}. \qquad (6.51d)$$

6.5.2 Geometrical interpretation of the sets of admissible states

To find the geometrical interpretation of the sets $\epsilon_0(\varphi_0(t))$ and $\epsilon_L(\varphi_L(t))$, we use Kružkov's entropy pairs.

Proposition 6.3 *The set $\epsilon_0(\varphi_0(t))$ of admissible states is characterized by the inequality*

$$\frac{f(\varphi,t) - f(k,t)}{\varphi - k} \leq 0 \quad \text{for all} \quad k \in I(\varphi_0(t), \varphi) \qquad (6.52)$$

and the set $\epsilon_L(\varphi_L(t))$ by the inequality

$$\frac{f(\varphi,t) - f(k,t)}{\varphi - k} \geq 0 \quad \text{for all} \quad k \in I(\varphi_L(t), \varphi). \qquad (6.53)$$

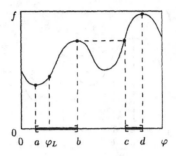

Figure 6.2: Graph of f vs φ showing the set $\epsilon_L(\varphi_L(t)) = [a, b] \cup [c, d]$.

Proof. Consider Kružkov's entropy pairs (α, β) defined in (6.37). Then inequalities (6.49) and (6.50) imply

$$(\operatorname{sgn}(\varphi - k) - \operatorname{sgn}(\varphi_0(t) - k))[f(\varphi, t) - f(k, t)] \leq 0 \text{ for all } k \in \mathbb{R}$$

and

$$(\operatorname{sgn}(\varphi - k) - \operatorname{sgn}(\varphi_L(t) - k))[f(\varphi, t) - f(k, t)] \geq 0 \text{ for all } k \in \mathbb{R}.$$

From Propositions 6.1 and 6.2 (replacing $\gamma\varphi$ by φ) we obtain inequalities (6.52) and (6.53) for all $k \in I(\varphi_0(t), \varphi)$ and $k \in I(\varphi_L(t), \varphi)$, respectively. Consider first condition (6.52) at the boundary $z = 0$. This condition indicates that, for an arbitrary function f, the chord that joins, for a given time, the points $(\varphi, f(\varphi, t))$ and $(k, f(k, t))$ in an f vs φ plot is *nonpositive*; see Figure 6.1. At $z = L$, condition (6.53) indicates that the chord joining the points $(\varphi, f(\varphi, t))$ and $(k, f(k, t))$ for a given time in an f vs φ plot is *nonnegative* as shown in Figure 6.2.

Chapter 7

Batch sedimentation of ideal suspensions

7.1 Initial value problem

As we saw in Chapter 2, the problem of batch sedimentation of ideal suspensions can be formulated as the following initial value problem:

$$\frac{\partial}{\partial t} + \frac{\partial f_b(\varphi)}{\partial z} = 0 \text{ for } z \in \mathbb{R}, \ t > 0, \tag{7.1a}$$

$$\varphi(z, 0) = \varphi_I(z) \text{ for } z \in \mathbb{R}, \tag{7.1b}$$

where $\varphi_I(z)$ is given by

$$\varphi_I(z) = \begin{cases} \varphi_L = 0 & \text{for } z \geq L, \\ \phi_0 & \text{for } 0 \leq z < L, \\ \varphi_\infty & \text{for } z < 0. \end{cases} \tag{7.2}$$

Recall that $\varphi_I(z)$ is the *initial data* and ϕ_0 is the *initial concentration*. Solutions to (7.1a) are in general discontinuous (see Chapters 4 and 5) so we work with entropy solutions, that is, generalized solutions that satisfy Oleinik's condition E. To construct the solutions, we start at $t = 0$, solving the two Riemann problems at $z = L$ between φ_L and ϕ_0 and at $z = 0$ between ϕ_0 and φ_∞. These solutions consist either of shocks or of rarefaction waves (see Chapter 5). Next, we intersect these two waves. It turns out that for flux density functions with exactly one or two inflection points it is sufficient to consider the interaction of two straight line shocks which give as a result another straight line shock or the interaction of a straight line shock with a rarefaction wave which generates a curved shock. Interactions between rarefaction waves do not occur for these flux density functions.

111

7.2   Modes of sedimentation

Definition 7.1 (Bustos and Concha 1988) *If $\varphi(z,t)$ is an entropy solution of problem* (7.1), *the pair* $(\varphi, f_b(\varphi))$ *constitutes a* Kynch Sedimentation Process *(KSP)*.

The KSP is completely defined when a constitutive equation is postulated for $f_b(\varphi)$.

Definition 7.2 Modes of Sedimentation (MS) *are the different possible types of KSP. We destinguish among the following seven types of concentration profiles existing in a sedimentation column for small times $t_0 > 0$ when proceeding downwards from $z = L$ to $z = 0$:*

MS-1 *The concentration jumps from zero to ϕ_0 and changes abruptly from ϕ_0 to φ_∞.*

MS-2 *The concentration jumps from zero to ϕ_0 and changes suddenly from ϕ_0 and then increases continuously to φ_∞.*

MS-3 *The concentration jumps from zero to ϕ_0 and increases continuously from ϕ_0 to φ_∞.*

MS-4 *The concentration jumps from zero to ϕ_0 and changes suddenly from ϕ_0 and then increases continuously, followed by an abrupt increase, to φ_∞.*

MS-5 *The concentration jumps from zero to ϕ_0, then increases continuously from ϕ_0 and finally jumps to φ_∞.*

MS-6 *The concentration jumps from zero to a value less than ϕ_0, then increases continuously to ϕ_0, followed by a jump to φ_∞.*

MS-7 *The concentration jumps from zero to a value less than ϕ_0, increases continuously and then jumps to ϕ_0, followed by a jump from ϕ_0 to φ_∞.*

The modes of sedimentation are entirely determined by the constitutive equation of the flux density function and by the initial concentration ϕ_0. Kynch analysed three types of flux density functions: a concave function, a function with one inflection point and a function with two inflection points between φ_L and φ_∞. Since a concave flux density function is purely theoretical because no such functions have been found by experiment, we consider only the last two types of flux density functions. It will turn out that the modes MS-2 and MS-3 exist only for flux density functions with one inflection point, while the modes MS-4 to MS-7 occur only for two inflection points. In the latter case, the range of possible modes of sedimentation depends on additional geometrical assumptions on the flux density function. Independently of the mode of sedimentation, the final state of all Kynch sedimentation processes consists of two regions: a sediment of concentration φ_∞ of height

$$z_c = \phi_0 L / \varphi_\infty \qquad (7.3)$$

and a layer of clear water on top. Mathematically, the horizontal line $z = z_c$ is a shock. The time at which this state is attained is called the *critical time* and is denoted by t_c. This is the natural final state of the batch sedimentation process, since the total mass of solid at the beginning of the sedimentation process, amounting to $\rho_s \phi_0 LS$ where S is the cross-sectional area of the sedimentation column, should be equal to the total mass of solid contained in the final sediment of concentration φ_∞ and height z_c, which is $\rho_s \varphi_\infty z_c S$, from which (7.3) follows immediately.

7.3 Construction of the global weak solution

In what follows, we construct the solution corresponding to each mode of sedimentation by considering the intersection of shocks and rarefaction waves. The solution is constructed (Cheng 1981) in such a way that it satisfies the initial condition (7.1b) and equation (7.1a) at the points of continuity. At the points of discontinuity it satisfies the Rankine-Hugoniot condition and Oleinik's condition E. This can be checked by direct substitution.

7.3.1 Flux density function with one inflection point

For a vast majority of sedimentation experiments, the constitutive equation of $f_b(\varphi)$ is a function with one inflection point. One such model is due to Richardson and Zaki (1954). However, these equations must be assessed carefully because they are seldom based on values of φ near φ_∞. The Richardson-Zaki equation, for example, predicts finite values of $u(\varphi)$ for $\varphi > \varphi_\infty$. Extrapolation from intermediate values is inaccurate because lubrication terms (which are unimportant there) dominate for values of φ very slightly less than φ_∞. This implies very small values of u for smooth particles, but less so for rough ones. We consider a flux density function with one inflection point at $\varphi = a$ between $\varphi_L = 0$ and φ_∞, which satisfies $f_b'' > 0$ for φ between 0 and a and $f_b'' < 0$ for φ between a and φ_∞. As we saw in Chapter 2, the flux density functions studied by Kynch have the following properties:

$$f_b(\varphi) < 0 \text{ for } \varphi \in (0, \varphi_\infty), \tag{7.4}$$

$$f_b(0) = f_b(\varphi_\infty) = 0, \tag{7.5}$$

$$f_b'(0) < 0 \text{ and } f_b'(\varphi_\infty) > 0. \tag{7.6}$$

From them, we deduce that

$$f_b'(\varphi) > 0 \quad \text{for} \quad a \leq \varphi \leq \varphi_\infty. \tag{7.7}$$

In an f_b vs φ plot, draw a tangent to the graph of f_b at $(\varphi_\infty, 0)$ and extend it backwards till it cuts f_b at the point $(\varphi_\infty^{**}, f_b(\varphi_\infty^{**}))$; see Figure 7.1. The superindex ** is introduced in Definition 5.3. Depending on the value of ϕ_0, there exist three modes of sedimentation.

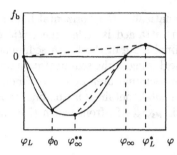 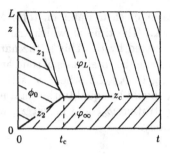

Figure 7.1: Graph of f_b vs φ and representation of the global weak solution $\varphi(z,t)$ for an MS-1 with $\phi_0 < \varphi_\infty^{**}$.

Theorem 7.1 *For a flux density function f_b with exactly one inflection point a, we have:*

(a) *If $\varphi_L = 0 < \phi_0 < \varphi_\infty^{**}$, the KSP is an MS-1 as shown in Figure 7.1.*

(b) *If $\varphi_\infty^{**} < \phi_0 < a$, the KSP is an MS-2 as shown in Figure 7.2.*

(c) *If $a < \phi_0$, the KSP is an MS-3 as shown in Figure 7.3.*

Proof.

(a) We have to solve two Riemann problems and then consider the interaction of the solutions. The solution of the Riemann problem at $z = L$, $t = 0$, which we already found in Chapter 5, is given by equation (5.15) with $\varphi_+ = \varphi_L = 0$ and $\varphi_- = \phi_0$. Then the solution consists of two constant states $\varphi_L = 0$ and ϕ_0 separated by the shock

$$z_1(t) = \sigma(\varphi_L, \phi_0)t + L. \tag{7.8}$$

This shock is the interface between the water and the suspension with constant concentration ϕ_0. Its slope $\sigma(\varphi_L, \phi_0) = f_b(\phi_0)/\phi_0$ is negative by property (2.14) of the flux density function, which corresponds to the fact that the particles move downward. To find the solution of the Riemann problem at $z = 0$, $t = 0$ we use Lemma 5.3 of Chapter 5 which says that $\phi_0 < \varphi_\infty^{**}$ implies that $\varphi_\infty < \phi_0^*$. Then for the Riemann problem at $z = 0$, $t = 0$ the solution is given by equation (5.15) with $\varphi_+ = \phi_0$ and $\varphi_- = \varphi_\infty$. Thus the solution consists of two constant states ϕ_0 and φ_∞ separated by the shock

$$z_2(t) = \sigma(\phi_0, \varphi_\infty)t. \tag{7.9}$$

This shock corresponds to the interface between the sediment and the suspension with constant initial concentration ϕ_0. Its slope

$$\sigma(\phi_0, \varphi_\infty) = \frac{f_b(\phi_0)}{\phi_0 - \varphi_\infty}$$

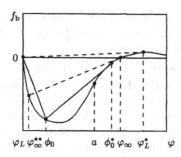 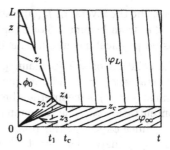

Figure 7.2: Graph of f_b vs φ and representation of the global weak solution $\varphi(z,t)$ for an MS-2 with $\varphi_\infty^{**} < \phi_0 < a$.

is positive which corresponds to the fact that the top of the sediment rises with time. Now we consider the intersection of these two shocks. Let us call (z_c, t_c) the point of intersection; then

$$z_c = z_1(t_c) = z_2(t_c) = \frac{\phi_0 - \varphi_L}{\varphi_\infty - \varphi_L} L, \quad t_c = \frac{L}{\sigma(\phi_0, \varphi_\infty) - \sigma(\varphi_L, \phi_0)}.$$

The value of z_c obtained is the final height of the sediment as given by equation (7.3) and therefore (z_c, t_c) is the critical point. The critical height z_c is positive and smaller than L. Since $\sigma(\phi_0, \varphi_\infty) > 0$ and $\sigma(\varphi_L, \phi_0) < 0$, t_c is positive, which ensures that the solution has a physical meaning.

The solution of problem (7.1) for $0 \le t < t_c$ is

$$\varphi(z,t) = \begin{cases} \varphi_L = 0 & \text{for } z_1(t) \le z, \\ \phi_0 & \text{for } z_2(t) \le z < z_1(t), \\ \varphi_\infty & \text{for } z < z_2(t). \end{cases} \tag{7.10}$$

Finally, we solve the Riemann problem at (z_c, t_c) between the states φ_L and φ_∞. The solution is given by equation (5.15) with $\varphi_+ = \varphi_L = 0$ and $\varphi_- = \varphi_\infty$. It consists of two constant states φ_L and φ_∞ separated by the horizontal shock $z_3(t) = z_c$. This shock corresponds to the water-sediment interface which, as shown in Figure 7.1, does not vary with time. The solution of problem (7.1) for $t_c \le t$ is

$$\varphi(z,t) = \begin{cases} \varphi_L = 0 & \text{for } z_c \le z, \\ \varphi_\infty & \text{for } z < z_c. \end{cases} \tag{7.11}$$

Now we consider a height z_0 such that $0 < z_0 < z_c$. The concentration $\varphi(z_0, t)$ changes suddenly from ϕ_0 to φ_∞ and therefore the KSP is an MS-1. This completes the proof of part (a) of the theorem. ∎

(b) Let h_2 be the inverse of $f'_b(\varphi)$ restricted to values of $\varphi > a$. Since $\varphi_L < \varphi_\infty^{**} < \phi_0 < a$, from Lemma 5.3 of Chapter 5, we have that $a < \phi_0^* < \varphi_\infty < \varphi_L^*$ (φ_L^* has no physical meaning). This is shown in Figure 7.2.

The solution of the Riemann problem at $z = L$, $t = 0$, consists of the two states $\varphi_L = 0$ and ϕ_0 separated by the shock $z_1(t) = \sigma(\varphi_L, \phi_0)t + L$. Similarly, $z_1(t)$ is the interface between the water and the suspension of initial concentration ϕ_0. The Riemann problem at $z = 0$, $t = 0$ is a jump from ϕ_0 to φ_∞. Since $\phi_0^* < \varphi_\infty$, this problem corresponds to Case II with $\varphi_+ = \phi_0$ and $\varphi_- = \varphi_\infty$ analyzed in Chapter 5 for a flux density function with one inflection point. The solution is given by equation (5.16) and consists of two constant states ϕ_0 and φ_∞ separated by the contact discontinuity $z_2(t) = f'_b(\phi_0^*)t$ and a rarefaction wave. Let (z_1, t_1) be the point of intersection of the shock $z_1(t)$ with the contact discontinuity $z_2(t)$. Then

$$z_1 = z_1(t_1) = z_2(t_1) = \frac{f'_b(\phi_0^*)L}{f'_b(\phi_0^*) - \sigma(\varphi_L, \phi_0)}, \quad t_1 = \frac{L}{f'_b(\phi_0^*) - \sigma(\varphi_L, \phi_0)}.$$

Since $a < \phi_0^*$, from property (7.7) we have that $f'_b(\phi_0^*) > 0$; moreover, $\sigma(\varphi_L, \phi_0) < 0$, hence the height z_1 is positive and smaller than L and t_1 is positive. The slope of the contact discontinuity $z_2(t) = f'_b(\phi_0^*)t$ is positive and represents the upward speed of the concentration ϕ_0^*. The line $z_3(t) = f'_b(\varphi_\infty)t$ is a line of continuity and its slope is the velocity with which the top of the sediment rises. The solution of problem (7.1) for $0 \le t < t_1$ is

$$\varphi(z, t) = \begin{cases} \varphi_L = 0 & \text{for } z_1(t) \le z, \\ \phi_0 & \text{for } z_2(t) \le z < z_1(t), \\ h_2(z/t) & \text{for } z_3(t) \le z < z_2(t), \\ \varphi_\infty & \text{for } z < z_3(t). \end{cases} \tag{7.12}$$

Next, we consider the interaction of the shock $z_1(t)$ with the rarefaction wave at (z_1, t_1). This consists of a sequence of Riemann problems: the concentration φ_L encounters consecutively the values of φ between ϕ_0^* and φ_∞. Each jump corresponds to Case I of Chapter 4 for a flux density function with one inflection point with $\varphi_+ = \varphi_L$ and $\varphi_- = \varphi$ with φ between ϕ_0^* and φ_∞, since $\varphi_- < \varphi_L^*$. This generates a curved shock $z_4(t)$ since the successive jumps produce shocks of decreasing slope till φ_- reaches the value φ_∞. The shock $z_4(t)$ is the interface between the fluid and the suspension. This shock satisfies the relation

$$z'_4(t) = \sigma(\varphi_L, h_2(z_4(t)/t)) \tag{7.13}$$

with $z_1(t_1) = z_2(t) = z_4(t_1)$. Taking the derivative of equation (7.13), we obtain

$$z''_4(t) = -\frac{[z'_4(t) - z_4(t)/t]^2}{t f''_b(h_2(z_4(t)/t)) [h_2(z_4(t)/t) - \varphi_L]}. \tag{7.14}$$

Since h_2 is restricted to $\varphi > a$, we have that $f_b'' \left(h_2 \left(z_4(t)/t \right) \right) < 0$. This implies that $z_4''(t) > 0$ for $t > t_1$, i.e. the shock $z_4(t)$ is concave for $t > t_1$. There is a finite time t_c when $z_4(t)$ meets $z_3(t)$: obviously,

$$z_4'(t) = \sigma \left(h_2 \left(z_4(t)/t \right), \varphi_L \right) = f_b \left(h_2 \left(z_4(t)/t \right) \right) / h_2 \left(z_4(t)/t \right) < 0$$

and therefore $z_4(t) < z_1$ for $t > t_1$. Consequently, $z_3(t)$ and $z_4(t)$ intersect at t_c before $z_3(t)$ reaches height z_1 at $t = z_1 / f_b'(\varphi_\infty)$. To determine the height z_c at which $z_4(t)$ meets $z_3(t)$, we use the function

$$\eta(t) = \int_{-\infty}^{A+\sigma(\varphi_L,\varphi_\infty)t} (\varphi(z,t) - \varphi_\infty) \, dz \tag{7.15}$$

defined by Liu (1978), which is time invariant when A is chosen large enough so that $\varphi(z,t) = \varphi_L = 0$ for $z > z(t) = A + \sigma(\varphi_L,\varphi_\infty)t$. For batch sedimentation $\sigma(\varphi_L,\varphi_\infty) = 0$. Obviously, $\eta(t)$ is time invariant for $A > L$. Then

$$\eta(t) = \int_{-\infty}^{A} (\varphi(z,t) - \varphi_\infty) dz = (\varphi_L - \varphi_\infty)(A - z_c) \quad \text{for} \quad t > t_c. \tag{7.16}$$

Moreover, evaluating (7.15) at $t = 0$ yields

$$\eta(0) = \int_{-\infty}^{A} (\varphi(z,0) - \varphi_\infty) dz = (\phi_0 - \varphi_\infty)L + (\varphi_L - \varphi_\infty)(A - L). \tag{7.17}$$

Since η is time invariant, the right-hand parts of (7.16) and (7.17) are equal, which implies

$$z_c = \frac{\phi_0 - \varphi_L}{\varphi_\infty - \varphi_L}L = \frac{\phi_0 L}{\varphi_\infty}, \tag{7.18}$$

which is the final height of the sediment (7.3). The value of t_c can be now determined from the intersection of $z_3(t)$ and z_c giving

$$t_c = \frac{\phi_0 L}{\varphi_\infty f_b'(\varphi_\infty)}. \tag{7.19}$$

The solution of problem (7.1) for $t_1 \leq t < t_c$ is

$$\varphi(z,t) = \begin{cases} \varphi_L = 0 & \text{for } z_4(t) \leq z, \\ h_2(z/t) & \text{for } z_3(t) \leq z < z_4(t), \\ \varphi_\infty & \text{for } z < z_3(t). \end{cases} \tag{7.20}$$

At (z_c, t_c), we solve the final Riemann problem between φ_L and φ_∞, which for $t \geq t_c$ is given by equation (5.15) with $\varphi_+ = \varphi_L = 0$ and $\varphi_- = \varphi_\infty$ and consists of two constant states φ_L and φ_∞ separated by the horizontal shock

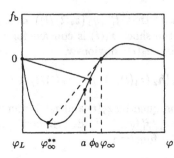 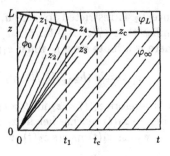

Figure 7.3: Graph of f_b vs φ and representation of the global weak solution $\varphi(z,t)$ for an MS-3 with $a < \phi_0$.

$z_5(t) = z_c$. This shock is the interface between the fluid and the sediment. The solution of problem (7.1) for $t \geq t_c$ is

$$\varphi(z,t) = \begin{cases} \varphi_L = 0 & \text{for } z_c \leq z, \\ \varphi_\infty & \text{for } z < z_c. \end{cases} \tag{7.21}$$

Finally, if we consider any time t_0 such that $0 < t_0 < t_1$, the concentration $\varphi(z_0,t)$ jumps from φ_L to ϕ_0 and then changes suddenly from ϕ_0 to ϕ_0^* and then continuously to φ_∞, which is an MS-2. This completes the proof of part (b) of the theorem. ∎

(c) The solution of the Riemann problem at $z = L$ and $t = 0$ consists, as before, of the two constant states $\varphi_L = 0$ and ϕ_0 separated by the shock $z_1(t) = \sigma(\varphi_L, \phi_0)t + L$. The solution of the Riemann problem at $z = 0$, $t = 0$ is given by equation (5.17) with $\varphi_+ = \phi_0$ and $\varphi_- = \varphi_\infty$. Here $z_2(t) = f_b'(\phi_0)t$ for $0 \leq t < t_1$ is a line of continuity and the states ϕ_0 and φ_∞ are separated only by a rarefaction wave. The point of intersection of $z_2(t)$ with $z_1(t)$ is

$$z_1 = z_1(t_1) = \frac{f_b'(\phi_0)L}{f_b'(\phi_0) - \sigma(\varphi_L, \phi_0)}, \quad t_1 = \frac{L}{f_b'(\phi_0) - \sigma(\varphi_L, \phi_0)}.$$

The height z_1 is positive and smaller than L and t_c is positive because $f_b'(\phi_0)$ is positive (see equation (7.7)) and $\sigma(\varphi_L, \phi_0)$ is negative. For $0 \leq t < t_1$, the solution is given by equation (7.12) with $z_2(t) = f_b'(\phi_0^*)t$ replaced by $z_2(t) = f_b'(\phi_0)t$. At the point (z_1, t_1), the interaction between the shock $z_1(t)$ and the rarefaction wave is identical to the one considered in part (b); therefore the solution for $t_1 \leq t < t_c$ is given by equation (7.20) with t_c as in equation (7.19). The final Riemann problem at (z_c, t_c) is identical with the one solved in part (b), so the solution for $t \geq t_c$ is given by equation (7.21). If we consider a time t_0 such that $0 < t_0 < t_1$, the concentration $\varphi(z,t_0)$ jumps from zero to ϕ_0 increases continuously from ϕ_0 to φ_∞, corresponding to an MS-3. This completes the proof of part (c) of the theorem. ∎

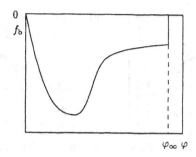

Figure 7.4: Cut flux density function with one inflection point.

7.3.2 Flux density function with two inflection points

A typical example of a flux density function with two inflection points at $\varphi = a$ and $\varphi = b$ is that which corresponds to a suspension of small and closely sized glass beads, such as that described by Shannon *et al.* (1963, 1964); see equation (7.59) and Figure 7.26. Some researchers (Fitch 1983, Kluwick 1977) have used the flux density function with one inflection point as a flux density function with two inflection points by cutting the function at a certain value of φ as plotted in Figure 7.4. The modes of sedimentation of this type of flux density function are similar to those of a flux density function with two inflection points. We assume that $a < b$ and that $f_b''(\varphi) > 0$ for $0 < \varphi < a$ and $b < \varphi < \varphi_\infty$ and that $f_b''(\varphi) < 0$ for $a < \varphi < b$.

Definition 7.3 *For $\varphi_L < a < b < \varphi_\infty$, let*

$$\gamma_1(\varphi_\infty) = \min\{\varphi_\infty\} \cup \{\varphi > a : f_b(a) + (\varphi - a)f_b'(a) \le f_b(\varphi)\},$$
$$\gamma_2(\varphi_L) = \max\{\varphi_L\} \cup \{\varphi < b : f_b(b) + (\varphi - b)f_b'(b) \le f_b(\varphi)\}.$$

There are four cases possible,

(i) $\gamma_1(\varphi_\infty) = \varphi_\infty$ and $\gamma_2(\varphi_L) = \varphi_L$,

(ii) $\gamma_1(\varphi_\infty) = \varphi_\infty$ and $\gamma_2(\varphi_L) > \varphi_L$,

(iii) $\gamma_1(\varphi_\infty) < \varphi_\infty$ and $\gamma_2(\varphi_L) = \varphi_L$ and

(iv) $\gamma_1(\varphi_\infty) < \varphi_\infty$ and $\gamma_2(\varphi_L) > \varphi_L$;

see Figure 7.5. As can be expected by comparing Definition 7.3 and Figure 7.5 with Definition 5.5, Case II of Theorem 5.3 and Figure 5.20, Case (i) corresponds to the existence of two tangential points $\tilde{\varphi}_t = \varphi_L^*$ and $\varphi_t = \varphi_\infty^*$ which allows rarefaction waves between φ_L and φ_∞ for appropriate values of ϕ_0. In Case (ii), only the tangential point $\varphi_t = \varphi_\infty^*$ exists and rarefaction waves are only possible between ϕ_0 and φ_∞; similarly, in Case (iii) only the tangential point $\tilde{\varphi}_t$ can be constructed

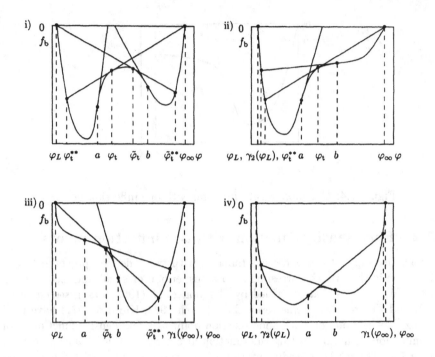

Figure 7.5: The four possible cases in Definition 7.3.

and rarefaction waves are possible only between φ_L and ϕ_0, while in Case (iv) there exists no tangential point, and rarefaction waves are excluded. Since the greatest variety of modes of sedimentation occurs for $\gamma_1(\varphi_\infty) = \varphi_\infty$ and $\gamma_2(\varphi_L) = \varphi_L$, the construction of the global weak solution of (7.1) will be presented in detail for this case. In the remaining cases, the construction is analogous and is summarized in Theorem 7.3.

We note that Spencer $et\ al.$ (1989) and Barton $et\ al.$ (1992) consider flux density functions of the type

$$f_b(\varphi) = v_0\varphi \left(1 - \frac{\varphi}{\varphi_\infty}\right)^C + v_1\varphi^2(\varphi_\infty - \varphi), \quad v_0, v_1 < 0, \quad C > 1, \qquad (7.22)$$

where $0 < \varphi_\infty \leq 1$ denotes the maximum solid concentration. By varying the parameters v_1, v_2 and C, any of the four situations of Definition 7.3 can be reproduced. This shows that the modes of sedimentation MS-6 and MS-7, which contain a rarefaction wave between zero and ϕ_0, although originally not considered by Bustos and Concha (1988), can indeed occur within a certain class of flux density functions, see Bürger and Tory (1999).

Lemma 7.1 $For\ a\ flux\ density\ function\ f_b\ with\ two\ inflection\ points\ a < b\ there$

exists a number $a < \tilde{\varphi}_t \le b$ satisfying

$$f'_b(\tilde{\varphi}_t) = \sigma(\varphi_L, \tilde{\varphi}_t) \qquad (7.23)$$

if and only if $\gamma_2(\varphi_L) = \varphi_L = 0$. Similarly, there exists a number $a \le \varphi_t < b$ satisfying

$$f'_b(\varphi_t) = \sigma(\varphi_\infty, \varphi_t) \qquad (7.24)$$

if and only if $\gamma_1(\varphi_\infty) = \varphi_\infty$. The numbers $\tilde{\varphi}_t$ and φ_t are unique.

Proof. Assume that $\gamma_2(\varphi_L) = \varphi_L$, then

$$f_b(b) + (\varphi - b)f'_b(b) > f_b(\varphi) \text{ for all } \varphi \in (\varphi_L, b)$$

and hence the function

$$g_2(\varphi) = f_b(\varphi) + (\varphi_L - \varphi)f'_b(\varphi)$$

satisfies $g_2(b) \ge f_b(\varphi_L)$. Moreover,

$$g'_2(\varphi) = (\varphi_L - \varphi)f''_b(\varphi) > 0 \text{ for all } \varphi \in (a, b), \qquad (7.25)$$

and, since $f''_b(\varphi) > 0$ for $\varphi < a$,

$$g_2(a) = f_b(a) + (\varphi_L - a)f'_b(a) < f_b(\varphi_L).$$

Consequently, there exists a unique number $a < \tilde{\varphi}_t \le b$ satisfying $g_2(\tilde{\varphi}_t) = f_b(\varphi_L)$ and therefore

$$f'_b(\tilde{\varphi}_t) = \frac{f_b(\varphi_L) - f_b(\tilde{\varphi}_t)}{\varphi_L - \tilde{\varphi}_t} = \sigma(\varphi_L, \tilde{\varphi}_t).$$

On the other hand, if $\gamma_2(\varphi_L) \ne \varphi_L$, i.e. $\gamma_2(\varphi_L) > \varphi_L$, we have $g_2(b) < f_b(\varphi_L)$. Therefore $g_2(\varphi) < f_b(\varphi_L)$ for $\varphi \in [a, b]$ and hence

$$f'_b(\varphi) > \frac{f_b(\varphi_L) - f_b(\varphi)}{\varphi_L - \varphi} = \sigma(\varphi_L, \varphi) \text{ for all } \varphi \in [a, b],$$

i.e. no tangential point $\tilde{\varphi}_t$ satisfying (7.23) exists. This proves the statement of Lemma 7.1 with respect to $\tilde{\varphi}_t$. The proof of the corresponding statement for φ_t is analogous, using $g_1(\varphi) = f_b(\varphi) + (\varphi_\infty - \varphi)f'_b(\varphi)$ instead of g_2. ∎

Extending the tangent to the graph of f through $(\varphi_L, 0)$ and $(\tilde{\varphi}_t, f_b(\tilde{\varphi}_t))$ leads to a third intersection point $(\tilde{\varphi}_t^{**}, f_b(\tilde{\varphi}_t^{**}))$. Similarly, extending the tangent through $(\varphi_t, f_b(\varphi_t))$ and $(\varphi_\infty, 0)$ backwards we obtain the intersection point $(\varphi_t^{**}, f_b(\varphi_t^{**}))$. The points φ_t, $\tilde{\varphi}_t$, φ_t^{**} and $\tilde{\varphi}_t^{**}$ are also displayed in Figure 7.5, wherever they exist. We now start constructing global weak solutions.

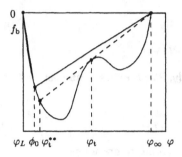 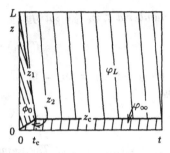

Figure 7.6: Graph of f_b vs φ and representation of the global weak solution $\varphi(z, t)$ for an MS-1 with $\gamma_1(\varphi_\infty) = \varphi_\infty$, $\gamma_2(\varphi_L) = \varphi_L$ and $\phi_0 < \varphi_t^{**}$.

Theorem 7.2 *For a flux density function f_b with two inflection points a and b which satisfies $\gamma_1(\varphi_\infty) = \varphi_\infty$ and $\gamma_2(\varphi_L) = \varphi_L$, we have:*

(a) *If $\phi_0 < \varphi_t^{**}$, the KSP is an MS-1 (see Figure 7.6).*

(b) *If $\varphi_t^{**} < \phi_0 < a$, the KSP is an MS-4 (see Figure 7.7).*

(c) *If $a < \phi_0 < \varphi_t$, the KSP is an MS-5 (see Figure 7.8).*

(d) *If $\varphi_t < \phi_0 < \tilde{\varphi}_t$, the KSP is an MS-1 (see Figure 7.9).*

(e) *If $\tilde{\varphi}_t < \phi_0 < b$, the KSP is an MS-6 (see Figure 7.10).*

(f) *If $b < \phi_0 < \tilde{\varphi}_t^{**}$, the KSP is an MS-7 (see Figure 7.11).*

(g) *If $\tilde{\varphi}_t^{**} < \phi_0 < \varphi_\infty$, the KSP is an MS-1 (see Figure 7.12).*

Proof.

(a) The solution of the Riemann problem at $z = L$, $t = 0$ is

$$\varphi(z, t) = \begin{cases} \varphi_L = 0 & \text{for } z_1(t) \leq z, \\ \varphi_\infty & \text{for } z < z_1(t), \end{cases} \qquad (7.26)$$

where $z_1(t) = \sigma(\varphi_L, \phi_0)t$. The states φ_L and ϕ_0 are separated by the discontinuity $z_1(t)$. This discontinuity is a shock since it satisfies Oleinik's condition E: the chord $\sigma(\varphi_L, \phi_0)$ lies completely on one side of the graph of f_b. This shock is the interface between the water and the suspension of constant concentration ϕ_0 and its slope $\sigma(\varphi_L, \phi_0) = f_b(\phi_0)/\phi_0$ is negative. The solution of the Riemann problem at $z = 0$, $t = 0$ corresponds to Case I for a flux density function with two inflection points which we saw in Chapter 5. The solution is given by equation (5.29) with $\varphi_+ = \phi_0$ and $\varphi_- = \varphi_\infty$ and consists of two constant states ϕ_0 and φ_∞ separated by the shock $z_2(t) = \sigma(\phi_0, \varphi_\infty)t$. This shock corresponds to the interface between

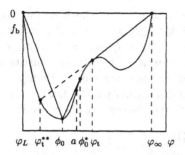 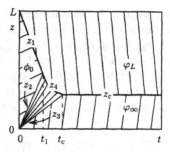

Figure 7.7: Graph of f_b vs φ and representation of the global weak solution $\varphi(z,t)$ for an MS-4 with $\gamma_1(\varphi_\infty) = \varphi_\infty$, $\gamma_2(\varphi_L) = \varphi_L$ and $\varphi_t^{**} < \phi_0 < a$.

the sediment and the suspension. Its slope $\sigma(\phi_0, \varphi_\infty) = f_b(\phi_0)/(\phi_0 - \varphi_\infty)$ is positive. The two shocks $z_1(t)$ and $z_2(t)$ intersect at the point $z = z_c$ and $t = t_c$:

$$z_c = z_1(t_c) = z_2(t_c) = \frac{\phi_0}{\varphi_\infty} L, \quad t_c = \frac{L}{\sigma(\phi_0, \varphi_\infty) - \sigma(\varphi_L, \phi_0)},$$

which is the critical point of the suspension. Here again t_c is positive and finite. The solution of problem (7.1) for $0 \leq t < t_c$ is

$$\varphi(z,t) = \begin{cases} \varphi_L = 0 & \text{for } z_1(t) \leq z, \\ \phi_0 & \text{for } z_2(t) \leq z < z_1(t), \\ \varphi_\infty & \text{for } z < z_2(t). \end{cases} \tag{7.27}$$

Finally, the solution of the Riemann problem at the point (z_c, t_c) is given by equation (5.29) with $\varphi_+ = \varphi_L$ and $\varphi_- = \varphi_\infty$. It consists of two constant states φ_L and φ_∞ separated by the horizontal shock $z_3(t) = z_c$. This shock is the water-sediment interface; see Figure 7.6. The solution of problem (7.1) for $t_c \leq t$ is

$$\varphi(z,t) = \begin{cases} \varphi_L = 0 & \text{for } z_c \leq z, \\ \varphi_\infty & \text{for } z < z_c. \end{cases} \tag{7.28}$$

At a time t_0 such that $0 < t_0 < t_c$, the concentration changes suddenly from φ_L to ϕ_0 and then from ϕ_0 to φ_∞; therefore the KSP is an MS-1. This completes the proof of part (a) of the theorem. ∎

(b) Let h_4 be the inverse of $f_b'(\varphi)$ restricted to values of φ between a and b. Since $\varphi_t^{**} < \phi_0 < a$ then by Lemma 5.3 of Chapter 5, we have that $a < \phi_0^* < \varphi_t$. The solution of the Riemann problem at $z = L, t = 0$ is given by (7.26) and consists of two constant states $\varphi_L = 0$ and ϕ_0 separated by the shock $z_1(t) =$

$\sigma(\varphi_L, \phi_0)t + L$ of negative slope. At $z = 0$, $t = 0$, the Riemann problem has the solution given by equation (5.30) with $\varphi_+ = \phi_0$ and $\varphi_- = \varphi_\infty$ and with $z_2(t) = f_b'(\phi_0^*)t = \sigma(\phi_0, \phi_0^*)t$ and $z_3(t) = f_b'(\varphi_t)t = \sigma(\varphi_t, \varphi_\infty)t$. The lines $z_2(t)$ and $z_3(t)$ are contact discontinuities. The solution consists of two constant states ϕ_0 and φ_∞ separated by the contact discontinuity $z_2(t)$, a rarefaction wave and the other contact discontinuity $z_3(t)$. Let (z_1, t_1) be the point of intersection of the shock $z_1(t)$ and the contact discontinuity $z_2(t)$:

$$z_1 = z_1(t_1) = z_2(t_1) = \frac{f_b'(\phi_0^*)L}{f_b'(\phi_0^*) - \sigma(\varphi_L, \phi_0)}, \quad t_1 = \frac{L}{f_b'(\phi_0^*) - \sigma(\varphi_L, \phi_0)}.$$

Since $a < \phi_0^* < \varphi_t$, $f_b''(\varphi) < 0$ for $a < \varphi < b$ and $f_b'(\varphi_t) > 0$, we have $f_b'(\phi_0^*) > 0$, and since $\sigma(\varphi_L, \phi_0) < 0$, the height z_1 is positive and smaller than L and t_1 is positive. The solution of problem (7.1) for $0 \le t < t_1$ is

$$\varphi(z, t) = \begin{cases} \varphi_L = 0 & \text{for } z_1(t) \le z, \\ \phi_0 & \text{for } z_2(t) \le z < z_1(t), \\ h_4(z/t) & \text{for } z_3(t) \le z < z_2(t), \\ \varphi_\infty & \text{for } z < z_3(t). \end{cases} \tag{7.29}$$

Next, we consider the interaction of the shock $z_1(t)$ with the rarefaction wave at the point (z_1, t_1), which is the curved shock $z_4(t)$ described in part (b) of Theorem 7.1 with $h_2(z/t)$ replaced by $h_4(z/t)$. The solution of problem (7.1) for $t_1 \le t < t_c$ is

$$\varphi(z, t) = \begin{cases} \varphi_L = 0 & \text{for } z_4(t) \le z, \\ h_4(z/t) & \text{for } z_3(t) \le z < z_4(t), \\ \varphi_\infty & \text{for } z < z_3(t) \end{cases} \tag{7.30}$$

with z_c and t_c given by

$$z_c = \frac{\phi_0}{\varphi_\infty}L, \quad t_c = \frac{\phi_0 L}{\varphi_\infty f_b'(\varphi_t)}. \tag{7.31}$$

The time t_c is positive as $f_b'(\varphi_t) > 0$. The solution of the last Riemann problem at (z_c, t_c) is given by equation (5.29) with $\varphi_+ = \varphi_L = 0$ and $\varphi_- = \varphi_\infty$ and consists of two constant states φ_L and φ_∞ separated by the horizontal shock $z_5(t) = z_c$. The solution of problem (7.1) is

$$\varphi(z, t) = \begin{cases} \varphi_L = 0 & \text{for } z_c \le z, \\ \varphi_\infty & \text{for } z < z_c. \end{cases} \tag{7.32}$$

If we consider a time t_0 such that $0 < t_0 < t_1$, the concentration $\varphi(z, t_0)$ jumps from φ_L to ϕ_0 and then from ϕ_0 to ϕ_0^*, then increases continuously to φ_t and finally jumps to φ_∞. This behavior corresponds to an MS-4. This completes the proof of part (b) of the theorem. ∎

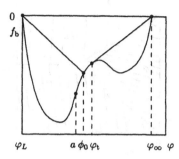
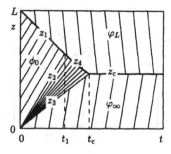

Figure 7.8: Graph of f_b vs φ and representation of the global weak solution $\varphi(z,t)$ for an MS-5 with $\gamma_1(\varphi_\infty) = \varphi_\infty$, $\gamma_2(\varphi_L) = \varphi_L$ and $a < \phi_0 < \varphi_t$.

(c) The solution of the initial Riemann problem at $z = L$ is given by (7.26) and consists of two constant states $\varphi_L = 0$ and ϕ_0 separated by the shock $z_1(t) = \sigma(\varphi_L, \phi_0)t + L$. At $z = 0$, $t = 0$, the solution of the Riemann problem is given in (5.31) with $\varphi_+ = \phi_0$ and $\varphi_- = \varphi_\infty$. The solution consists of two constant states ϕ_0 and φ_∞ separated by a rarefaction wave and the contact discontinuity $z_3(t) = f_b'(\varphi_t)t = \sigma(\varphi_t, \varphi_\infty)t$. The line $z_2(t) = f_b'(\phi_0^*)t$ is a line of continuity. The point of intersection of the shock $z_1(t)$ with the characteristic $z_2(t)$ is

$$z_1 = z_1(t_1) = z_2(t_1) = \frac{f_b'(\phi_0)L}{f_b'(\phi_0) - \sigma(\varphi_L, \phi_0)}, \quad t_1 = \frac{L}{f_b'(\phi_0) - \sigma(\varphi_L, \phi_0)}.$$

Since $f_b'(\phi_0) > 0$ as $f_b'(\varphi) > 0$ for $a < \varphi \le \varphi_t$ and since $\sigma(\varphi_L, \phi_0)$ is negative, t_1 is positive and z_1 is positive and smaller than L. The solution of problem (7.1) for $0 \le t < t_1$ is given by equation (7.29). At $z = z_1$ and $t = t_1$, the interaction of the shock $z_1(t)$ with the rarefaction wave is the curved shock $z_4(t)$ described in part (b) of Theorem 7.1 with $h_2(z/t)$ replaced by $h_4(z/t)$. The solution of problem (7.1) for $t_1 \le t < t_c$ is given by equation (7.30) with z_c and t_c as given by (7.31). The solution of the last Riemann problem at $z = z_c$ and $t = t_c$ is given by equation (5.29) with $\varphi_+ = \varphi_L = 0$ and $\varphi_- = \varphi_\infty$ and consists of two constant states φ_L and φ_∞ separated by the horizontal shock $z_5(t) = z_c$. The solution of problem (7.1) for $t \ge t_c$ is given by (7.32). For a time $0 < t_0 < t_1$, the concentration $\varphi(z, t_0)$ jumps from φ_L to ϕ_0, then increases continuously from ϕ_0 to φ_t and finally jumps to φ_∞. This behavior corresponds to an MS-5. This completes the proof of part (c) of the theorem. ■

(d) The solution of the initial Riemann problem at $z = L$ is given by (7.26) and consists of two constant states $\varphi_L = 0$ and ϕ_0 separated by the shock $z_1(t) = \sigma(\varphi_L, \phi_0)t + L$. At $z = 0$, $t = 0$, the solution of the Riemann problem is given by equation (5.32) with $\varphi_+ = \phi_0$ and $\varphi_- = \varphi_\infty$. It consists of two

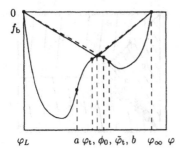
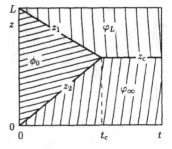

Figure 7.9: Graph of f_b vs φ and representation of the global weak solution $\varphi(z,t)$ for an MS-1 with $\gamma_1(\varphi_\infty) = \varphi_\infty$, $\gamma_2(\varphi_L) = \varphi_L$ and $\varphi_t < \phi_0 < \tilde{\varphi}_t$.

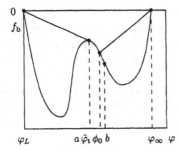
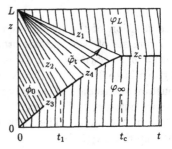

Figure 7.10: Graph of f_b vs φ and representation of the global weak solution $\varphi(z,t)$ for an MS-6 with $\gamma_1(\varphi_\infty) = \varphi_\infty$, $\gamma_2(\varphi_L) = \varphi_L$ and $\tilde{\varphi}_t < \phi_0 < b$.

constant states ϕ_0 and φ_∞ separated by the shock $z_2(t) = \sigma(\phi_0, \varphi_\infty)t$. The two shocks $z_1(t)$ and $z_2(t)$ intersect at the point

$$z_1(t_c) = z_2(t_c) = z_c = \frac{\phi_0 L}{\varphi_\infty}, \quad t_c = \frac{L}{\sigma(\phi_0, \varphi_\infty) - \sigma(\varphi_L, \varphi_\infty)},$$

which is the critical point of the suspension. Here again t_c is positive and finite. The solution of problem (7.1) for $0 \le t < t_c$ is given by equation (7.27). Finally, the solution of the Riemann problem at $z = z_c$ and $t = t_c$ is given by equation (5.29) with $\varphi_+ = \varphi_L$ and $\varphi_- = \varphi_\infty$. It consists of two constant states φ_L and φ_∞ separated by the horizontal shocks $z_3(t) = z_c$. The solution of problem (7.1) for $t \ge t_c$ is given by equation (7.28). At a time t_0 such that $0 < t_0 < t_c$, the concentration changes suddenly from φ_L to ϕ_0 and then ϕ_0 to φ_∞. Therefore the KSP is an MS-1. This completes the proof of part (d) of the theorem. ∎

(e) Let h_4 denote the inverse of f_b' for $\varphi \in [a, b]$. From Case II of Theorem 5.2 we conclude that the solution of the Riemann problem at $z = L$, $t = 0$ is given by

$$\varphi(z,t) = \begin{cases} \varphi_L = 0 & \text{for} \quad z \geq z_1(t), \\ h_4((z-L)/t) & \text{for} \quad z_2(t) \leq z < z_1(t), \\ \phi_0 & \text{for} \quad z_3(t) \leq z < z_2(t), \end{cases} \tag{7.33}$$

where the lines $z_1(t) = L + tf_b'(\tilde{\varphi}_t)$ and $z_2(t) = L + tf_b'(\phi_0)$ are contact discontinuities. Moreover, the solution of the Riemann problem at $z = 0$, $t = 0$ is

$$\varphi(z,t) = \begin{cases} \phi_0 & \text{for} \quad z \geq z_3(t), \\ \varphi_\infty & \text{for} \quad 0 \leq z < z_3(t), \end{cases} \tag{7.34}$$

which follows from the analogue of Case I of Theorem 5.2 obtained by interchanging the convexity assumptions stated at the beginning of Section 5.3. The shock $z_3(t)$ and the contact discontinuity $z_2(t)$ meet at the point (z_1, t_1) where

$$z_1 = z_2(t_1) = \frac{\sigma(\varphi_\infty, \phi_0)L}{\sigma(\varphi_\infty, \phi_0) - f_b'(\phi_0)}, \quad t_1 = \frac{L}{\sigma(\phi_0, \varphi_\infty) - f_b'(\phi_0)}.$$

The height z_1 is positive and smaller than L and t_1 is positive: from $f''(\varphi) < 0$ for $a < \varphi < b$ we obtain $f_b'(\phi_0) < f_b'(\tilde{\varphi}_t) = \sigma(0, \tilde{\varphi}_t)$, and $f_b(\phi_0) < f_b(\varphi_\infty)$ yields $\sigma(\phi_0, \varphi_\infty) > 0$ and hence $t_1 > 0$. Obviously,

$$0 < \frac{\sigma(\varphi_\infty, \phi_0)}{\sigma(\varphi_\infty, \phi_0) - f_b'(\phi_0)} < 1$$

and therefore $0 < z_1 < L$. At $t = t_1$ the shock $z_3(t)$ and the rarefaction wave start to interact. Similar to the interaction considered previously for the case of one inflection point only, the concentration value φ_∞ now meets values $\varphi \in [\tilde{\varphi}_t, \phi_0]$. Each jump corresponds to the analogue of Case I of Theorem 5.2 mentioned above, where $\varphi_- = \varphi_\infty$ and φ_+ varies from $\varphi_+ = \phi_0$ to $\varphi_+ = \tilde{\varphi}_t$. Here we obtain a curved shock $z_4(t)$ since the successive jumps produce shocks of decreasing slopes till φ_+ reaches the value $\tilde{\varphi}_t$. Consequently, the shock $z_4(t)$ satisfies the relation

$$z_4'(t) = \sigma(\varphi_\infty, h_4((z_4(t) - L)/t)) = \frac{f_b(\varphi_\infty) - f_b(h_4((z_4(t) - L)/t))}{\varphi_\infty - h_4((z_4(t) - L)/t)}. \tag{7.35}$$

Differentiating equation (7.35) yields

$$z_4''(t) = \frac{(z_4'(t) - (z_4(t) - L)/t)^2}{tf_b''(h_4((z_4(t) - L)/h))(\varphi_\infty - h_4((z_4(t) - L)/t))}. \tag{7.36}$$

Since $f_b''(\varphi) < 0$ for $a < \varphi < b$, we obtain $z_4''(t) < 0$, i.e., the shock $z_4(t)$ is convex for $t > t_1$. Next, we prove that $z_4(t)$ and $z_1(t)$ intersect. Again

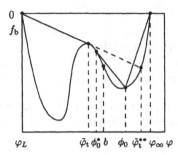 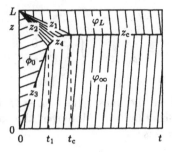

Figure 7.11: Graph of f_b vs φ and representation of the global weak solution $\varphi(z,t)$ for an MS-7 with $\gamma_1(\varphi_\infty) = \varphi_\infty$, $\gamma_2(\varphi_L) = \varphi_L$ and $b < \phi_0 < \tilde{\varphi}_t^{**}$.

the intersection point is denoted by (z_c, t_c), since it turns out that the intersection point is the critical point of the suspension. By the assumptions on the function f_b, we have $z_4'(t) > 0$ and hence $z_4(t) > z_1$; moreover, $z_1'(t) = f_b'(\tilde{\varphi}_t) < 0$. Since $z_1(0) = L > z_1$, it is evident that $z_4(t)$ and $z_1(t)$ intersect. Repeating the arguments that led to (7.18), we obtain that this intersection takes place at

$$z_c = \frac{\phi_0 L}{\varphi_\infty}, \quad t_c = \frac{\phi_0 L}{\varphi_\infty f_b'(\tilde{\varphi}_t)}.$$

The solution of problem (7.1) for $0 \le t < t_c$ is

$$\varphi(z,t) = \begin{cases} \varphi_L = 0 & \text{for} \quad z \ge z_1(t), \\ h_4((z-L)/t) & \text{for} \quad z_2(t) \le z < z_1(t), \\ \phi_0 & \text{for} \quad z_3(t) \le z < z_2(t), \\ \varphi_\infty & \text{for} \quad z < z_3(t) \end{cases} \tag{7.37}$$

for $0 \le t < t_1$ and

$$\varphi(z,t) = \begin{cases} \varphi_L = 0 & \text{for} \quad z \ge z_1(t), \\ h_4((z-L)/t) & \text{for} \quad z_4(t) \le z < z_1(t), \\ \varphi_\infty & \text{for} \quad z < z_4(t). \end{cases} \tag{7.38}$$

for $t_1 \le t < t_c$. Finally, the solution of the Riemann problem at $z = z_c$ and $t = t_c$ is given by equation (5.29) with $\varphi_+ = \varphi_L$ and $\varphi_- = \varphi_\infty$. It consists of two constant states φ_L and φ_∞ separated by the horizontal shock $z_3(t) = z_c$. The solution of problem (7.1) for $t \ge t_c$ is given by equation (7.28). At a time t_0 such that $0 < t_0 < t_1$, the concentration changes suddenly from φ_L to $\tilde{\varphi}_t$ and then continuously to ϕ_0. Next, it changes abruptly to φ_∞. Therefore the KSP is an MS-6. This completes the proof of part (e) of the theorem. ∎

(f) Let h_4 denote the inverse of f'_b for $\varphi \in [a, b]$. From Case II of Theorem 5.3 we conclude that the solution of the Riemann problem at $z = L$, $t = 0$ is given by

$$\varphi(z, t) = \begin{cases} \varphi_L = 0 & \text{for} \quad z \geq z_1(t), \\ h_4((z - L)/t) & \text{for} \quad z_2(t) \leq z < z_1(t), \\ \phi_0 & \text{for} \quad z_3(t) \leq z < z_2(t), \end{cases} \tag{7.39}$$

where the lines $z_1(t) = L + t f'_b(\bar{\varphi}_t)$ and $z_2(t) = L + t f'_b(\phi_0^*)$ are contact discontinuities. Moreover, the solution of the Riemann problem at $z = 0$, $t = 0$ is given by (7.34). The shock $z_3(t)$ and the contact discontinuity $z_2(t)$ meet at the point (z_1, t_1) where

$$z_1 = z_2(t_1) = \frac{\sigma(\varphi_\infty, \phi_0)L}{\sigma(\varphi_\infty, \phi_0) - f'_b(\phi_0^*)}, \quad t_1 = \frac{L}{\sigma(\phi_0, \varphi_\infty) - f'_b(\phi_0^*)}.$$

As in part (e), it is easy to see that the height z_1 is positive and smaller than L and that t_1 is positive. At $t = t_1$ the shock $z_3(t)$ and the rarefaction wave start to interact, and we obtain a curved shock $z_4(t)$ satisfying (7.35), which is convex as in part (e). Repeating the arguments of part (e), we see that $z_4(t)$ and $z_1(t)$ intersect at

$$z_c = \frac{\phi_0 L}{\varphi_\infty}, \quad t_c = \frac{\phi_0 L}{\varphi_\infty f'_b(\bar{\varphi}_t)}.$$

The solution of problem (7.1) for $0 \leq t < t_c$ is

$$\varphi(z, t) = \begin{cases} \varphi_L = 0 & \text{for} \quad z \geq z_1(t), \\ h_4((z - L)/t) & \text{for} \quad z_2(t) \leq z < z_1(t), \\ \phi_0 & \text{for} \quad z_3(t) \leq z < z_2(t), \\ \varphi_\infty & \text{for} \quad z < z_3(t) \end{cases} \tag{7.40}$$

for $0 \leq t < t_1$ and

$$\varphi(z, t) = \begin{cases} \varphi_L = 0 & \text{for} \quad z \geq z_1(t), \\ h_4((z - L)/t) & \text{for} \quad z_4(t) \leq z < z_1(t), \\ \varphi_\infty & \text{for} \quad z < z_4(t). \end{cases} \tag{7.41}$$

for $t_1 \leq t < t_c$. The solution of problem (7.1) for $t \geq t_c$ is given by equation (7.28), as in part (e). At a time t_0 such that $0 < t_0 < t_1$, the concentration changes suddenly from φ_L to $\bar{\varphi}_t$, then continuously to ϕ_0^* and then suddenly again to ϕ_0. Next, it changes abruptly to φ_∞. Therefore the KSP is an MS-7. This completes the proof of part (f) of the theorem. ∎

(g) This part of the theorem is proved exactly as part (a), except that here the Riemann problem whose solution corresponds to Case I for a flux density function with two inflection points in Chapter 5 is the one at $z = L$, $t = 0$.

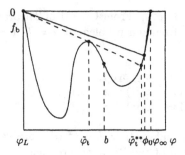 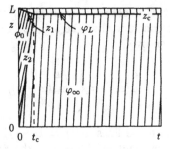

Figure 7.12: Graph of f_b vs φ and representation of the global weak solution $\varphi(z,t)$ for an MS-1 with $\gamma_1(\varphi_\infty) = \varphi_\infty$, $\gamma_2(\varphi_L) = \varphi_L$ and $\tilde{\varphi}_t^{**} < \phi_0 < \varphi_\infty$.

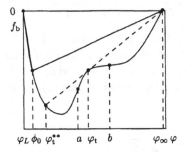 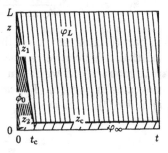

Figure 7.13: Graph of f_b vs φ and representation of the global weak solution $\varphi(z,t)$ for an MS-1 with $\gamma_1(\varphi_\infty) = \varphi_\infty$, $\gamma_2(\varphi_L) > \varphi_L$ and $\phi_0 < \tilde{\varphi}_t$.

We now assume $\gamma_1(\varphi_\infty) < \varphi_\infty$ or $\gamma_2(\varphi_L) > \varphi_L$. Since, by Lemma 7.1, at least one of the tangent points φ_t and $\tilde{\varphi}_t$ does not exist then, fewer cases for the location of ϕ_0 have to be distinguished. In all cases, the proofs coincide with the proofs presented for Theorem 7.2, and will not be repeated here. However, illustrations are presented.

Theorem 7.3 *Consider a flux density function f_b with two inflection points a and b. For $\gamma_1(\varphi_\infty) = \varphi_\infty$ and $\gamma_2(\varphi_L) > \varphi_L$, we have:*

(a) *If $\phi_0 < \varphi_t^{**}$, the KSP is an MS-1 (see Figure 7.13).*

(b) *If $\varphi_t^{**} < \phi_0 < a$, the KSP is an MS-4 (see Figure 7.14).*

(c) *If $a < \phi_0 < \varphi_t$, the KSP is an MS-5 (see Figure 7.15).*

(d) *If $\varphi_t < \phi_0 < \varphi_\infty$, the KSP is an MS-1 (see Figure 7.16).*

For $\gamma_1(\varphi_\infty) < \varphi_\infty$ and $\gamma_2(\varphi_L) = \varphi_L$, we have:

(e) *If $\phi_0 < \tilde{\varphi}_t$, the KSP is an MS-1 (see Figure 7.17).*

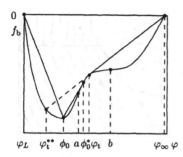
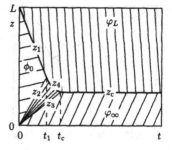

Figure 7.14: Graph of f_b vs φ and representation of the global weak solution $\varphi(z,t)$ for an MS-4 with $\gamma_1(\varphi_\infty) = \varphi_\infty$, $\gamma_2(\varphi_L) > \varphi_L$ and $\varphi_t^{**} < \phi_0 < a$.

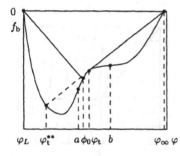
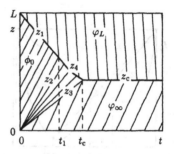

Figure 7.15: Graph of f_b vs φ and representation of the global weak solution $\varphi(z,t)$ for an MS-5 with $\gamma_1(\varphi_\infty) = \varphi_\infty$, $\gamma_2(\varphi_L) > \varphi_L$ and $a < \phi_0 < \varphi_t$.

(f) If $\tilde{\varphi}_t < \phi_0 < b$, the KSP is an MS-6 (see Figure 7.18).

(g) If $b < \phi_0 < \tilde{\varphi}_t^{**}$, the KSP is an MS-7 (see Figure 7.19).

(h) If $\tilde{\varphi}_t^{**} < \phi_0 < \varphi_\infty$, the KSP is an MS-1 (see Figure 7.20).

For $\gamma_1(\varphi_\infty) < \varphi_\infty$ and $\gamma_2(\varphi_L) > \varphi_L$, only one mode of sedimentation is possible:

(i) If $\gamma_1(\varphi_\infty) < \varphi_\infty$ and $\gamma_2(\varphi_L) > \varphi_L$, the KSP is an MS-1 for all values of ϕ_0 (see Figure 7.21).

Proof.

(a)–(c) The solution is the same as in the respective parts of Theorem 7.2. ∎

(d) Since the point $\tilde{\varphi}_t$ does not exist here, the solution of the initial Riemann problem at $z = L$ is always a shock. Therefore the proof of part (d) of Theorem 7.2 applies for the whole range $\varphi_t < \phi_0 < \varphi_\infty$ considered here. ∎

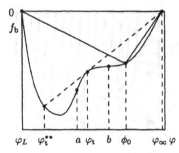 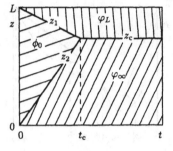

Figure 7.16: Graph of f_b vs φ and representation of the global weak solution $\varphi(z,t)$ for an MS-1 with $\gamma_1(\varphi_\infty) = \varphi_\infty$, $\gamma_2(\varphi_L) > \varphi_L$ and $\varphi_t < \phi_0 < \varphi_\infty$.

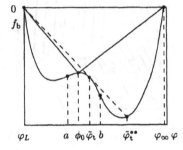 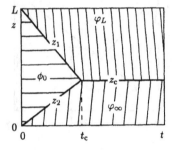

Figure 7.17: Graph of f_b vs φ and representation of the global weak solution $\varphi(z,t)$ for an MS-1 with $\gamma_1(\varphi_\infty) < \varphi_\infty$, $\gamma_2(\varphi_L) = \varphi_L$ and $\phi_0 < \bar{\varphi}_t$.

(e) Here, the point φ_t does not exist. Therefore the solution of the initial Riemann problem at $z = 0$ is always a shock, and again the proof of part (d) of Theorem 7.2 applies for all values of ϕ_0 between φ_L and $\bar{\varphi}_t$ here. ∎

(f)–(h) The solution is the same as in the parts (e), (f) and (g), respectively, of Theorem 7.2. ∎

(i) Since neither φ_t nor $\bar{\varphi}_t$ exists, solutions of both initial Riemann problems at $z = 0$ and $z = L$ are always shocks. Hence for all values of ϕ_0, the solution is the same as in part (a) of Theorem 7.2. ∎

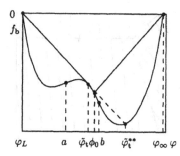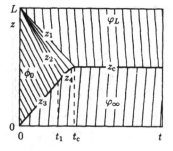

Figure 7.18: Graph of f_b vs φ and representation of the global weak solution $\varphi(z,t)$ for an MS-6 with $\gamma_1(\varphi_\infty) < \varphi_\infty$, $\gamma_2(\varphi_L) = \varphi_L$ and $\tilde{\varphi}_t < \phi_0 < b$.

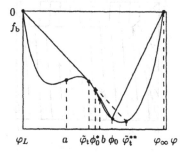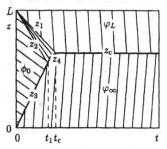

Figure 7.19: Graph of f_b vs φ and representation of the global weak solution $\varphi(z,t)$ for an MS-7 with $\gamma_1(\varphi_\infty) < \varphi_\infty$, $\gamma_2(\varphi_L) = \varphi_L$ and $b < \phi_0 < \tilde{\varphi}_t^{**}$.

7.4 Non-homogeneous initial concentration

Now we consider the case when the initial concentration is no longer constant but decreases monotonically with z between 0 and L:

$$\phi_0(z) = \begin{cases} \alpha_1 & \text{for } 2L/3 \leq z < L, \\ \alpha_2 & \text{for } L/3 \leq z < 2L/3, \\ \alpha_3 & \text{for } 0 \leq z < L/3. \end{cases} \qquad (7.42)$$

To present this case, we take a flux density function with two inflection points with $\gamma_1(\varphi_\infty) = \varphi_\infty$, $\gamma_2(\varphi_L) > \varphi_L$, $0 < \alpha_1 < \varphi_t^{**}$, $\varphi_t^{**} < \alpha_2 < a$, $a < \alpha_3 < \varphi_t$ and $\alpha_2^* < \alpha_3$. See Figure 7.22. This concentration can be interpreted physically as an initial condition of an intermediate state of a continuous sedimentation process. Of course, the solution we will obtain does not correspond to any of the modes of sedimentation considered so far, which appear only for a suspension with a

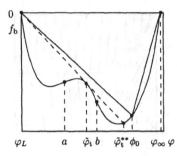 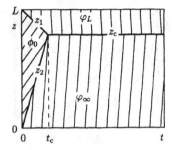

Figure 7.20: Graph of f_b vs φ and representation of the global weak solution $\varphi(z,t)$ for an MS-1 with $\gamma_1(\varphi_\infty) < \varphi_\infty$, $\gamma_2(\varphi_L) = \varphi_L$ and $\tilde{\varphi}_t^{**} < \phi_0 < \varphi_\infty$.

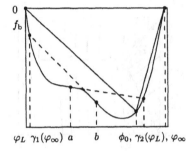 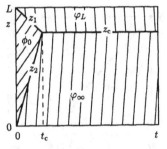

Figure 7.21: Graph of f_b vs φ and representation of the global weak solution $\varphi(z,t)$ for an MS-1 with $\gamma_1(\varphi_\infty) < \varphi_\infty$, $\gamma_2(\varphi_L) > \varphi_L$ and $\varphi_L < \phi_0 < \varphi_\infty$.

constant initial concentration. We consider the initial value problem

$$\frac{\partial \varphi}{\partial t} + \frac{\partial f_b(\varphi)}{\partial z} = 0 \quad \text{for } z \in \mathbb{R},\ t > 0, \tag{7.43a}$$

$$\varphi(z,0) = \varphi_I(z) \quad \text{for } z \in \mathbb{R} \tag{7.43b}$$

with

$$\varphi_I(z) = \begin{cases} \varphi_L = 0 & \text{for } z \geq L, \\ \phi_0(z) & \text{for } 0 \leq z < L, \\ \varphi_\infty & \text{for } z < 0. \end{cases} \tag{7.44}$$

Not to make the construction of the global weak solution of problem (7.43) too tedious, we give only the results and leave the details to the reader.

The initial Riemann problems

The solution of the Riemann problem at $z = L$, $t = 0$ consists of two constant states φ_L and α_1 separated by the shock $z_1(t) = L + \sigma(\varphi_L, \alpha_1)t$. The solution

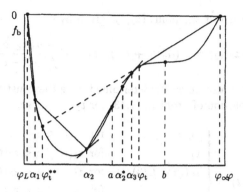

Figure 7.22: Display of the initial values.

of the Riemann problem at $z = 2L/3$, $t = 0$ consists also of two constant states α_1 and α_2 separated by the shock $z_2(t) = 2L/3 + \sigma(\alpha_1, \alpha_2)t$. The solution of the Riemann problem at $z = L/3$, $t = 0$ consists of two constant states α_2 and α_3 separated by the contact discontinuity $z_3(t) = L/3 + f_b'(\alpha_2^*)t = L/3 + \sigma(\alpha_2, \alpha_2^*)t$ and a rarefaction wave. The solution of the Riemann problem at $z = 0$, $t = 0$ consists of two constant states α_3 and φ_∞ separated by a rarefaction wave and a contact discontinuity $z_6(t) = \sigma(\varphi_t, \varphi_\infty)t = f_b'(\varphi_t)t$. The two shocks $z_1(t)$ and $z_2(t)$ intersect at the point (z_1, t_1) with

$$z_1 = z_1(t_1) = z_2(t_1) = \frac{\sigma(\alpha_1, \alpha_2) - \frac{2}{3}\sigma(\varphi_L, \alpha_1)}{\sigma(\alpha_1, \alpha_2) - \sigma(\varphi_L, \alpha_1)}L, \quad t_1 = \frac{L/3}{\sigma(\alpha_1, \alpha_2) - \sigma(\varphi_L, \alpha_1)}.$$

In some cases, the shocks $z_2(t)$ and $z_3(t)$ may intersect first. This can be checked by computing the two times of intersection and choosing the smaller one, which in this example is t_1. The point (z_1, t_1) can be determined once f_b is known. The solution of problem (7.43) for $0 \leq t < t_1$ is

$$\varphi(z, t) = \begin{cases} \varphi_L = 0 & \text{for } z_1(t) \leq z, \\ \alpha_1 & \text{for } z_2(t) \leq z < z_1(t), \\ \alpha_2 & \text{for } z_3(t) \leq z < z_2(t), \\ h_4\left((z - L/3)/t\right) & \text{for } z_4(t) \leq z < z_3(t), \\ \alpha_3 & \text{for } z_5(t) \leq z < z_4(t), \\ h_4(z/t) & \text{for } z_6(t) \leq z < z_5(t), \\ \varphi_\infty & \text{for } z < z_6(t). \end{cases}$$

Here $z_4(t) = (L/3) + f_b'(\alpha_3)t$ and $z_5(t) = f_b'(\alpha_3)t$ are lines of continuity.

The Riemann problem at the point (z_1, t_1)

The solution of the Riemann problem at (z_1, t_1) consists of two constant states $\varphi_L = 0$ and α_2 separated by the shock $z_7(t) = z_1 + \sigma(\varphi_L, \alpha_2)(t - t_1)$. The shock

$z_7(t)$ and $z_3(t)$ intersect at the point (z_2, t_2) given by

$$z_2 = z_7(t_2) = z_3(t_2) = f_b'(\alpha_2^*)t_2 + L/3, \quad t_2 = \frac{z_1 - \sigma(\varphi_L, \alpha_2)t_1 - L/3}{f_b'(\alpha_2^*) - \sigma(\varphi_L, \alpha_2)}.$$

The point (z_2, t_2) can also be easily determined once the constitutive equation of f_b is known. The solution of problem (7.43) for $t_1 \leq t < t_2$ is

$$\varphi(z,t) = \begin{cases} \varphi_L = 0 & \text{for } z_7(t) \leq z, \\ \alpha_2 & \text{for } z_3(t) \leq z < z_7(t), \\ h_4\left((z - L/3)/t\right) & \text{for } z_4(t) \leq z < z_3(t), \\ \alpha_3 & \text{for } z_5(t) \leq z < z_4(t), \\ h_4(z/t) & \text{for } z_6(t) \leq z < z_5(t), \\ \varphi_\infty & \text{for } z < z_6(t). \end{cases}$$

The Riemann problem at the point (z_2, t_2)

The interaction of the shock $z_7(t)$ with the rarefaction wave between $z_3(t)$ and $z_4(t)$ is the curved shock $z_8(t)$ that satisfies

$$z_8'(t) = \sigma\left(\varphi_L, h_4\left((z_8(t) - L/3)/t\right)\right)$$

for $t_2 \leq t \leq t_3$ with $z_7(t_2) = z_3(t_2) = z_8(t_2)$. The shock $z_8(t)$ and the line $z_4(t)$ intersect at the point (z_3, t_3) which cannot be determined in closed form. This shows the shortcomings of the method of characteristics for more general problems and justifies the need for a numerical method which we present in the next section. Supposing that the point (z_3, t_3) has been determined by an approximate method, the solution of the problem (7.43) for $t_2 \leq t < t_3$ is

$$\varphi(z,t) = \begin{cases} \varphi_L = 0 & \text{for } z_8(t) \leq z, \\ h_4\left((z - L/3)/t\right) & \text{for } z_4(t) \leq z < z_8(t), \\ \alpha_3 & \text{for } z_5(t) \leq z < z_4(t), \\ h_4(z/t) & \text{for } z_6(t) \leq z < z_5(t), \\ \varphi_\infty & \text{for } z < z_6(t). \end{cases}$$

The Riemann problem at the point (z_3, t_3)

The solution of the Riemann problem at the point (z_3, t_3) consists of two constant states φ_L and α_3 separated by the shock

$$z_9(t) = z_3 + \sigma(\varphi_L, \alpha_3)(t - t_3).$$

The shock $z_9(t)$ and the line $z_5(t)$ intersect at the point (z_4, t_4) given by

$$z_4 = z_5(t_4) = z_9(t_4) = f_b'(\alpha_3)t_4, \quad t_4 = \frac{z_3 - \sigma(\varphi_L, \alpha_3)t_3}{f_b'(\alpha_3) - \sigma(\varphi_L, \alpha_3)}.$$

The solution of problem (7.43) for $t_3 \leq t < t_4$ is

$$\varphi(z,t) = \begin{cases} \varphi_L = 0 & \text{for } z_9(t) \leq z, \\ \alpha_3 & \text{for } z_5(t) \leq z < z_9(t), \\ h_4(z/t) & \text{for } z_6(t) \leq z < z_5(t), \\ \varphi_\infty & \text{for } z < z_6(t). \end{cases}$$

The Riemann problem at the point (z_4, t_4)

The interaction of the shock $z_9(t)$ with the rarefaction wave between $z_5(t)$ and $z_6(t)$ is the curved shock $z_{10}(t)$ that satisfies

$$z_{10}'(t) = \sigma\left(\varphi_L, h_4\left(z_{10}(t)/t\right)\right)$$

for $t_4 \leq t \leq t_c$ with $z_9(t_4) = z_5(t_4) = z_{10}(t_4)$. The shock $z_{10}(t)$ and the contact discontinuity intersect $z_6(t)$ intersect at the point (z_c, t_c) given by

$$z_c = z_6(t_c) = \frac{\alpha_1 + \alpha_2 + \alpha_3}{3\varphi_\infty} L, \quad t_c = \frac{z_c}{\sigma(\varphi_t, \varphi_\infty)}. \tag{7.45}$$

The solution of the problem (7.43) for $t_4 \leq t < t_c$ is

$$\varphi(z,t) = \begin{cases} \varphi_L = 0 & \text{for } z_{10}(t) \leq z, \\ h_4(z/t) & \text{for } z_6(t) \leq z < z_{10}(t), \\ \varphi_\infty & \text{for } z < z_6(t). \end{cases}$$

The Riemann problem at the point (z_c, t_c)

The solution of this last Riemann problem consists of two constant states φ_L and φ_∞ separated by the horizontal shock $z_{11}(t) = z_c$. The solution of problem (7.43) for $t_c \leq t$ is

$$\varphi(z,t) = \begin{cases} \varphi_L = 0 & \text{for } z_c \leq z, \\ \varphi_\infty & \text{for } z < z_c. \end{cases}$$

This completes the construction of the global weak solution of problem (7.43). A sketch of the solution is shown in Figure 7.23.

7.5 Numerical computation of curved trajectories

We require $z(t)$ for those regions where φ and hence

$$u(\varphi) = \sigma(\varphi_L, \varphi) = f_b(\varphi)/\varphi, \tag{7.46}$$

which is needed to compute the curved shock $z(t) = z_4(t)$ of an MS-2, MS-3, MS-4 or MS-5 described by the differential equation (7.13), change continuously.

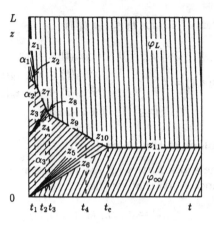

Figure 7.23: Sketch of the weak solution for monotonically increasing initial concentration.

Analogous considerations apply to the curved shock $z(t) = z_4(t)$ of an MS-6 or MS-7 expressed by equation (7.35) and, of course, to curved shocks occurring with non-homogeneous initial concentrations, such as z_8 or z_{10} in Sect. 7.4.

Starting from given values of z_{i-1}, t_{i-1} and φ_{i-1}, we obtain an initial estimate of z_i from

$$z_i^0 = z_{i-1} + u(\varphi_{i-1})\Delta t,$$

where

$$\Delta t = t_i - t_{i-1}. \tag{7.47}$$

We know that (in the simple examples) $f_b'(\varphi_i) = z_i/t_i$, but is generally impossible or at least inefficient to obtain φ_i directly. Fortunately, Newton-Raphson iteration is very efficient because φ changes very slowly. Thus

$$\varphi_i^0(0) = \varphi_{i-1}, \tag{7.48}$$

$$\varphi_i^m(n+1) = \varphi_i^m(n) - \frac{f_b'(\varphi_i^m(n)) - z_i^m/t_i}{f_b''(\varphi_i^m(n))}, \tag{7.49}$$

where m is the index of the outer iteration and n of the inner one. Then

$$z_i^{m+1} = z_{i-1} + \frac{1}{2}[u(\varphi_{i-1}) + u(\varphi_i^m)]\Delta t \tag{7.50}$$

yields an improved estimate. We can refine this estimate by iterating (7.49) and (7.50), but this may not be necessary. Alternatively, we could base our estimates on

$$z_{i-1/2}^0 = z_{i-1} + u(\varphi_{i-1})\frac{\Delta t}{2} \quad \text{and} \quad z_i = z_{i-1} + u(\varphi_{i-1/2})\Delta t.$$

If refinement is necessary, it is better to use an efficient method such as a fourth-order Runge-Kutta or a predictor-corrector method. The only difference from standard schemes is the use of φ to yield $dz/dt = u(\varphi)$. A detailed example of a simple method is given by Shannon *et al.* (1964).

7.6 Dafermos' polygonal approximation method

7.6.1 Polygonal flux-density function

Dafermos (1972) was the first to consider the problem

$$\frac{\partial \varphi}{\partial t} + \frac{\partial \Pi}{\partial z} = 0 \ \text{ for } z \in \mathbb{R}, \ t > 0, \tag{7.51a}$$

$$\varphi(0, z) = \varphi_{\mathrm{I}}(z) \ \text{ for } z \in \mathbb{R} \tag{7.51b}$$

for real polygons Π defined in the interval $[a_1, a_2]$ and step functions φ_{I}. The method of characteristics cannot be applied since the function Π is not continuous. Nevertheless, Dafermos showed that the initial value problem (7.51) has a solution for any polygon Π and every step function φ_{I}. This solution is piecewise constant and consists of straight-line shocks only, which satisfy the Rankine-Hugoniot condition (4.3) and Oleinik's condition E (4.7).

Dafermos' method has become known as the front tracking method, which has been used as an efficient computational tool by many authors. In particular, Holden and Holden (1988) and Holden *et al.* (1988) proved that the method is well-defined and developed it into a numerical method. Front tracking was later extended to systems of equations by Risebro (1993), see also Risebro and Tveito (1991, 1992). We refer to Holden and Risebro (1997) for an excellent introduction to front tracking techniques, see also Espedal and Karlsen (1999) for their application to problems of multiphase flow in porous media.

The Riemann problem

We consider first two examples of Riemann problems for polygons Π and construct their weak solutions.

Example. Consider the real polygonal function Π in the interval $[0, 0.5]$ shown in Figure 7.24:

$$\Pi(\varphi) = \begin{cases} -4\varphi & \text{for } 0.00 \leq \varphi \leq 0.05, \\ 0.5\varphi - 0.225 & \text{for } 0.05 \leq \varphi \leq 0.15, \\ -5\varphi + 0.6 & \text{for } 0.15 \leq \varphi \leq 0.20, \\ 2\varphi - 0.8 & \text{for } 0.20 \leq \varphi \leq 0.30, \\ -0.5\varphi - 0.05 & \text{for } 0.30 \leq \varphi \leq 0.40, \\ 1.25\varphi - 0.75 & \text{for } 0.40 \leq \varphi \leq 0.50, \end{cases}$$

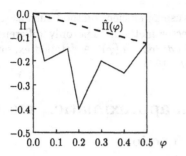 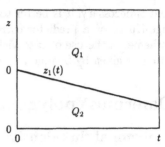

Figure 7.24: Polygonal flux density function $\Pi(\varphi)$ with the convex hull $\hat{\Pi}$ (dashed line) and the corresponding Riemann solution.

and solve the Riemann problem

$$\frac{\partial \varphi}{\partial t} + \frac{\partial \Pi(\varphi)}{\partial z} = 0 \quad \text{for } z \in \mathbb{R}, \quad t > 0,$$

$$\varphi_{\mathrm{I}} = \varphi(z, 0) = \begin{cases} 0 & \text{for } 0 \le z, \\ 0.5 & \text{for } z < 0. \end{cases}$$

Denote by $\hat{\Pi}$ a polygon constructed in the following way: starting at $\varphi = 0$, line segments are drawn connecting the origin with the rest of the corner points $(\varphi, \Pi(\varphi))$ of Π in a Π vs φ plot, and select the line segment with greatest slope. Then, proceed starting with the other vertices till the end point of the interval is reached. The set of these segments so obtained constitutes the polygon $\hat{\Pi}$ also called the *concave hull of* Π. For the polygon Π calculate the slope of the line segments drawn from the origin to the points $(0.05, -0.2)$, $(0.15, -0.15)$, $(0.2, -0.4)$, $(0.3, -0.2)$, $(0.4, -0.25)$ and $(0.5, -0.125)$. The line segment from the origin to the end point $(0.5, -0.125)$ has the greatest slope which is equal to -0.25. Then, the concave hull $\hat{\Pi}$ of Π consists of the line segment $\hat{\Pi}(\varphi) = -0.25\varphi$ only. The solution of the Riemann problem is shown in Figure 7.24 and consists of two constant states corresponding to the φ coordinate of the corner points of the concave hull denoted by: $Q_1 = 0$ and $Q_2 = 0.5$. These two states are separated by the shock $z_1(t) = -0.25t$.

In general, the number of segments of the concave hull is equal to the *number of shocks of the solution* (Kunik 1993), the *slopes of the shocks* are given by the slopes of the segments of the concave hull and the *values of the constant states* separated by shocks are given by the values of φ at the corner points of the concave hull.

Example. Consider now the following polygonal function

$$\Pi_-(\varphi) = -\Pi(\varphi), \tag{7.52}$$

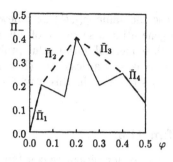 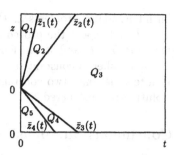

Figure 7.25: Polygonal flux density function $\Pi_-(\varphi)$ with the segments $\overline{\Pi}_i(\varphi)$, $i = 1, \ldots, 4$, of the convex hull (dashed lines) and the corresponding Riemann solution.

where Π was given in the preceding example. The corresponding Riemann problem is

$$\frac{\partial \varphi}{\partial t} + \frac{\partial \Pi_-(\varphi)}{\partial z} = 0 \text{ for } z \in \mathbb{R}, \quad t > 0$$

$$\varphi_I = \varphi(z, 0) = \begin{cases} 0 & \text{for } 0 \leq z, \\ 0.5 & \text{for } z < 0. \end{cases}$$

To build the concave hull $\hat{\Pi}_-$ of Π_-, draw the line segments that connect the origin with all corner points of Π_-. The line segment with the greatest slope is the first one: $\overline{\Pi}_1(\varphi) = 4\varphi$ for $0 \leq \varphi \leq 0.05$ and therefore it is the first segment of the concave hull $\hat{\Pi}_-$. The line that connects the second corner point with the fourth corner point has the next greatest slope, that is: $\overline{\Pi}_2 = 1.\overline{3}\varphi + 0.1\overline{3}$ for $0.05 \leq \varphi \leq 0.2$. The line segment with the next greatest slope is $\overline{\Pi}_3 = -0.75\varphi + 0.55$ for $0.2 \leq \varphi \leq 0.4$. The last segment of the concave hull $\hat{\Pi}_-$ is the segment $\overline{\Pi}_4 = -1.25\varphi + 0.75$ for $0.4 \leq \varphi \leq 0.5$. Then, the Riemann solution consists of five constant states separated by four shocks: $\overline{z}_1(t) = 4t$, $\overline{z}_2(t) = 1.\overline{3}t$, $\overline{z}_3(t) = -0.75t$, $\overline{z}_4(t) = -1.25t$ for $t > 0$. The five constant states correspond to the φ coordinate of the corner values of the concave hull $\hat{\Pi}_-$: $Q_1 = 0$, $Q_2 = 0.05$, $Q_3 = 0.2$, $Q_4 = 0.4$, $Q_5 = 0.5$. This solution is shown in Figure 7.25.

General step function $\varphi_I(z)$

Consider now an increasing (or decreasing) step function $\varphi_I(z)$, with jumps at the points z_k, $k = 1, \ldots, m + 1$, on the z-axis with $m \in \mathbb{N}$, as initial data and a polygonal flux density function $\Pi(\varphi)$. The solution φ can be obtained as a sequence of Riemann problems in time. For sufficiently small and positive t, determine the solution by the independent construction of local Riemann problems at the points z_k. From the shock interactions of two neighbouring Riemann solutions select the smallest intersection time and denote it by $t = t^1$. Evaluate the solution obtained from the initial Riemann problems at t^1, use the value $\varphi(\cdot, t^1)$ as a new initial datum at $t = t^1$ and repeat the procedure generating a sequence of initial value problems at $t = t^2$, $t = t^3$ and so on. For each sequential initial value problem t^i, $i = 1, 2, \ldots$, the number of constant states decreases by at least one (Kunik 1993) so, after a finite time, only two constant states remain. In this way, the global polygonal solution is constructed.

7.6.2 Continuous flux-density function

The polygonal approximation method of Dafermos for piecewise constant monotonically decreasing initial data and a continuous flux-density function consists in approximating the continuous flux-density function $f(\varphi)$ by a piecewise linear function $\Pi(\varphi)$. Consider a real function $f(\varphi)$ defined in the interval $[0, \varphi_\infty]$ and the following monotonically decreasing step function $\varphi_I(z)$ as initial data,

$$
\varphi_I(z) = \begin{cases} \alpha_0 = 0 & \text{for } z_1 = L \leq z, \\ \vdots & \\ \alpha_k & \text{for } z_{k+1} \leq z < z_k, \\ \vdots & \\ \alpha_{m+1} = \varphi_\infty & \text{for } z < z_{m+1} = 0, \end{cases} \tag{7.53}
$$

where φ_∞ and L are given constants and

$$
z_k = L - (k-1)L/m \text{ for } k = 1, \ldots, m+1, \tag{7.54}
$$

$\alpha_k < \alpha_{k+1}$ for $k = 0, \ldots, m$. Let M be a positive constant and divide each interval $[\alpha_k, \alpha_{k+1}]$, $k = 0, \ldots, m$ into M equal parts. Denote by φ_j, $j = 1, \ldots, (m+1)M - 1$ the values of φ so obtained with

$$
\varphi_0 = 0 < \varphi_1 < \ldots < \varphi_{(m+1)M} = \varphi_\infty \tag{7.55}
$$

and determine the vertices $(\varphi_j, f(\varphi_j))$, $j = 0, \ldots, (m+1)M$ of the polygon $\Pi(\varphi)$. Once the continuous flux-density function $f(\varphi)$ has been approximated by the polygon $\Pi(\varphi)$, the method described for polygonal flux-density functions applies.

Error estimate

For numerical solutions, it is very important to have an a priori error estimate such that the user knows beforehand how good his approximate solution is. For our problem, the following a priori error estimate for the absolute error (Kunik 1993) is available:

$$
\epsilon = \int_0^L |\psi - \varphi| dz \leq \frac{t\varphi_\infty^2}{(m+1)M} \max_{0 \leq \varphi \leq \varphi_\infty} |f''(\varphi)|, \tag{7.56}
$$

where ψ is the exact solution and φ is the approximate solution.

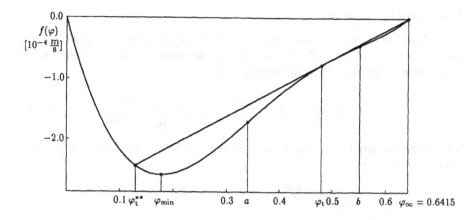

Figure 7.26: Flux density function after Shannon *et al.* (1963, 1964).

7.6.3 Application to batch sedimentation

The initial value problem

Consider the batch sedimentation of an ideal suspension in a settling column (Concha and Bustos 1991) of height $L = 1[\text{m}]$. We take a more general initial concentration $\phi_0(z)$ for $0 < z < L$. At $z = 0$, $\varphi = \varphi_\infty$ and at $z = L$, $\varphi = \varphi_L = 0$. Therefore we formulate the batch sedimentation process as

$$\frac{\partial \varphi}{\partial t} + \frac{\partial f(\varphi)}{\partial z} = 0 \qquad \text{for } z \in \mathbb{R}, \ t > 0, \qquad (7.57a)$$

$$\varphi(z,0) = \varphi_\text{I}(z) \qquad \text{for } z \in \mathbb{R} \qquad (7.57b)$$

with $\varphi_\text{I}(z)$ given by

$$\varphi_\text{I}(z) = \begin{cases} \alpha_0 = 0 & \text{for } z \geq L, \\ \alpha_k = \phi_0(z), \ k = 1, \dots, m & \text{for } 0 \leq z < L, \\ \alpha_{m+1} = \varphi_\infty & \text{for } z < 0. \end{cases} \qquad (7.58)$$

To complete the description of the process, a constitutive equation must be chosen for the solid flux density function. Consider the function $f(\varphi)$ in the interval $[0, \varphi_\infty = 0.642]$ having two inflection points at $a = 0.3391$ and $b = 0.5515$, $\varphi_t^{**} = 0.1294$, $\varphi_t = 0.4796$, and a minimum at $\varphi_{\min} = 0.1787$ described by Shannon *et al.* (1963, 1964), see Figure 7.26:

$$f(\varphi) = \left(-0.33843\varphi + 1.37672\varphi^2 - 1.62275\varphi^3\right.$$
$$\left. -0.11264\varphi^4 + 0.902253\varphi^5\right) \times 10^{-2}[\text{m/s}]. \quad (7.59)$$

Note that $\gamma_1(\varphi_\infty) = \varphi_\infty$ and $\gamma_2(\varphi_L) > \varphi_L$ for this flux density function.

Approximate solution

We wish to obtain a numerical solution of the initial value problem (7.57) with a relative error of less than 5%. The relationship between the absolute error ϵ given by equation (7.56) and the relative error ϵ_r is

$$\epsilon_r = \frac{\epsilon}{\int_0^L \phi_0(z)dz} \leq \frac{t\varphi_\infty^2}{(m+1)M} \max_{0\leq\varphi\leq\varphi_\infty} \frac{|f''(\varphi)|}{\int_0^L \phi_0(z)dz}. \qquad (7.60)$$

For the flux density function given by equation (7.59), the maximum value of the second derivative is

$$\max_{0\leq\varphi\leq\varphi_\infty} |f''(\varphi)| = 2.7534 \times 10^{-2}[\text{m/s}].$$

Replacing in (7.60), with $\varphi_\infty = 0.642$, we obtain

$$\epsilon_r \leq \frac{0.01135t}{(m+1)M \int_0^L \phi_0(z)dz}. \qquad (7.61)$$

The number M of divisions of each interval $[\alpha_k, \alpha_{k+1}]$, $k = 0,\ldots,m$ can be determined from (7.61) so as to obtain a relative error of less than 5%:

$$M \geq \frac{0.2269t}{(m+1) \int_0^L \phi_0(z)dz}. \qquad (7.62)$$

As it will become evident later, the numerical method is much more precise than is estimated a priori and experience shows that the value of M can be chosen as a small fraction of the analytical value.

Algorithm

We implemented Dafermos' polygonal approximation method in Turbo Basic with the following general algorithm:

1. Determine the points φ_j defined in (7.55) in the interval $[\alpha_0, \alpha_{m+1}]$ and the polygon $\Pi(\varphi)$.

2. Build the concave hull $\hat{\Pi}$ of $\Pi(\varphi)$ in each interval and determine the values of Q_i.

3. Define $n = 0$ and solve the Riemann problems for $t = t^n = 0$.

4. Find the intersection times of the shocks for $t = t^n$.

5. Call the smallest of the intersection times t^{n+1}.

6. Find the general equation of the shocks for $t = t^{n+1}$.

7. Take t^{n+1} as the initial time, find the new initial data and increase n by one.

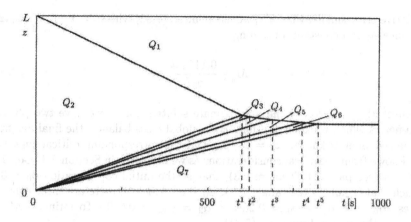

Figure 7.27: The approximate global solution with $\varphi_t^{**} < \phi_0 = 0.25 < a$.

8. Repeat steps 4 to 7 until only the two constant states 0 and φ_∞ remain.

The input to the program is: the extreme values of the interval $\varphi = 0$ and $\varphi_\infty = 0.642$, the initial height of the mixture $L = 1[\mathrm{m}]$, the number m of partitions of the interval $[0, \varphi_\infty]$, the values α_k, $k = 1, \dots, m$ and the number M of partitions of each subinterval $[\alpha_k, \alpha_{k+1}]$, $k = 0, \dots, m$. The output of the program is:

a) The φ coordinate Q_i of the corner points of the concave hull.

b) The minimum times t^i of intersection.

c) The corresponding heights z_i.

d) A settling plot of z vs t, where the global solution is shown.

Homogeneous initial concentration

It is interesting to solve the problem of sedimentation of an initially homogeneous ideal suspension, since it provides a way to compare the approximate with the exact solution obtained by the method of characteristics. Consider the case when the initial concentration ϕ_0 of the mixture is constant. The initial data for this case are

$$\varphi_I(z) = \begin{cases} \alpha_0 = 0 & \text{for } z \geq L, \\ \alpha_k = \phi_0 & \text{for } 0 \leq z < L, \\ \alpha_{m+1} = \varphi_\infty & \text{for } z < 0. \end{cases} \tag{7.63}$$

Comparing this definition of $\varphi_I(z)$ with (7.58), we conclude that $m = 1$ and therefore the interval is divided into $m + 1 = 2$ subintervals, $[0, \phi_0]$ and $[\phi_0, \varphi_\infty]$. To determine the minimum number M of subdivisions of each subinterval to obtain a relative error smaller than 5%, we use relation (7.62) with $m = 1$ and $\int_0^L \phi_0 dz = \phi_0$, since ϕ_0 is a constant, obtaining

$$M \geq \frac{0.11345t}{\phi_0 L}. \qquad (7.64)$$

To check the precision of the approximate solution for φ, we have two physical parameters which we know a priori. By a global mass balance, the final height z_c of the sediment is known: $z_c = \phi_0/\varphi_\infty$ and the corresponding critical time t_c is also known from analytical considerations as we have seen in Section 7.3. Consider an MS-4 (see part (b) Theorem 7.3); that is, the initial concentration $\phi_0 = 0.25$ is such that it falls in the range $\varphi_t^{**} < \phi_0 < a$. At $t = 0$, there are three constant states: $\alpha_0 = 0$, $\alpha_1 = \phi_0 = 0.25$ and $\alpha_2 = \varphi_\infty = 0.642$. To estimate M, we calculate the critical time from (7.31),

$$t_c = z_c/f'(\varphi_t) = 823[s],$$

where $z_c = 0.25/0.642 = 0.389[m]$. Introducing the value of t_c, L and ϕ_0 in (7.64), we get

$$M \geq \frac{0.11345 \times 822.642}{0.25} = 373.$$

Again different values of M were tried and it was found that $M = 11$ gave the approximate values $z_c^a = 0.389[m]$ and $t_c^a = 823[s]$, correct to three significant digits. Following the procedure of the numerical method, divide each interval $[0, \phi_0]$ and $[\phi_0, \varphi_\infty]$ into eleven parts obtaining the values φ_j, $j = 0, \ldots, 22$ of (7.55), and use them to construct the polygon $\Pi(\varphi)$. Then construct the concave hull $\hat{\Pi}(\varphi)$ of $\Pi(\varphi)$ on each of the subintervals $[0, \phi_0]$ and $[\phi_0, \varphi_\infty]$. On the first interval, the concave hull $\hat{\Pi}(\varphi)$ consists of only one line segment that joins the origin with the point $(\phi_0, f(\phi_0))$. On the interval $[\phi_0, \varphi_\infty]$, the concave hull consists of six line segments joining the points $(Q_i, f(Q_i))$ to $(Q_{i+1}, f(Q_{i+1}))$ with $i = 2, \ldots, 7$. Then proceed to solve the two Riemann problems at $t = 0$. The jump from $\varphi = 0$ to $\varphi = 0.25$ at $z = 1$ generates a shock with negative slope. The jump from 0.25 to 0.642 at $z = 0$ generates six constant states separated by five shocks all emanating from the origin and having positive slope. Then, intersect the first shock with these five and find that $t^1 = 601[s]$ is the time of intersection of the shock between $Q_1 = 0$ and $Q_2 = 0.25$ with the shock between $Q_2 = 0.25$ and $Q_3 = 0.393$. Now take $t = t^1$ as the new initial time, recalling that the number of constant states has decreased by one, since the state 0.25 no longer appears. Continue until the number of constant states is reduced to two. In Figure 7.27 we show the approximate global solution of problem (7.57) and (7.63). The solution consists of seven constant states separated by shocks. The values of the constant states are the values of the concentration at the corner points of the concave hull: $Q_1 = 0$, $Q_2 = 0.25$, $Q_3 = 0.393$, $Q_4 = 0.428$, $Q_5 = 0.464$, $Q_6 = 0.499$, $Q_7 = 0.642$.

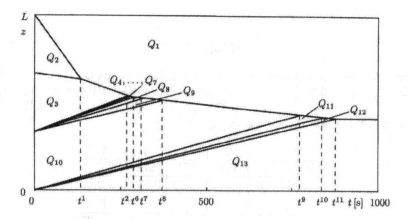

Figure 7.28: The approximate global solution for a monotonically decreasing initial concentration.

The intersection times and the corresponding heights are:

i	1	2	3	4	5
t^i [s]	601	625	689	777	823
z_i [m]	0.436	0.428	0.412	0.396	0.389

Note that it is much easier to evaluate the solution given here than the solutions given by equations (7.29), (7.30) and (7.32).

Monotonically decreasing initial concentration

Consider as our last example an initial concentration $\phi_0(z)$ that decreases monotonically with z between $z = 0$ and $z = L$:

$$\phi_0(z) = \begin{cases} \alpha_1 = 0.05 & \text{for } 2L/3 \le z < L, \\ \alpha_2 = 0.30 & \text{for } L/3 \le z < 2L/3, \\ \alpha_3 = 0.45 & \text{for } 0 \le z < L/3. \end{cases} \tag{7.65}$$

This is the problem we could not solve before by the method of characteristics. If the initial concentration is a continuous monotonically decreasing continuous function of z, it can also be approximated by a step function and the method of Dafermos applies. The final height of the sediment with $L = 1$ is obtained from (7.45) and is $z_c = 0.415$ [m]. The corresponding critical time is $t_c = 878$ [s]. In Figure 7.28, we present the approximate solution. The number M of subdivisions of each interval to obtain a relative error of less than 5% is obtained from inequality (7.62) with $t = 887$ [s], $m = 3$ and $\int_0^L \phi_0(z)dz = 0.267$, $L = 1$, giving $M \ge 150.8$. This bound again is too large since for $M = 8$ we obtain the following approximate values:

$$z_c^a = 0.415 \, [\text{m}], \quad t_c^a = 878 \, [\text{s}]. \tag{7.66}$$

We get thirteen corner points for the concave hull which are $Q_1 = 0$, $Q_2 = 0.05$, $Q_3 = 0.3$, $Q_4 = 0.356$, $Q_5 = 0.375$, $Q_6 = 0.394$, $Q_7 = 0.413$, $Q_8 = 0.431$, $Q_9 = 0.45$, $Q_{10} = 0.45$, $Q_{11} = 0.474$, $Q_{12} = 0.498$ and $Q_{13} = 0.642$. The intersection times and the corresponding heights are:

i	1	2	3	4	5
t^i [s]	135	270	272	278	287
z_i [m]	0.632	0.541	0.540	0.538	0.535

i	6	7	8	9	10	11
t^i [s]	300	316	372	775	842	878
z_i [m]	0.531	0.527	0.515	0.432	0.420	0.415

We already pointed out that this problem cannot be solved by the method of characteristics, which shows the advantage of the approximate method described in this chapter. The relative errors in the approximate solutions are much smaller than the a priori estimates; for this reason, the number of subdivisions of the concentration intervals can be chosen one order of magnitude smaller than that calculated by that method.

Chapter 8

Continuous sedimentation of ideal suspensions

8.1 Mathematical model for continuous sedimentation

We formulate (Bustos, Concha and Wendland 1990, Bustos, Paiva and Wendland 1990, Concha and Bustos 1992) the mathematical model for continuous sedimentation of an ideal suspension in an ideal continuous thickener (ICT) as an initial-boundary value problem (see Chapter 6):

$$\frac{\partial \varphi}{\partial t} + \frac{\partial f(\varphi, t)}{\partial z} = 0 \ \text{ for } (z, t) \in Q_T, \tag{8.1a}$$

$$\varphi(z, 0) = \varphi_{\mathrm{I}}(z) \ \text{ for } z \in \Omega, \tag{8.1b}$$

$$\gamma \varphi(0, t) \in \epsilon_0(\varphi_\infty) \ \text{ for } t \in \mathcal{T}, \tag{8.1c}$$

$$\gamma \varphi(L, t) \in \epsilon_L(\varphi_L(t)) \ \text{ for } t \in \mathcal{T}. \tag{8.1d}$$

The solid flux density function $f = f(\varphi, t)$ is given in (2.18) and the sets ϵ_0, ϵ_L, Q_T, Ω and $\mathcal{T}$ were defined in Chapter 6. The solution φ of the quasilinear hyperbolic equation (8.1a) is constant along the characteristic lines with slope

$$\frac{dz}{dt} = \frac{\partial f}{\partial \varphi}(\varphi, t) = q(t) + f'_b(\varphi), \tag{8.2}$$

which shows that the propagation speed of a solution value is in general a function of time. Solutions of problem (8.1) are usually discontinuous, so we consider entropy solutions (see Chapter 6).

In the next two sections we analyse the different behaviors of the transient states of an ICT which is initially filled to a height A with sediment of concentration φ_∞ and from A to L with a suspension of concentration φ_L independent of time.

The thickener is fed at $z = L$ with a mixture of constant feed solid flux density $f_F = f(\varphi_L)$. The bottom of the vessel is open, producing a constant average velocity q. This phenomenon is modeled by the following initial-boundary value problem:

$$\frac{\partial \varphi}{\partial t} + \frac{\partial f(\varphi)}{\partial z} = 0 \quad \text{for } (z, t) \in Q_T,$$ (8.3a)

$$\varphi(z, 0) = \varphi_I(z) = \begin{cases} \varphi_L & \text{for } A \leq z \leq L, \\ \varphi_\infty & \text{for } 0 \leq z < A, \end{cases}$$ (8.3b)

$$\gamma\varphi(0, t) \in \epsilon_0(\varphi_\infty) \quad \text{for } t \in \mathcal{T},$$ (8.3c)

$$\gamma\varphi(L, t) \in \epsilon_L(\varphi_L) \quad \text{for } t \in \mathcal{T}.$$ (8.3d)

In contrast to the general discussion in Chapter 6, we assume throughout this section that the boundary datum at $z = 0$, φ_∞, is constant. In Sections 8.2 and 8.3, φ_L is also independent of time. In Section 8.4, which corresponds to the control of the ICT, we consider the more general case that both $\varphi_L = \varphi_L(t)$ and $q = q(t)$ are dependent on t. The initial-boundary value problem (8.3) is Riemann-like in the sense that the initial data consist of only two constant states and the solution in general will consist of elementary waves: shocks, rarefaction waves and contact discontinuities. Note that, from a mathematical point of view, it is easier to determine exact solutions to the problem (8.3) than to solve the problem of batch sedimentation as in Chapter 7, since only *one* Riemann problem and, in particular, no interaction between solutions of Riemann problems has to be considered.

8.2   Modes of continuous sedimentation

Definition 8.1 *Let $\varphi(z, t)$ be an entropy solution of problem* (8.3). *Then the pair* $(\varphi, f(\varphi))$ *constitutes a* continuous Kynch sedimentation process (CKSP).

Definition 8.2 Modes of continuous sedimentation (MCS) *are the different possible types of continuous Kynch sedimentation processes.*

The modes of continuous sedimentation are entirely determined by the constitutive equation of the flux density function, the initial data, the boundary conditions and the average velocity q_0. We consider flux density functions $f(\varphi)$ with one and with two inflection points. If φ_∞ denotes the final concentration for batch sedimentation, then we assume for $f(\varphi)$ the following properties:

$$f(0) = 0 \quad \text{and} \quad f'(0) < 0,$$ (8.4)
$$f(\varphi) \leq 0 \quad \text{for } 0 \leq \varphi \leq \varphi_\infty.$$ (8.5)

For such functions, five modes of continuous sedimentation exist.

Definition 8.3 *A CKSP is:*

(1) An MCS-1 *if it consists of two constant states separated by a shock.*

(2) An MCS-2 *if it consists of two constant states separated by a contact discontinuity and a rarefaction wave.*

(3) An MCS-3 *if it is continuous and consists of two constant states separated by a rarefaction wave.*

(4) An MCS-4 *if it consists of two constant states separated by a contact discontinuity, a rarefaction wave and another contact discontinuity.*

(5) An MCS-5 *if it consists of two constant states separated by a rarefaction wave and a contact discontinuity.*

In what follows, we consider different values for the feed flux density f_F and we calculate the corresponding values of φ_L from equation (2.27). Then we construct the global weak solutions of problem (8.3) so that they satisfy the initial condition (8.3b), the generalized boundary conditions (8.3c) and (8.3d), and the differential equation (8.3a) at the points of continuity. At discontinuities they satisfy the Rankine-Hugoniot condition and Oleinik's condition E. This can be checked by direct substitution. In the proof of each of the following theorems, we consider only those MCS that are distinct. The other proofs are similar.

8.3 Flux density function with one inflection point

Let a be the inflection point of $f(\varphi)$. Then from properties (8.4) and (8.5) we have

$$f''(\varphi) > 0 \text{ for } 0 \le \varphi < a, \tag{8.6}$$
$$f''(\varphi) < 0 \text{ for } a \le \varphi \le \varphi_\infty. \tag{8.7}$$

Lemma 8.1 *If $\varphi < \varphi_\infty^{**} < \varphi_\infty$, then $\sigma(\varphi, \varphi_\infty)$ is an increasing function of φ.*

Proof. Fix φ_∞, then

$$\frac{d\sigma(\varphi, \varphi_\infty)}{d\varphi} = \frac{f'(\varphi) - \sigma(\varphi, \varphi_\infty)}{\varphi - \varphi_\infty}.$$

But $\varphi < \varphi_\infty^{**}$ implies that $\varphi_\infty < \varphi^*$ by Lemma 5.3. Since $\varphi_\infty \in (\varphi, \varphi^*)$ for $\varphi < a$, then by Lemma 5.1, we get $f'(\varphi) < \sigma(\varphi, \varphi_\infty)$. Then $d\sigma/d\varphi > 0$. This completes the proof. ∎

8.3.1 Case I: Both $f'(a)$ and $f'(\varphi_\infty)$ are positive

We assume that $|q|$ is small enough so that both $f'(a) > 0$ and $f'(\varphi_\infty) > 0$. Due to the concavity of f between a and φ_∞, we have

$$f'(\varphi) > 0 \text{ for } \varphi \in [a, \varphi_\infty]. \tag{8.8}$$

Physically, this selection of q means that, most of the time, the thickener overflows since there is not enough capacity to discharge the thickened slurry. Let φ_s be such that

$$\sigma(\varphi_s, \varphi_\infty) = 0, \quad 0 < \varphi_s \le \varphi_\infty^{**}. \tag{8.9}$$

Corollary 8.1 *The function $\sigma(\varphi, \varphi_\infty)$ satisfies*

$$\sigma(\varphi, \varphi_\infty) < 0 \text{ for } 0 < \varphi < \varphi_s, \tag{8.10}$$
$$\sigma(\varphi, \varphi_\infty) > 0 \text{ for } \varphi_s < \varphi < \varphi_\infty^{**}. \tag{8.11}$$

Proof. Corollary 8.1 follows immediately from Lemma 8.1.   ∎

We start with a dilute suspension (φ_L small) and consider the different behaviors of the suspension when the feed concentration is increased.

Theorem 8.1 *If both $f'(a) > 0$ and $f'(\varphi_\infty) > 0$, then for the solution of problem (8.3) we have:*

(a) *For $\varphi_L < \varphi_s$ there exists a time t_1 such that for $0 \le t < t_1$ the CKSP is an MCS-1 and afterwards the ICT empties. The boundary conditions are satisfied classically at $z = L$ and in a generalized sense at $z = 0$.*

(b) *For $\varphi_L = \varphi_s$ the CKSP is an MCS-1 for all t and the ICT attains a steady state with $\varphi_D = \varphi_\infty$. Both boundary conditions are satisfied classically.*

(c) *For $\varphi_s < \varphi_L \le \varphi_\infty^{**}$ there exists a time t_1 such that for $0 \le t < t_1$ the CKSP is an MCS-1 and afterwards the ICT overflows. The boundary conditions are satisfied in a generalized sense at $z = L$ and classically at $z = 0$.*

(d) *For $\varphi_\infty^{**} < \varphi_L < a$ there exists a time t_1 such that for $0 \le t < t_1$ the CKSP is an MCS-2 and afterwards the ICT overflows. The boundary conditions are satisfied in a generalized sense at $z = L$ and classically at $z = 0$.*

(e) *For $a \le \varphi_L < \varphi_\infty$ there exists a time t_1 such that for $0 \le t < t_1$ the CKSP is an MCS-3 and afterwards the ICT overflows. The boundary conditions are satisfied in a generalized sense at $z = L$ and classically at $z = 0$.*

Proof. The set of admissible states at $z = 0$ for all cases considered in this theorem is

$$\epsilon_0(\varphi_\infty) = [0, \varphi_s] \cup \{\varphi_\infty\}. \tag{8.12}$$

Remark The values of $f'(\varphi)$ are negative for all $\varphi \in \epsilon_0(\varphi_\infty)$, which is a necessary but not sufficient condition for a concentration to belong to the set $\epsilon_0(\varphi_\infty)$. Petty (1975) recognized this problem and, by purely physical arguments, indicates that at $z = 0$, $f'(\varphi)$ must be negative or equal to zero. The set of admissible states at the boundary $z = L$ is considered separately.

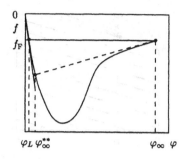 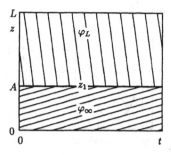

Figure 8.1: Graph of f vs φ and representation of the global weak solution $\varphi(z,t)$ for an MCS-1 with $\varphi_L = \varphi_s$.

(b) For $\varphi_L = \varphi_s$, as shown in Figure 8.1, the global weak solution of problem (8.3) for $t \geq 0$ is given by

$$\varphi(z,t) = \begin{cases} \varphi_L & \text{for } z_1(t) \leq z \leq L, \\ \varphi_\infty & \text{for } 0 \leq z < z_1(t). \end{cases} \tag{8.13}$$

The shock $z_1(t) = A$ is horizontal since $\sigma(\varphi_L, \varphi_\infty) = 0$. The solution consists of two constant states φ_L and φ_∞ separated by the shock $z_1(t)$, hence the CKSP is an MCS-1 for all t. At $z = L$, $\epsilon_L(\varphi_L) = \{\varphi_L\} \cup \{\varphi_\infty\}$ and $\gamma\varphi(L,t) = \varphi_L$. At $z = 0$, the discharge concentration $\gamma\varphi(0,t) = \varphi_D(t) = \varphi_\infty$ for all t and the ICT is at steady state. Both boundary conditions are satisfied classically. This proves part (b) of the theorem. ∎

(d) For $\varphi_\infty^{**} < \varphi_L < a$ (see Figure 8.2) the global weak solution of problem (8.3) for $0 \leq t < t_1$ is given by

$$\varphi(z,t) = \begin{cases} \varphi_L & \text{for } z_1(t) \leq z \leq L, \\ h_2\left((z-A)/t\right) & \text{for } z_2(t) \leq z < z_1(t), \\ \varphi_\infty & \text{for } 0 \leq z < z_2(t). \end{cases}$$

For $t_1 \leq t < t_2$ the solution is

$$\varphi(z,t) = \begin{cases} h_2\left((z-A)/t\right) & \text{for } z_2(t) \leq z < L, \\ \varphi_\infty & \text{for } 0 \leq z < z_2(t), \end{cases}$$

and for $t \geq t_2$ is

$$\varphi(z,t) = \varphi_\infty \text{ for } 0 \leq z \leq L.$$

Here, as in Chapter 7, h_2 is the inverse of $f'(\varphi)$ restricted to values of φ greater than a. Since $\varphi_L < a$, we have $f'(\varphi_L) < \sigma(\varphi_L, \varphi_L^*) = f'(\varphi_L^*)$ by Corollary 5.2 and Lemma 5.2, and therefore the discontinuity

$$z_1(t) = A + f'(\varphi_L^*)t = A + \sigma(\varphi_L, \varphi_L^*)t$$

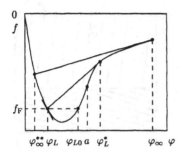
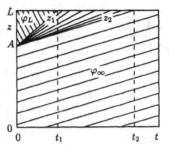

Figure 8.2: Graph of f vs φ and representation of the global weak solution $\varphi(z,t)$ for an MCS-2 with $\varphi_\infty^{**} < \varphi_L < a$.

is a contact discontinuity. The line $z_2(t) = A + f'(\varphi_\infty)t$ is a line of continuity. The contact discontinuity $z_1(t)$ cuts the line $z = L$ at the time

$$t_1 = \frac{L - A}{\sigma(\varphi_L, \varphi_L^*)} = \frac{L - A}{f'(\varphi_L^*)}.$$

From Lemma 5.3, $a < \varphi_L^* < \varphi_\infty$ and therefore from relation (8.8), we deduce that $f'(\varphi_L^*) > 0$ and t_1 is a positive and finite time. For $0 \le t < t_1$ the solution consists of two constant states φ_L and φ_∞ separated by the contact discontinuity $z_1(t)$ and a rarefaction wave; then the CKSP is an MCS-2. The line $z_2(t)$ cuts $z = L$ at the time

$$t_2 = \frac{L - A}{f'(\varphi_\infty)}$$

which is positive. Since $f'' < 0$ for $\varphi > a$, then $f'(\varphi_L^*) > f'(\varphi_\infty)$ and therefore $t_2 > t_1$. At $z = L$ the solution is

$$\gamma\varphi(L,t) = \begin{cases} \varphi_L & \text{for } 0 \le z < t_1, \\ h_2\left((L - A)/t\right) & \text{for } t_1 \le t < t_2, \\ \varphi_\infty & \text{for } t_2 \le t. \end{cases}$$

Let φ_m be the point at which f attains its minimum in $[0, \varphi_\infty]$. Then, for $\varphi_L < \varphi_m$, we define (see also Figure 8.2)

$$\varphi_{L0} = \min\left\{\varphi > \varphi_m | \sigma(\varphi_L, \varphi_{L0}) = 0\right\}. \tag{8.14}$$

Then, the set $\epsilon_L(\varphi_L) = \{\varphi_L\} \cup [\varphi_{L0}, \varphi_\infty]$ or $\epsilon_L(\varphi_L) = [\varphi_m, \varphi_\infty]$ in the case that $\varphi_L \ge \varphi_m$. Since φ_L belongs to $\epsilon_L(\varphi_L)$ and $h_2((L - A)/t)$ for $t_1 \le t < t_2$ assumes values between φ_L^* and φ_∞ which also belong to $\epsilon_L(\varphi_L)$, the entropy boundary condition is satisfied at $z = L$. The thickener overflows for $t \ge t_1$. At $z = 0$, the discharge concentration $\varphi_D(t) = \gamma\varphi(0,t) = \varphi_\infty$ for all t and

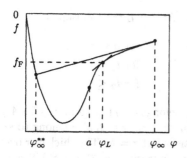
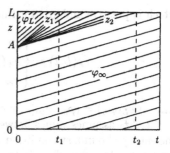

Figure 8.3: Graph of f vs φ and representation of the global weak solution $\varphi(z,t)$ for an MCS-3 with $a < \varphi_L < \varphi_\infty$.

the boundary condition is satisfied classically. This proves part (d) of the theorem. ∎

(e) For $a \le \varphi_L < \varphi_\infty$, the global weak solution of problem (8.3) for $0 \le t < t_1$ is given by (see Figure 8.3)

$$\varphi(z,t) = \begin{cases} \varphi_L & \text{for } z_1(t) \le z \le L, \\ h_2\left((z-A)/t\right) & \text{for } z_2(t) \le z < z_1(t), \\ \varphi_\infty & \text{for } 0 \le z < z_2(t). \end{cases}$$

For $t_1 \le t < t_2$ the solution is

$$\varphi(z,t) = \begin{cases} h_2\left((z-A)/t\right) & \text{for } z_2(t) \le z < L, \\ \varphi_\infty & \text{for } 0 \le z < z_2(t), \end{cases}$$

and for $t \ge t_2$ is

$$\varphi(z,t) = \varphi_\infty \text{ for } 0 \le z \le L.$$

The lines $z_1(t) = A + f'(\varphi_L)t$ and $z_2(t) = A + f'(\varphi_\infty)t$ are both lines of continuity. The line $z_1(t)$ cuts $z = L$ at the time

$$t_1 = \frac{L-A}{f'(\varphi_L)},$$

which is positive and finite since $f'(\varphi) > 0$ for $\varphi > a$. For $0 \le t < t_1$ the solution consists of two constant states φ_L and φ_∞ separated by a rarefaction wave; then the CKSP is an MCS-3. The line $z_2(t)$ cuts $z = L$ at the time

$$t_2 = \frac{L-A}{f'(\varphi_\infty)},$$

which is positive and greater than t_1. At $z = L$ the solution is

$$\gamma\varphi(L,t) = \begin{cases} \varphi_L & \text{for } 0 \leq z < t_1, \\ h_2\left((L-A)/t\right) & \text{for } t_1 \leq t < t_2, \\ \varphi_\infty & \text{for } t_2 \leq t. \end{cases}$$

The set $\epsilon_L(\varphi_L) = [\varphi_m, \varphi_\infty]$. Since φ_L belongs to $\epsilon_L(\varphi_L)$ and $h_2((L-A)/t)$ for $t_1 \leq t < t_2$ assumes values between φ_L and φ_∞ which also belong to $\epsilon_L(\varphi_L)$, the entropy boundary condition is satisfied at $z = L$. The thickener overflows for $t \geq t_1$. At $z = 0$, the discharge concentration $\varphi_D(t) = \gamma\varphi(0,t) = \varphi_\infty$ for all t and the boundary condition is satisfied classically. This proves part (e) and finally the theorem (since the proof of parts (a) and (c) are trivial). ∎

8.3.2   Case II: $f'(a)$ is positive and $f'(\varphi_\infty)$ is negative

We increase $|q|$ in such a way that $f'(\varphi_\infty) < 0$ while $f'(a)$ remains positive. This implies that there exists a relative maximum φ_M of f, that is, $f'(\varphi_M) = 0$ and $f''(\varphi_M) < 0$. Now that Q_D has been further increased, we expect more cases in which the thickener empties.

Theorem 8.2 *If $f'(a) > 0$ and $f'(\varphi_\infty) < 0$, then for the solution of problem (8.3) we have:*

(a) *For $0 < \varphi_L \leq \varphi_\infty^{**}$ there exists a time t_1 such that for $0 \leq t < t_1$ the CKSP is an MCS-1 and afterwards the ICT empties. The boundary conditions are satisfied classically at $z = L$ and in a generalized sense at $z = 0$.*

(b) *For $\varphi_\infty^{**} < \varphi_L < \varphi_M^{**}$ there exists a time t_2 such that for $0 \leq t < t_2$ the CKSP is an MCS-2. Also there exists a time $t_1 > t_2$ after which the ICT empties. The boundary conditions are satisfied classically at $z = L$ and in a generalized sense at $z = 0$.*

(c) *For $\varphi_L = \varphi_M^{**}$ there exists a time t_2 such that for $0 \leq t < t_2$ the CKSP is an MCS-2 and the ICT attains a steady state with $\varphi_D = \varphi_M$. The boundary conditions are satisfied classically at $z = L$ and in a generalized sense at $z = 0$.*

(d) *For $\varphi_M^{**} < \varphi_L < a$ there exist times t_1 and t_2 such that for $0 \leq t < \min(t_1, t_2)$, the CKSP is an MCS-2. For $t > t_1$ the ICT overflows. The boundary conditions are satisfied in a generalized sense on both boundaries.*

(e) *For $a \leq \varphi_L < \varphi_M$ there exist times t_1 and t_2 such that for $0 \leq t < \min(t_1, t_2)$, the CKSP is an MCS-3. For $t \geq t_1$ the ICT overflows. The boundary conditions are satisfied in a generalized sense on both boundaries.*

(f) *For $\varphi_L = \varphi_M$ there exists a time t_2 such that for $0 \leq t < t_2$ the CKSP is an MCS-3. The ICT empties and the boundary conditions are satisfied classically at $z = L$ and in a generalized sense at $z = 0$.*

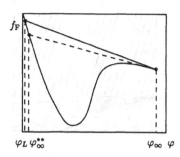
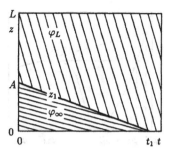

Figure 8.4: Graph of f vs φ and representation of the global weak solution $\varphi(z,t)$ for an MCS-1 with $0 < \varphi_L < \varphi_\infty^{**}$.

(g) For $\varphi_M < \varphi_L < \varphi_\infty$ there exists a time t_2 such that for $0 \leq t < t_2$ the CKSP is an MCS-3. Also there exists a time $t_1 > t_2$ after which the ICT empties. The boundary conditions are satisfied classically at $z = L$ and in a generalized sense at $z = 0$.

Proof. The set of admissible states at $z = 0$ for all cases considered in this theorem is

$$\epsilon_0(\varphi_\infty) = [0, \varphi_M^{**}] \cup [\varphi_M, \varphi_\infty]. \tag{8.15}$$

The set of admissible states at the boundary $z = L$ is considered separately.

(a) For $0 < \varphi_L \leq \varphi_\infty^{**} < \varphi_M^{**}$, as shown in Figure 8.4, the global weak solution of problem (8.3) for $0 \leq t < t_1$ is given by

$$\varphi(z,t) = \begin{cases} \varphi_L & \text{for } z_1(t) \leq z \leq L, \\ \varphi_\infty & \text{for } 0 \leq z < z_1(t), \end{cases}$$

and for $t \geq t_1$ by

$$\varphi(z,t) = \varphi_L \text{ for } 0 \leq z \leq L.$$

The shock $z_1(t) = A + \sigma(\varphi_L, \varphi_\infty)t$ intersects $z = 0$ at the time $t_1 = A/\sigma(\varphi_L, \varphi_\infty)$. In Lemma 8.1, we proved that $\sigma(\varphi, \varphi_\infty)$ is an increasing function of φ for $\varphi < \varphi_\infty^{**}$. Since $\sigma(\varphi_\infty^{**}, \varphi_\infty) = f'(\varphi_\infty) < 0$ and since $\varphi_L < \varphi_\infty^{**}$, then also $\sigma(\varphi_L, \varphi_\infty) < 0$ and t_1 is positive and finite. For $0 \leq t < t_1$ the solution consists of two constant states φ_L and φ_∞ separated by the shock $z_1(t)$; then the CKSP is an MCS-1. At $z = L$, $\epsilon_L(\varphi_L) = \{\varphi_L\}$ and, since $\gamma\varphi(L, T) = \varphi_L$, the boundary condition is satisfied in a classical sense. At $z = 0$, the discharge concentration $\varphi_D(t) = \gamma\varphi(0, t)$ is

$$\varphi_D(t) = \begin{cases} \varphi_\infty & \text{for } 0 \leq t < t_1, \\ \varphi_L & \text{for } t \geq t_1. \end{cases}$$

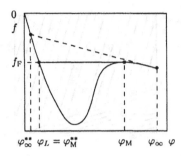
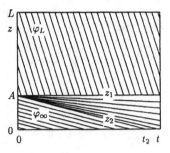

Figure 8.5: Graph of f vs φ and representation of the global weak solution $\varphi(z,t)$ for an MCS-2 with $\varphi_L = \varphi_M^{**}$.

Since both φ_L and φ_∞ belong to $\epsilon_0(\varphi_\infty)$, the solution satisfies the boundary condition in a generalized sense. Since $\varphi_L < \varphi_F$, the ICT empties for $t > t_1$. This proves part (a) of the theorem. ∎

(c) For $\varphi_L = \varphi_M^{**}$, as shown in Figure 8.5, the global weak solution of problem (8.3) for $0 \le t < t_2$ is given by

$$\varphi(z,t) = \begin{cases} \varphi_L & \text{for } z_1(t) \le z \le L, \\ h_2\left((z - A)/t\right) & \text{for } z_2(t) \le z < z_1(t), \\ \varphi_\infty & \text{for } 0 \le z < z_2(t). \end{cases}$$

For $t \ge t_2$, the solution is

$$\varphi(z,t) = \begin{cases} \varphi_L & \text{for } z_1(t) \le z < L, \\ h_2\left((z - A)/t\right) & \text{for } 0 \le z < z_1(t). \end{cases}$$

The line $z_2(t) = A + f'(\varphi_\infty)t$ is a line of continuity and cuts the t-axis at $t_2 = -A/f'(\varphi_\infty)$ which is positive and finite. Since $\varphi_L < a$, we have that $f'(\varphi_L) < \sigma(\varphi_L, \varphi_L^*) = f'(\varphi_L^*)$ by Corollary 5.2 and Lemma 5.2, and therefore the discontinuity

$$z_1(t) = A + f'(\varphi_L^*)t = A + \sigma(\varphi_L, \varphi_L^*)t$$

is a contact discontinuity and horizontal. For $0 \le t < t_2$ the solution consists of two constant states φ_L and φ_∞ separated by the contact discontinuity $z_1(t)$ and a rarefaction wave, hence the CKSP is an MCS-2. At $z = L$, the set $\epsilon_L(\varphi_L) = \{\varphi_L\} \cup \{\varphi_M\}$ and since $\gamma\varphi(L,t) = \varphi_L$ for all t, the boundary condition is satisfied classically. At $z = 0$, the discharge concentration $\varphi_D(t) = \gamma\varphi(0,t)$ is

$$\varphi_D(t) = \begin{cases} \varphi_\infty & \text{for } 0 \le z < t_2, \\ h_2(-A/t) & \text{for } t \ge t_2. \end{cases} \tag{8.16}$$

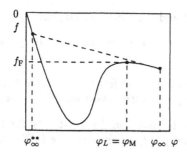 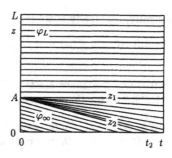

Figure 8.6: Graph of f vs φ and representation of the global weak solution $\varphi(z,t)$ for an MCS-3 with $\varphi_L = \varphi_M$.

Since $h_2(-A/t)$ for $t_2 \leq t$ assumes values between φ_M and φ_∞ which belongs to $\epsilon_0(\varphi_\infty)$, the entropy boundary condition is satisfied at $z = 0$. Here the thickener attains a steady state with $\lim_{t\to\infty} \varphi_D(t) = \varphi_M$. This proves part (c) of the theorem. ■

(f) For $\varphi_L = \varphi_M$, as depicted in Figure 8.6, the global weak solution of problem (8.3) for $0 \leq t < t_2$ is given by

$$\varphi(z,t) = \begin{cases} \varphi_L & \text{for } z_1(t) \leq z \leq L, \\ h_2((z-A)/t) & \text{for } z_2(t) \leq z < z_1(t), \\ \varphi_\infty & \text{for } 0 \leq z < z_2(t), \end{cases}$$

and for $t \geq t_2$ is

$$\varphi(z,t) = \begin{cases} \varphi_L & \text{for } z_1(t) \leq z < L, \\ h_2((z-A)/t) & \text{for } 0 \leq z < z_1(t). \end{cases}$$

The horizontal line $z_1(t) = A$ and $z_2(t) = A + f'(\varphi_\infty)t$ are lines of continuity. The line $z_2(t)$ cuts the t-axis at $t_2 = -A/f'(\varphi_\infty)$ which is positive and finite. For $0 \leq t < t_2$ the solution consists of two constant states φ_L and φ_∞ separated by a rarefaction wave; thus the CKSP is an MCS-3. At $z = L$, we have that $\epsilon_L(\varphi_L) = [\varphi_m, \varphi_M = \varphi_L]$ and $\gamma\varphi(L,t) = \varphi_L$ and the boundary condition is satisfied classically. At $z = 0$, the discharge concentration $\varphi_D(t) = \gamma\varphi(0,t)$ is given by (8.16). Then the entropy condition is satisfied at $z = 0$. The ICT attains a steady state with $\lim_{t\to\infty} \varphi_D(t) = \varphi_M$. Since $\varphi_D = \varphi_M = \varphi_L = \varphi_F$, the ICT empties. This proves part (f) of the theorem. ■

8.3.3 Case III: Both $f'(a)$ and $f'(\varphi_\infty)$ are negative

We increase $|q|$ again in such a way that $f'(a) < 0$ and $f'(\varphi_\infty)$. This means that now f has no relative maximum nor minimum and $f'(\varphi) < 0$ for φ between 0 and

φ_∞. Physically, no steady state is attained and the thickener empties.

Theorem 8.3 *If both $f'(a)$ and $f'(\varphi_\infty)$ are negative, then the solution of problem (8.3) is:*

(a) *For $0 < \varphi_L < a$ there exists a time t_2 such that for $0 \leq t < t_2$ the CKSP is an MCS-2. Also there exists a time $t_1 > t_2$ after which the ICT empties.*

(b) *For $a \leq \varphi_L < \varphi_\infty$ there exists a time t_2 such that for $0 \leq t < t_2$ the CKSP is an MCS-3. Also there exists a time $t_1 > t_2$ after which the ICT empties.*

The boundary conditions are satisfied classically at $z = L$ and in a generalized sense at $z = 0$.

Proof. For all cases considered in this theorem, the sets of admissible states are $\epsilon_L(\varphi_L) = \{\varphi_L\}$ and $\epsilon_0(\varphi_\infty) = [0, \varphi_\infty]$ at $z = L$ and $z = 0$, respectively. Let h_2 be the inverse of $f'(\varphi)$ restricted to values of φ greater than a. Moreover, $f'(\varphi) < 0$ for $0 \leq \varphi \leq \varphi_\infty$.

(a) For $0 < \varphi_L < a$, as shown in Figure 8.7, the global weak solution of problem (8.3) for $0 \leq t < t_2$ is given by

$$\varphi(z,t) = \begin{cases} \varphi_L & \text{for } z_1(t) \leq z \leq L, \\ h_2((z - A)/t) & \text{for } z_2(t) \leq z < z_1(t), \\ \varphi_\infty & \text{for } 0 \leq z < z_2(t). \end{cases}$$

For $t_2 \leq t < t_1$, the solution is

$$\varphi(z,t) = \begin{cases} \varphi_L & \text{for } z_1(t) \leq z < L, \\ h_2((z - A)/t) & \text{for } 0 \leq z < z_1(t), \end{cases}$$

and for $t_1 \leq t$ is

$$\varphi(z,t) = \varphi_L \text{ for } 0 \leq z \leq L.$$

Since $\varphi_L < a$, we have $f'(\varphi_L) < \sigma(\varphi_L, \varphi_L^*) = f'(\varphi_L^*)$ by Corollary 5.2 and Lemma 5.2, and therefore the discontinuity

$$z_1(t) = A + f'(\varphi_L^*)t = A + \sigma(\varphi_L, \varphi_L^*)t$$

is a contact discontinuity. The line $z_2(t) = A + f'(\varphi_\infty)t$ is a line of continuity. The line $z_2(t)$ cuts $z = 0$ at the time $t_2 = -A/f'(\varphi_\infty)$, which is positive and finite. For $0 \leq t < t_2$ the solution consists of two constant states φ_L and φ_∞ separated by the contact discontinuity $z_1(t)$ and a rarefaction wave, then the CKSP is an MCS-2. The contact discontinuity $z_1(t)$ cuts the line $z = 0$ at the time

$$t_1 = -\frac{A}{\sigma(\varphi_L, \varphi_L^*)} = -\frac{A}{f'(\varphi_L^*)},$$

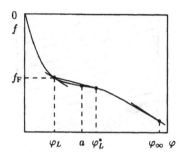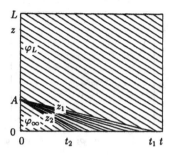

Figure 8.7: Graph of f vs φ and representation of the global weak solution $\varphi(z,t)$ for an MCS-2 with $0 < \varphi_L < a$.

which is positive and finite. Moreover, since $\varphi_L^* < \varphi_\infty$ due to the fact that f'' is negative for $\varphi > a$, then $f'(\varphi_\infty) < f'(\varphi_L^*)$ which shows that $t_1 > t_2$. At $z = L$, $\gamma\varphi(L,t) = \varphi_L$ for all t; the boundary condition is satisfied classically. At $z = 0$, the discharge concentration $\varphi_D(t) = \gamma\varphi(0,t)$ is

$$\varphi_D(t) = \begin{cases} \varphi_\infty & \text{for } 0 \leq z < t_2, \\ h_2(-A/t) & \text{for } t_2 \leq t < t_1, \\ \varphi_L & \text{for } t_1 \leq t. \end{cases} \qquad (8.17)$$

Here $h_2(-A/t)$ assumes, for $t_2 \leq t < t_1$, values between φ_L^* and φ_∞ which also belong to $\epsilon_0(\varphi_\infty)$; the entropy boundary condition is satisfied at $z = 0$. Since $\varphi_L < \varphi_F$, the thickener empties for $t \geq t_1$. This proves part (a) of the theorem. ∎

(b) For $a \leq \varphi_L < \varphi_\infty$, as shown in Figure 8.8, the global weak solution of problem (8.3) for $0 \leq t < t_2$ is given by

$$\varphi(z,t) = \begin{cases} \varphi_L & \text{for } z_1(t) \leq z \leq L, \\ h_2((z-A)/t) & \text{for } z_2(t) \leq z < z_1(t), \\ \varphi_\infty & \text{for } 0 \leq z < z_2(t). \end{cases}$$

For $t_2 \leq t < t_1$ the solution is

$$\varphi(z,t) = \begin{cases} \varphi_L & \text{for } z_1(t) \leq z < L, \\ h_2((z-A)/t) & \text{for } 0 \leq z < z_1(t), \end{cases}$$

and for $t_1 \leq t$, the solution is

$$\varphi(z,t) = \varphi_L \text{ for } 0 \leq z \leq L.$$

Both $z_1(t) = A + f'(\varphi_L)t$ and $z_2(t) = A + f'(\varphi_\infty)t$ are lines of continuity. The line $z_1(t)$ cuts the t-axis at $t_1 = -A/f'(\varphi_L)$, which is positive and finite.

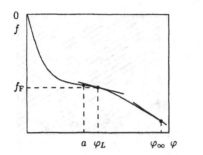 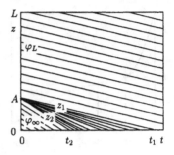

Figure 8.8: Graph of f vs φ and representation of the global weak solution $\varphi(z,t)$ for an MCS-3 with $a < \varphi_L < \varphi_\infty$.

The line $z_2(t)$ cuts the t-axis at $t_2 = -A/f'(\varphi_\infty)$ which is also positive and finite. Moreover, since $\varphi_L < \varphi_\infty$, $f'(\varphi_L) > f'(\varphi_\infty)$ and hence $t_1 > t_2$. For $0 \le t < t_2$ the solution consists of two constant states φ_L and φ_∞ separated by a rarefaction wave; therefore the CKSP is an MCS-3. At $z = L$, the solution is $\gamma\varphi(L,t) = \varphi_L$, then the boundary condition is satisfied classically. At $z = 0$, the discharge concentration $\varphi_D(t) = \gamma\varphi(0,t)$ is given by equation (8.17). For $t_2 \le t < t_1$, $h_2(-A/t)$ assumes values between φ_L and φ_∞ and therefore the entropy boundary condition is satisfied. Since $\varphi_L < \varphi_F$ the thickener empties for $t \ge t_1$. This proves part (b) and finally the theorem. ∎

8.4 Flux density function with two inflection points

Let a and b be the two inflection points of f and let us assume that $a < b$ and $f''(\varphi) > 0$ for $0 < \varphi < a$ and $b < \varphi < \varphi_\infty$ and $f''(\varphi) < 0$ for $a < \varphi < b$. To exclude the degenerate case (which is equivalent to a function with no inflection points), we further assume that

$$\frac{f(\varphi_\infty) - f(a)}{\varphi_\infty - a} < f'(a). \qquad (8.18)$$

From the point $(\varphi_\infty, f(\varphi_\infty))$ in the f vs φ plot, draw a tangent to the graph and extend it till it cuts the graph. Call such a point $(\varphi_t^{**}, f(\varphi_t^{**}))$ and φ_t the coordinate of the point of tangency. See Figure 8.9.

Lemma 8.2 *If $\varphi < \varphi_t^{**} < \varphi_\infty$, then $\sigma(\varphi, \varphi_\infty)$ is an increasing function of φ.*

Proof. Fix φ_∞, then

$$\frac{d\sigma}{d\varphi} = \frac{f'(\varphi) - \sigma(\varphi, \varphi_\infty)}{\varphi - \varphi_\infty}. \qquad (8.19)$$

The chord that joins the point $(\varphi, f(\varphi))$ with the point $(\varphi_\infty, f(\varphi_\infty))$ in the flux density plot is completely on one side of the graph of f. Therefore

$$\sigma(\varphi, \varphi_\infty) > \sigma(\varphi, u) \text{ for all } u \text{ between } \varphi \text{ and } \varphi_\infty. \tag{8.20}$$

Taking the limit of the above relation when $u \to \varphi$, we get $\sigma(\varphi, \varphi_\infty) > f'(\varphi)$. This last relation proves that $d\sigma/d\varphi > 0$ in (8.19). This completes the proof. ∎

In the three theorems which follow, we present all the cases which can occur with the experimental flux curve determined by Shannon *et al.* (1963, 1964); see Figure 7.26 in Chapter 7. Other cases, which may arise with different flux curves, are discussed in the corollaries.

8.4.1 Case I: $f'(a)$, $f'(b)$ and $f'(\varphi_\infty)$ are positive

We assume that $|q|$ is small enough so that $f'(a) > 0$, $f'(b) > 0$ and $f'(\varphi_\infty) > 0$. This implies that

$$f'(\varphi) > 0 \text{ for } \varphi \in [a, \varphi_t]. \tag{8.21}$$

Physically, this selection of q means that most of the time the thickener overflows. Let φ_s be such that

$$\sigma(\varphi_s, \varphi_\infty) = 0, \ 0 < \varphi_s < \varphi_t^{**}, \tag{8.22}$$

then; as a consequence of Lemma 8.2, we have the following corollary:

Corollary 8.2 *The function $\sigma(\varphi, \varphi_\infty)$ satisfies $\sigma(\varphi, \varphi_\infty) < 0$ for $0 < \varphi < \varphi_s$ and $\sigma(\varphi, \varphi_\infty) > 0$ for $\varphi_s < \varphi < \varphi_t^{**}$.*

Depending on the feed solid flux density f_F, we have different CKSP. In the next theorem we analyse all possible cases, considering first a small $|f_F|$ for which naturally the thickener empties. Then a larger $|f_F|$ is used attaining a steady state. In the next cases for $|f_F|$ even larger, the thickener overflows.

Theorem 8.4 *If $f'(a) > 0$, $f'(b) > 0$ and $f'(\varphi_\infty) > 0$, then for the solution of problem (8.3) we have:*

(a) *For $\varphi_L < \varphi_s$ there exists a time t_1 such that for $0 \le t < t_1$ the CKSP is an MCS-1 and afterwards the ICT empties. The boundary conditions are satisfied classically at $z = L$ and in a generalized sense at $z = 0$.*

(b) *For $\varphi_L = \varphi_s$ the CKSP is an MCS-1 for all t and the ICT attains a steady state with $\varphi_D = \varphi_\infty$. Both boundary conditions are satisfied classically.*

(c) *For $\varphi_s < \varphi_L \le \varphi_t^{**}$ there exists a time t_1 such that for $0 \le t < t_1$ the CKSP is an MCS-1 and afterwards the ICT overflows. The boundary conditions are satisfied in a generalized sense at $z = L$ and classically at $z = 0$.*

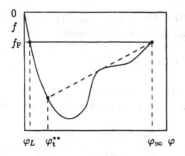
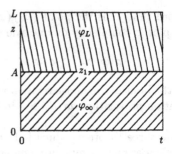

Figure 8.9: Graph of f vs φ and representation of the global weak solution $\varphi(z,t)$ for an MCS-1 with $\varphi_L = \varphi_s$.

(d) For $\varphi_t^{**} < \varphi_L < a$ there exists a time t_1 such that for $0 \le t < t_1$ the CKSP is an MCS-4 and afterwards the ICT overflows. The boundary conditions are satisfied in a generalized sense at $z = L$ and classically at $z = 0$.

(e) For $a \le \varphi_L < \varphi_t$ there exists a time t_1 such that for $0 \le t < t_1$ the CKSP is an MCS-5 and afterwards the ICT overflows. The boundary conditions are satisfied in a generalized sense at $z = L$ and classically at $z = 0$.

(f) For $\varphi_t \le \varphi_L < \varphi_\infty$ there exists a time t_1 such that for $0 \le t < t_1$ the CKSP is an MCS-1 and afterwards the ICT overflows. The boundary conditions are satisfied in a generalized sense at $z = L$ and classically at $z = 0$.

Proof. For all cases considered in this theorem, we have $\epsilon_0(\varphi_\infty) = [0, \varphi_s] \cup \{\varphi_\infty\}$.

(b) For $\varphi_L = \varphi_s$, as shown in Figure 8.9, the global weak solution of problem (8.3) for $t \ge 0$ is given by

$$\varphi(z,t) = \begin{cases} \varphi_L & \text{for } z_1(t) \le z \le L, \\ \varphi_\infty & \text{for } 0 \le z < z_1(t). \end{cases}$$

The shock $z_1(t) = A$ is horizontal since $\sigma(\varphi_L, \varphi_\infty) = 0$. The solution consists of two constant states φ_L and φ_∞ separated by the shock $z_1(t)$, hence the CKSP is an MCS-1 for all t. At $z = L$, $\epsilon_L(\varphi_L) = \{\varphi_L\} \cup \{\varphi_\infty\}$ and $\gamma\varphi(L,t) = \varphi_L$. At $z = 0$, the discharge concentration $\gamma\varphi(0,t) = \varphi_D(t) = \varphi_\infty$ for all t and the ICT is at steady state. Both boundary conditions are satisfied classically. This proves part (b) of the theorem. ∎

(d) For $\varphi_t^{**} < \varphi_L < a$, see Figure 8.10, the global weak solution of problem (8.3) for $0 \le t < t_1$ is given by

$$\varphi(z,t) = \begin{cases} \varphi_L & \text{for } z_1(t) \le z \le L, \\ h_4((z - A)/t) & \text{for } z_2(t) \le z < z_1(t), \\ \varphi_\infty & \text{for } 0 \le z < z_2(t). \end{cases}$$

For $t_1 \leq t < t_2$, the solution is

$$\varphi(z,t) = \begin{cases} h_4((z-A)/t) & \text{for } z_2(t) \leq z < L, \\ \varphi_\infty & \text{for } 0 \leq z < z_2(t), \end{cases}$$

and for $t \geq t_2$ is

$$\varphi(z,t) = \varphi_\infty \text{ for } 0 \leq z \leq L.$$

Here, as in Chapter 7, h_4 is the inverse of $f'(\varphi)$ restricted to values of φ between a and b. Since $\varphi_L < a$, we have $f'(\varphi_L) < \sigma(\varphi_L, \varphi_L^*) = f'(\varphi_L^*)$ by Corollary 5.2 and Lemma 5.2, and therefore the discontinuity

$$z_1(t) = A + f'(\varphi_L^*)t = A + \sigma(\varphi_L, \varphi_L^*)t$$

is a contact discontinuity. This contact discontinuity $z_1(t)$ cuts the line $z = L$ at the time

$$t_1 = \frac{L-A}{\sigma(\varphi_L, \varphi_L^*)} = \frac{L-A}{f'(\varphi_L^*)}.$$

Since $f'(\varphi_L^*) > 0$ by property (8.21), t_1 is a positive and finite time. For the discontinuity $z_2(t) = A + f'(\varphi_t)t$ we have from Lemma 5.6 that $f'(\varphi_t) = \sigma(\varphi_t, \varphi_\infty) < f'(\varphi_\infty)$ and therefore $z_2(t)$ is a contact discontinuity. This contact discontinuity cuts the line $z = L$ at the time

$$t_2 = \frac{L-A}{f'(\varphi_t)},$$

which is positive and finite. Since $f''(\varphi) < 0$ for φ between a and b, we have that $f'(\varphi_L^*) > f(\varphi_t)$ and hence $t_2 > t_1$. For $0 \leq t < t_1$ the solution consists of two constant states φ_L and φ_∞ separated by the contact discontinuity $z_1(t)$, a rarefaction wave and the other contact discontinuity $z_2(t)$; thus the CKSP is an MCS-4. At $z = L$ the solution is

$$\gamma\varphi(L,t) = \begin{cases} \varphi_L & \text{for } 0 \leq z < t_1, \\ h_4((L-A)/t) & \text{for } t_1 \leq t < t_2, \\ \varphi_\infty & \text{for } t_2 \leq t. \end{cases}$$

The set $\epsilon_L(\varphi_L) = \{\varphi_L\} \cup [\varphi_{L0}, \varphi_\infty]$ or $\epsilon_L(\varphi_L) = [\varphi_m, \varphi_\infty]$ in the case that $\varphi_L \geq \varphi_m$, where φ_{L0} was defined in (8.14) and φ_m is the point at which f attains its minimum in $[0, \varphi_\infty]$. Since φ_L belongs to $\epsilon_L(\varphi_L)$ and $h_4((L-A)/t)$ for $t_1 \leq t < t_2$ assumes values between φ_L^* and φ_∞ which also belong to $\epsilon_L(\varphi_L)$, the entropy boundary condition is satisfied at $z = L$. The thickener overflows for $t \geq t_1$. At $z = 0$, the discharge concentration $\varphi_D(t) = \gamma\varphi(0,t) = \varphi_\infty$ for all t and the boundary condition is satisfied classically. This proves part (d) of the theorem. ∎

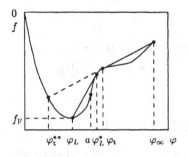 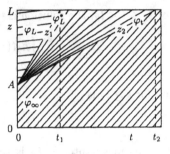

Figure 8.10: Graph of f vs φ and representation of the global weak solution $\varphi(z,t)$ for an MCS-4 with $\varphi_t^{**} < \varphi_L < a$.

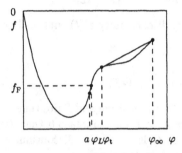 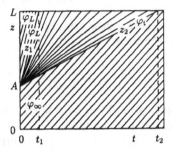

Figure 8.11: Graph of f vs φ and representation of the global weak solution $\varphi(z,t)$ for an MCS-5 with $a < \varphi_L < \varphi_t$.

(e) For $a \leq \varphi_L < \varphi_t$, the global weak solution of problem (8.3) for $0 \leq t < t_1$ is given by (see Figure 8.11)

$$\varphi(z,t) = \begin{cases} \varphi_L & \text{for } z_1(t) \leq z \leq L, \\ h_4((z-A)/t) & \text{for } z_2(t) \leq z < z_1(t), \\ \varphi_\infty & \text{for } 0 \leq z < z_2(t). \end{cases}$$

For $t_1 \leq t < t_2$ the solution is

$$\varphi(z,t) = \begin{cases} h_4((z-A)/t) & \text{for } z_2(t) \leq z < L, \\ \varphi_\infty & \text{for } 0 \leq z < z_2(t), \end{cases}$$

and for $t_2 \leq t$ is

$$\varphi(z,t) = \varphi_\infty \text{ for } 0 \leq z \leq L.$$

The line $z_1(t) = A + f'(\varphi_L)t$ is a line of continuity and the line $z_2(t) = A + f'(\varphi_t)t$, the same as in part (d) of the theorem, is a contact discontinuity.

The line $z_1(t)$ cuts $z = L$ at the time

$$t_1 = \frac{L - A}{f'(\varphi_L)},$$

which is positive and finite since $f'(\varphi_L) > 0$ by property (8.21). For $0 \leq t < t_1$ the solution consists of two constant states φ_L and φ_∞ separated by a rarefaction wave and a contact discontinuity; therefore the CKSP is an MCS-5. The line $z_2(t)$ cuts $z = L$ at the time

$$t_2 = \frac{L - A}{f'(\varphi_\infty)},$$

which is positive. Since $f'' < 0$ for φ between a and b, we have $f'(\varphi_L) > f'(\varphi_t)$ and thus $t_2 > t_1$. At $z = L$ the solution is

$$\gamma\varphi(L, t) = \begin{cases} \varphi_L & \text{for } 0 \leq z < t_1, \\ h_4((L - A)/t) & \text{for } t_1 \leq t < t_2, \\ \varphi_\infty & \text{for } t \geq t_2. \end{cases}$$

The set $\epsilon_L(\varphi_L) = [\varphi_m, \varphi_\infty]$. Since φ_L belongs to $\epsilon_L(\varphi_L)$ and $h_4((L-A)/t)$ for $t_1 \leq t < t_2$ assumes values between φ_L and φ_∞ which also belong to $\epsilon_L(\varphi_L)$, the entropy boundary condition is satisfied at $z = L$. The thickener overflows for $t \geq t_1$. At $z = 0$, the discharge concentration $\varphi_D(t) = \gamma\varphi(0, t) = \varphi_\infty$ for all t and the boundary condition is satisfied classically. This proves part (e) of the theorem. ∎

Corollary 8.3 *If $f'(a) > 0$, $f'(b) < 0$ and $f'(\varphi_\infty) > 0$ with $f(\varphi_\infty) > f(\varphi_M)$, then the solution of problem (8.3) is given by cases (a) to (f) of Theorem 8.4.*

Proof. It follows from these conditions and the definitions of a and b that there is a local maximum, $f(\varphi_M)$, and a local minimum, $f(\varphi_m)$, such that $f(\varphi_m) < f(\varphi_M) < f(\varphi_\infty)$ and $a < \varphi_M < b < \varphi_m < \varphi_\infty$ as shown in Figure 8.12. It is clear

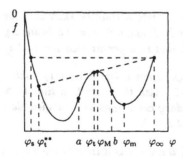

Figure 8.12: Graphs of f vs φ showing the relevant coordinates in Corollary 8.3.

that $0 < \varphi_s < \varphi_t^{**} < a < \varphi_t < \varphi_\infty$ (just as in Theorem 8.4) and that no other values (including φ_M and φ_m) are significant. Also, $f(\varphi_t^{**}) < f(\varphi_t) < f(\varphi_\infty)$ as in Theorem 8.4. Therefore, all the conditions in all parts of Theorem 8.4 apply and the corollary is proved.

8.4.2 Case II: $f'(a)$ is positive and $f'(b)$ and $f'(\varphi_\infty)$ are negative

We increase $|q|$ in such a way that $f'(\varphi) < 0$ for $\varphi_t \leq \varphi \leq \varphi_\infty$ while $f'(a)$ remains positive. This implies that there exists a $\varphi_M > a$ at which f has a relative maximum, that is $f'(\varphi_M) = 0$ and $f''(\varphi_M) < 0$.

Theorem 8.5 *If $f'(a) > 0$, $f'(b) < 0$, $f'(\varphi_\infty) < 0$ and φ_t^{**} does not exist, then for the solution of problem (8.3) we have:*

(a) *For $0 < \varphi_L < \varphi_M^{**}$ there exists a time t_2 such that for $0 \leq t < t_2$ the CKSP is an MCS-4. Also there exists a time $t_1 > t_2$ after which the ICT empties. The boundary conditions are satisfied classically at $z = L$ and in a generalized sense at $z = 0$.*

(b) *For $\varphi_L = \varphi_M^{**}$ there exists a time t_2 such that for $0 \leq t < t_2$ the CKSP is an MCS-4 and the ICT attains a steady state with $\varphi_D = \varphi_M$. The boundary conditions are satisfied classically at $z = L$ and in a generalized sense at $z = 0$.*

(c) *For $\varphi_M^{**} < \varphi_L < a$ there exist times t_1 and t_2 such that for $0 \leq t < t_0$ with $t_0 = min(t_1, t_2)$, the CKSP is an MCS-4 and afterwards the ICT overflows. The boundary conditions are satisfied in a generalized sense on both boundaries.*

(d) *For $a \leq \varphi_L < \varphi_M$ there exist times t_1 and t_2 such that for $0 \leq t < t_0$ with $t_0 = min(t_1, t_2)$, the CKSP is an MCS-5. For $t_0 \leq t$ the ICT overflows. The boundary conditions are satisfied in a generalized sense on both boundaries.*

(e) *For $\varphi_L = \varphi_M$ there exists a time t_2 such that for $0 \leq t < t_2$ the CKSP is an MCS-5. The ICT empties and the boundary conditions are satisfied classically at $z = L$ and in a generalized sense at $z = 0$.*

(f) *For $\varphi_M < \varphi_L < \varphi_t$ there exists a time t_2 such that for $0 \leq t < t_2$ the CKSP is an MCS-5. Also there exists a time $t_1 > t_2$ after which the ICT empties. The boundary conditions are satisfied classically at $z = L$ and in a generalized sense at $z = 0$.*

(g) *For $\varphi_t \leq \varphi_L < \varphi_\infty$ there exists a time t_2 such that for $0 \leq t < t_2$ the CKSP is an MCS-1. Afterwards the ICT empties. The boundary conditions are satisfied classically at $z = L$ and in a generalized sense at $z = 0$.*

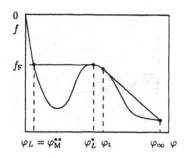 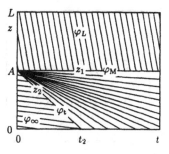

Figure 8.13: Graph of f vs φ and representation of the global weak solution $\varphi(z,t)$ for an MCS-4 with $\varphi_L = \varphi_M^{**}$.

Proof. The set of admissible states at $z = 0$ for all cases considered in this theorem is $\epsilon_0(\varphi_\infty) = [0, \varphi_M^{**}] \cup [\varphi_M, \varphi_\infty]$. The set of admissible states at the boundary $z = L$ is considered separately. Let h_4 be the inverse of $f'(\varphi)$ restricted to values of φ between a and b.

(b) For $\varphi_L = \varphi_M^{**}$, as shown in Figure 8.13, the global weak solution of problem (8.3a) to (8.3d) for $0 \leq t < t_2$ is given by

$$\varphi(z,t) = \begin{cases} \varphi_L & \text{for } z_1(t) \leq z \leq L, \\ h_4((z-A)/t) & \text{for } z_2(t) \leq z < z_1(t), \\ \varphi_\infty & \text{for } 0 \leq z < z_2(t). \end{cases}$$

For $t \geq t_2$ the solution is

$$\varphi(z,t) = \begin{cases} \varphi_L & \text{for } z_1(t) \leq z < L, \\ h_4((z-A)/t) & \text{for } 0 \leq z < z_1(t). \end{cases}$$

The lines

$$z_1(t) = A + f'(\varphi_L^*)t = A + \sigma(\varphi_L, \varphi_L^*)t$$

and

$$z_2(t) = A + f'(\varphi_t)t = A + \sigma(\varphi_t, \varphi_\infty)t$$

are shocks. Since $\varphi_L < a$, we have $f'(\varphi_L) < \sigma(\varphi_L, \varphi_L^*) = f'(\varphi_L^*)$ by Corollary 5.2 and Lemma 5.2, and therefore $z_1(t)$ is a contact discontinuity. Also from Lemma 5.6, $f'(\varphi_t) = \sigma(\varphi_t, \varphi_\infty) < f'(\varphi_\infty)$ and so $z_2(t)$ is also a contact discontinuity. The line $z_2(t)$ cuts the t-axis at $t_2 = -A/f'(\varphi_t)$ which is positive and finite. The contact discontinuity $z_1(t)$ is horizontal. For $0 \leq t < t_2$ the solution consists of two constant states φ_L and φ_∞ separated by the contact discontinuity $z_1(t)$, a rarefaction wave and the other

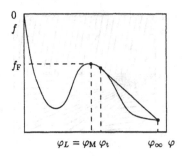
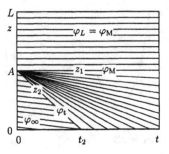

Figure 8.14: Graph of f vs φ and representation of the global weak solution $\varphi(z,t)$ for an MCS-5 with $\varphi_L = \varphi_M$.

contact discontinuity $z_2(t)$; hence the CKSP is an MCS-4. At $z = L$, the set $\epsilon_L(\varphi_L) = \{\varphi_L\} \cup \{\varphi_M\}$ and since $\gamma\varphi(L,t) = \varphi_L$ for all t, the boundary condition is satisfied classically. At $z = 0$, the discharge concentration $\varphi_D(t) = \gamma\varphi(0,t)$ is

$$\varphi_D(t) = \begin{cases} \varphi_\infty & \text{for } 0 \le z < t_2, \\ h_4(-A/t) & \text{for } t \ge t_2. \end{cases} \qquad (8.23)$$

Since $h_4(-A/t)$ for $t \ge t_2$ assumes values between φ_M and φ_t which belong to $\epsilon_0(\varphi_\infty)$, the entropy boundary condition is satisfied at $z = 0$. Here the thickener attains a steady state with $\lim_{t\to\infty} \varphi_D(t) = \varphi_M$. This proves part (b) of the theorem. $\blacksquare$

(e) For $\varphi_L = \varphi_M$, as depicted in Figure 8.14, the global weak solution of problem (8.3) for $0 \le t < t_2$ is given by

$$\varphi(z,t) = \begin{cases} \varphi_L & \text{for } z_1(t) \le z \le L, \\ h_4((z-A)/t) & \text{for } z_2(t) \le z < z_1(t), \\ \varphi_\infty & \text{for } 0 \le z < z_2(t), \end{cases}$$

and for $t \ge t_2$ is

$$\varphi(z,t) = \begin{cases} \varphi_L & \text{for } z_1(t) \le z < L, \\ h_4((z-A)/t) & \text{for } 0 \le z < z_1(t). \end{cases}$$

The horizontal line $z_1(t) = A$ is a line of continuity and

$$z_2(t) = A + f'(\varphi_t)t = A + \sigma(\varphi_t, \varphi_\infty)t$$

is a contact discontinuity. The line $z_2(t)$ cuts the t-axis at $t_2 = -A/f'(\varphi_t)$ which is positive and finite. For $0 \le t < t_2$ the solution consists of two

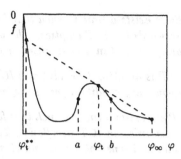

Figure 8.15: Graph of f vs φ showing the relevant coordinates in Corollary 8.4.

constant states φ_L and φ_∞ separated by a rarefaction wave and the contact discontinuity $z_2(t)$; therefore the CKSP is an MCS-5. At $z = L$, we have that $\epsilon_L(\varphi_L) = [\varphi_m, \varphi_M = \varphi_L]$ and $\gamma\varphi(L, t) = \varphi_L$ and the boundary condition is satisfied classically. At $z = 0$, the discharge concentration $\varphi_D(t) = \gamma\varphi(0, t)$ is given by (8.23). Then the entropy condition is satisfied at $z = 0$. The ICT attains a steady state with $\lim_{t\to\infty} \varphi_D(t) = \varphi_M$. Since $\varphi_D = \varphi_M = \varphi_L \le \varphi_F$, the ICT empties. This proves part (e) of the theorem. ■

Corollary 8.4 *If $f'(a) > 0$, $f'(b) < 0$, $f'(\varphi_\infty) < 0$ and φ_t^{**} exists, as shown in Figure 8.15, then case (a) in Theorem 8.5 is replaced by*

(i) *For $0 < \varphi_L \le \varphi_t^{**}$ there exists a time t_1 such that for $0 \le t < t_1$ the CKSP is an MCS-1 and afterwards the ICT empties. The boundary conditions are satisfied classically at $z = L$ and in a generalized sense at $z = 0$.*

(ii) *For $\varphi_t^{**} < \varphi_L < \varphi_M^{**}$ there exists a time t_2 such that for $0 \le t < t_2$ the CKSP is an MCS-4. Also there exists a time $t_1 > t_2$ after which the ICT empties. The boundary conditions are satisfied classically at $z = L$ and in a generalized sense at $z = 0$.*

and the rest of the cases remain the same.

Proof. The conditions imply that $\varphi_M < \varphi_t < \varphi_\infty$ and hence $0 < \varphi_t^{**} < \varphi_M^{**}$. Case (i) is similar to case (a) of Theorem 8.2 in that no intermediate concentrations are constructed. Case (ii) is similar to case (a) of Theorem 8.5. For the other cases, note that $\varphi_M^{**} < a < \varphi_M < \varphi_t < \varphi_\infty$ and $f(\varphi_M^{**}) = f(\varphi_M) > f(\varphi_t) > f(\varphi_\infty)$ just as in Theorem 8.5. ■

Corollary 8.5 *If $f'(a) > 0$, $f'(b) < 0$, $f'(\varphi_\infty) > 0$ and $f(\varphi_\infty) < f(\varphi_M)$, as shown in Figure 8.16, then the solution of problem (8.3) is given either by cases (a) to (f) of Theorem 8.5 or by cases (i) and (ii) of Corollary 8.4 together with cases (b) to (f) of Theorem 8.5 and in both instances case (g) is replaced by*

(iii) For $\varphi_t < \varphi_L < \varphi_s$, there exists a time t_1 such that for $0 \leq t < t_1$ the CKSP is an MCS-1 and afterwards the ICT empties. The boundary conditions are satisfied classically at $z = L$ and in a generalized sense at $z = 0$.

(iv) For $\varphi_L = \varphi_s$, the CKSP is an MCS-1 for all t and ICT attains a steady state with $\varphi_D = \varphi_\infty$; both boundary conditions are satisfied classically.

(v) For $\varphi_s < \varphi_L < \varphi_\infty$, there exists a time t_1 such that for $0 \leq t < t_1$ the CKSP is an MCS-1 and afterwards the ICT overflows. The boundary conditions are satisfied in a generalized sense at $z = L$ and classically at $z = 0$.

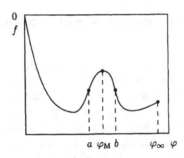

Figure 8.16: Graph of f vs φ showing the relevant coordinates in Corollary 8.5.

Proof. The conditions imply that $0 < \varphi_M < \varphi_t < \varphi_\infty$ and $f(\varphi_M) > f(\varphi_t) > f(\varphi_\infty)$, so φ_t^{**} may or may not exist. If it does, then the cases (i) and (ii) obviously apply. If not, then case (a). In cases (b) to (f), note that $\varphi_M^{**} < a < \varphi_M < \varphi_t$ and $f(\varphi_M^{**}) = f(\varphi_M) > f(\varphi_t) > f(\varphi_\infty)$. For (iii) and (iv), note that the conditions imply a local minimum, $f(\varphi_m)$, such that $\varphi_M < \varphi_m < \varphi_\infty$ and hence φ_s such that $\varphi_M < \varphi_s < \varphi_m$ and $f(\varphi_s) = f(\varphi_\infty)$. The proof of (iii) is similar to that of (g) of Theorem 8.5. The proofs of (iv) and (v) are obvious. ∎

8.4.3   Case III: $f'(a)$, $f'(b)$ and $f'(\varphi_\infty)$ are negative

We increase $|q|$ again in such a way that $f'(a)$, $f'(b)$ and $f'(\varphi_\infty)$ are negative. This means that now f has no relative maximum nor minimum between a and b and $f'(\varphi) < 0$ for $0 \leq \varphi \leq \varphi_\infty$. Physically, no steady state is attained and the thickener empties.

Theorem 8.6 *If $f'(a)$, $f'(b)$ and $f'(\varphi_\infty)$ are negative and φ_t^{**} does not exist, then for the solution of problem (8.3) we have:*

(a) For $0 < \varphi_L < a$ there exists a time t_2 such that for $0 \leq t < t_2$ the CKSP is an MCS-4. Also there exists a time $t_1 > t_2$ after which the ICT empties.

(b) For $a \leq \varphi_L < \varphi_t$ there exists a time t_2 such that for $0 \leq t < t_2$ the CKSP is an MCS-5. Also there exists a time $t_1 > t_2$ after which the ICT empties.

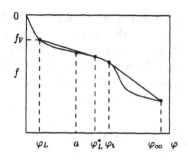 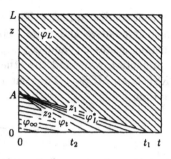

Figure 8.17: Graph of f vs φ and representation of the global weak solution $\varphi(z,t)$ for an MCS-4 with $0 < \varphi_L < a$.

(c) For $\varphi_t \leq \varphi_L < \varphi_\infty$ there exists a time t_1 such that for $0 \leq t < t_1$ the CKSP is an MCS-1 and afterwards the ICT empties.

In all three cases the boundary conditions are satisfied classically at $z = L$ and in a generalized sense at $z = 0$.

Proof. For all cases considered in this theorem, the set of admissible states are $\epsilon_L(\varphi_L) = \{\varphi_L\}$ and $\epsilon_0(\varphi_\infty) = [0, \varphi_\infty]$ at $z = L$ and $z = 0$, respectively. Let h_4 be the inverse of $f'(\varphi)$ restricted to values of φ between a and b.

(a) For $0 < \varphi_L < a$, as shown in Figure 8.17, the global weak solution of problem (8.3) for $0 \leq t < t_2$ is given by

$$\varphi(z,t) = \begin{cases} \varphi_L & \text{for } z_1(t) \leq z \leq L, \\ h_4((z-A)/t) & \text{for } z_2(t) \leq z < z_1(t), \\ \varphi_\infty & \text{for } 0 \leq z < z_2(t). \end{cases}$$

For $t_2 \leq t < t_1$ the solution is

$$\varphi(z,t) = \begin{cases} \varphi_L & \text{for } z_1(t) \leq z < L, \\ h_4((z-A)/t) & \text{for } 0 \leq z < z_1(t), \end{cases}$$

and for $t_1 \leq t$ is

$$\varphi(z,t) = \varphi_L \text{ for } 0 \leq z \leq L.$$

Since $\varphi_L < a$, we have $f'(\varphi_L) < \sigma(\varphi_L, \varphi_L^*) = f'(\varphi_L^*)$ by Corollary 5.2 and Lemma 5.2, and therefore the discontinuity

$$z_1(t) = A + f'(\varphi_L^*)t = A + \sigma(\varphi_L, \varphi_L^*)t$$

is a contact discontinuity. Also from Lemma 5.6 of Chapter 5 we have that $f'(\varphi_t) = \sigma(\varphi_t, \varphi_\infty) < f'(\varphi_\infty)$, so the line

$$z_2(t) = A + f'(\varphi_t)t = A + \sigma(\varphi_t, \varphi_\infty)$$

is a contact discontinuity. This line cuts $z = 0$ at the time $t_2 = -A/f'(\varphi_t)$, which is positive and finite. For $0 \le t < t_2$ the solution consists of two constant states φ_L and φ_∞ separated by the contact discontinuity $z_1(t)$, a rarefaction wave and the other contact discontinuity $z_2(t)$; therefore the CKSP is an MCS-4. The contact discontinuity $z_1(t)$ cuts the line $z = 0$ at the time

$$t_1 = -\frac{A}{\sigma(\varphi_L, \varphi_L^*)} = -\frac{A}{f'(\varphi_L^*)},$$

which is positive and finite. Moreover, since $\varphi_L^* < \varphi_t$ due to the fact that f'' is negative between a and φ_t, we have $t_1 > t_2$. At $z = L$, $\gamma\varphi(L, t) = \varphi_L$ for all t, so the boundary condition is satisfied classically. At $z = 0$, the discharge concentration $\varphi_D(t) = \gamma\varphi(0, t)$ is

$$\varphi_D(t) = \begin{cases} \varphi_\infty & \text{for } 0 \le z < t_2, \\ h_4(-A/t) & \text{for } t_2 \le t < t_1, \\ \varphi_L & \text{for } t_1 \le t. \end{cases} \tag{8.24}$$

Here $h_4(-A/t)$ for $t_2 \le t < t_1$ assumes values between φ_L^* and φ_t which belong to $\epsilon_0(\varphi_\infty)$, so the entropy boundary condition is satisfied at $z = 0$. Since $\varphi_L < \varphi_F$, the thickener empties for $t \ge t_1$. This proves part (a) of the theorem. ■

(b) For $a \le \varphi_L < \varphi_t$, as shown in Figure 8.18, the global weak solution of problem (8.3) for $0 \le t < t_2$ is given by

$$\varphi(z, t) = \begin{cases} \varphi_L & \text{for } z_1(t) \le z \le L, \\ h_4((z - A)/t) & \text{for } z_2(t) \le z < z_1(t), \\ \varphi_\infty & \text{for } 0 \le z < z_2(t). \end{cases}$$

For $t_2 \le t < t_1$ the solution is

$$\varphi(z, t) = \begin{cases} \varphi_L & \text{for } z_1(t) \le z < L, \\ h_4((z - A)/t) & \text{for } 0 \le z < z_1(t), \end{cases}$$

and for $t_1 \le t$, the solution is

$$\varphi(z, t) = \varphi_L \text{ for } 0 \le z \le L.$$

As before, the line

$$z_2(t) = A + f'(\varphi_t)t = A + \sigma(\varphi_t, \varphi_\infty)$$

is a contact discontinuity that cuts the t-axis at $t_2 = -A/f'(\varphi_\infty)$ which is also positive and finite. For $0 \le t < t_2$ the solution consists of two constant states φ_L and φ_∞ separated by a rarefaction wave and the contact

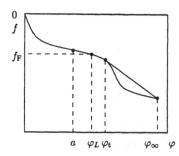 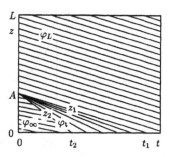

Figure 8.18: Graph of f vs φ and representation of the global weak solution $\varphi(z,t)$ for an MCS-5 with $a < \varphi_L < \varphi_t$.

discontinuity $z_2(t)$; hence the CKSP is an MCS-5. Moreover, since $\varphi_L < \varphi_\infty$, $f'(\varphi_L) > f'(\varphi_\infty)$ and therefore $t_1 > t_2$. The line of continuity $z_1(t) = A + f'(\varphi_L)t$ cuts the t-axis at $t_1 = -A/f'(\varphi_L)$, which is positive and finite. Since $f'' < 0$ between φ_L and φ_t, then $f'(\varphi_L) > f'(\varphi_t)$ and therefore $t_1 > t_2$. At $z = L$, the solution is $\gamma\varphi(L,t) = \varphi_L$, so the boundary condition is satisfied classically. At $z = 0$, the discharge concentration $\varphi_D(t) = \gamma\varphi(0,t)$ is given by equation (8.24). For $t_2 \leq t < t_1$, $h_4(-A/t)$ assumes values between φ_L and φ_t and therefore the entropy boundary condition is satisfied. Since $\varphi_L = \varphi_F$ the thickener empties for $t \geq t_1$. This proves part (b) of the theorem. ∎

(c) For $\varphi_t \leq \varphi_L < \varphi_\infty$, the solution is obvious. This completes the proof of the theorem. ∎

Corollary 8.6 *If $f'(a)$, $f'(b)$ and $f'(\infty)$ are negative and φ_t^{**} exists, as shown in Figure 8.19, then case (a) in Theorem 8.6 is replaced by:*

(i) *For $0 < \varphi_L \leq \varphi_t^{**}$, there exists a time t_1 such that for $0 \leq t < t_1$, the CKSP is an MCS-1 and afterwards the ICT empties. The boundary conditions are satisfied classically at $z = L$ and in a generalized sense at $z = 0$.*

(ii) *For $\varphi_t^{**} < \varphi_L < a$, there exists a time t_2 such that for $0 \leq t < t_2$, the CKSP is an MCS-4. Also, there exists a time $t_1 > t_2$ after which the ICT empties. The boundary conditions are satisfied classically at $z = L$ and in a generalized sense at $z = 0$.*

Proof. Similar to those given earlier. ∎

Corollary 8.7 *If $f'(a) < 0$, $f'(b) < 0$ and $f'(\varphi_\infty) > 0$ as in Figure 8.20, then for the solution of problem (8.3a)–(8.3d) we have (i) and (ii) of Corollary 8.6, (b) of Theorem 8.6 and (iii), (iv) and (v) of Corollary 8.5.*

Proof. Similar to those given earlier. ∎

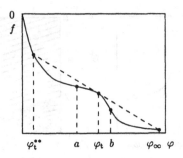

Figure 8.19: Graph of f vs φ showing the relevant coordinates in Corollary 8.6.

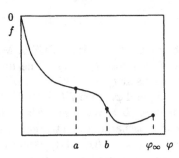

Figure 8.20: Graph of f vs φ showing the relevant coordinates in Corollary 8.7.

8.5   Control of continuous sedimentation

In this section, we use Kynch's theory and the method of characteristics together with the interaction of shocks and rarefaction waves to control the operation of an ICT. Let an ICT be operating at steady state under the conditions stated in part (b) of Theorem 8.1. Let

$$\varphi(z,0) = \begin{cases} \varphi_{L0} & \text{for } A \leq z \leq L, \\ \varphi_\infty & \text{for } 0 \leq z < A. \end{cases} \tag{8.25}$$

Let f_{F1} and q_1 be the feed solid flux density and the volume average velocity of the suspension at the steady state. At $t = t_1$, the feed solid flux density decreases in absolute value from $|f_{F1}|$ to $|f_{F2}|$. See Figure 8.21. Such a perturbation will cause the ICT to empty, so a control action is needed to recover the steady state. This control action consists in reducing the volume average velocity in absolute value from $|q_1|$ to $|q_2|$ such that

$$q_2 = f_{F2}/\varphi_\infty. \tag{8.26}$$

This control action is performed at $t = t_2$. This time must be small enough so

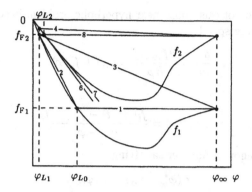

Figure 8.21: Graph of f vs φ showing f_1, f_2 and $\varphi_L(t)$. The numbers $i = 1, 2, 3, 4, 6, 7, 8$ correspond to the straight lines $z_i(t)$ in Figure 8.22.

that the ICT does not empty. At $z = L$, the concentration $\varphi_L(t)$ is

$$\varphi_L(t) = \begin{cases} \varphi_{L0} & \text{for } 0 \leq t < t_1, \\ \varphi_{L1} & \text{for } t_1 \leq t < t_2, \\ \varphi_{L2} & \text{for } t_2 \leq t. \end{cases} \tag{8.27}$$

Here φ_{L0}, φ_{L1} and φ_{L2} are obtained from

$$f_{F1} = q_1 \varphi_{L0} + f_b(\varphi_{L0}), \tag{8.28}$$
$$f_{F2} = q_1 \varphi_{L1} + f_b(\varphi_{L1}), \tag{8.29}$$
$$f_{F2} = q_2 \varphi_{L2} + f_b(\varphi_{L2}). \tag{8.30}$$

At $z = 0$ the concentration remains equal to φ_∞.

8.5.1 Model of the control problem

We model the control problem just described as the following initial-boundary value problem:

$$\frac{\partial \varphi}{\partial t} + \frac{\partial f(\varphi, t)}{\partial z} = 0 \text{ for } (z, t) \in Q_T, \tag{8.31a}$$
$$\varphi(z, 0) = \varphi_I(z) \text{ for } z \in \Omega, \tag{8.31b}$$
$$\gamma \varphi(0, t) \in \epsilon_0(\varphi_\infty) \text{ for } t \in \mathcal{T}, \tag{8.31c}$$
$$\gamma \varphi(L, t) \in \epsilon_L(\varphi_L) \text{ for } t \in \mathcal{T} \tag{8.31d}$$

with the flux density function

$$f(\varphi, t) = \begin{cases} f_1(\varphi) = q_1 \varphi + f_b(\varphi) & \text{for } 0 \leq t \leq t_1, \\ f_2(\varphi) = q_2 \varphi + f_b(\varphi) & \text{for } t_2 \leq t. \end{cases} \tag{8.32}$$

The smooth solution of this quasilinear hyperbolic problem is constant along the characteristics having speed

$$\frac{dz}{dt} = \frac{\partial f(\varphi, t)}{\partial \varphi} = q(t) + f_b'(\varphi). \tag{8.33}$$

Define

$$\sigma_i(\varphi^+, \varphi^-) = \frac{f_i(\varphi^+, t) - f_i(\varphi^-, t)}{\varphi^+ - \varphi^-}, \tag{8.34}$$

let $\varphi_{\infty i}^{**}$, $i = 1, 2$ denote a number satisfying

$$\sigma_i(\varphi_{\infty i}^{**}, \varphi_\infty) = f_i'(\varphi_\infty), \tag{8.35}$$

and let φ_{si}, $i = 1, 2$, with $0 < \varphi_{si} < \varphi_{\infty i}^{**}$ be defined as in (8.9):

$$\sigma_i(\varphi_{si}, \varphi_\infty) = 0. \tag{8.36}$$

Since the ICT initially operated at the steady state defined in part (b) of Theorem 8.1, we have that

$$\varphi_{L0} = \varphi_{s1} \quad \text{and} \quad \varphi_{L0} < \varphi_{\infty 1}^{**} \quad \text{for} \quad 0 \le t < t_1. \tag{8.37}$$

Lemma 8.3 *The concentration values φ_{L0} and φ_{L1} satisfy $\varphi_{L1} < \varphi_{L0}$.*

Proof. We know that $\sigma_1(\varphi_{L0}, \varphi_\infty) = 0$ and

$$\sigma_1(\varphi_{L1}, \varphi_\infty) = \frac{f_1(\varphi_{L1}) - f_1(\varphi_\infty)}{\varphi_{L1} - \varphi_\infty} = \frac{f_{F2} - q_1\varphi_\infty}{\varphi_{L1} - \varphi_\infty} = \frac{f_{F2} - f_{F1}}{\varphi_{L1} - \varphi_\infty} < 0,$$

and from Corollary 8.1 we deduce that $\varphi_{L1} < \varphi_{L0}$. This completes the proof. ■

Finally, we assume that φ_{L2} obtained in (8.30) is such that $\varphi_{L2} < \varphi_{\infty 2}^{**}$. In this way we can recover the same type of steady state we started with.

8.5.2 Construction of the entropy solution

We construct the entropy solution of problem (8.31) so that it satisfies the initial condition (8.31b), the differential equation (8.31a) at the points of continuity, the Rankine-Hugoniot (6.18) and Oleinik's condition E (6.20) at discontinuities, and the boundary entropy conditions (8.31c) and (8.31d).

Solution for $0 \le t < t_1$

For $0 \le t < t_1$ (see Figure 8.22) the ICT is at steady state and the solution is given by (8.13) with φ_L replaced by φ_{L0}:

$$\varphi(z, t) = \begin{cases} \varphi_{L0} & \text{for } z_1(t) \le z \le L, \\ \varphi_\infty & \text{for } 0 \le z < z_1(t), \end{cases}$$

with $z_1(t) = A + \sigma_1(\varphi_{L0}, \varphi_\infty)t = A$ for $0 \le t < t_1$.

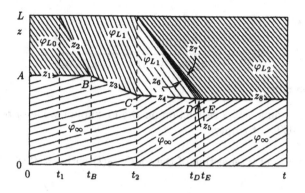

Figure 8.22: Representation of the global weak solution $\varphi(z,t)$ for the control problem of an ICT.

Solution for $t_1 \leq t < t_2$

At $t = t_1$ (see Figures 8.21 and 8.22) f_{F_1} changes to f_{F_2}. This causes the concentration at $z = L$ to change from φ_{L0} to φ_{L1}. Then at $t = t_1$, we solve an initial-boundary value problem. The jump from φ_{L1} to φ_{L0} is the shock $z_2(t) = L + \sigma_1(\varphi_{L0}, \varphi_{L1})(t - t_1)$ which intersects the shock $z_1(t)$ at the point (B, t_B) with coordinates

$$B = A, \quad t_B = t_1 + \frac{A - L}{\sigma_1(\varphi_{L1}, \varphi_{L0})}. \tag{8.38}$$

Lemma 8.4 *The time t_B is greater than t_1.*

Proof. Since $\varphi_{L1} < \varphi_{L0} = \varphi_{s1}$, from Corollary 8.1 we have that $\sigma_1(\varphi_{L1}, \varphi_{L0}) < 0$ and then $t_B > t_1$ by equation (8.38). This completes the proof. ∎

At $t = t_B$, the interaction of the shocks $z_1(t)$ and $z_2(t)$ produces a jump from φ_{L1} to φ_∞ that is the shock $z_3(t) = B + \sigma_1(\varphi_{L1}, \varphi_\infty)(t - t_B)$. The solution for $t_1 \leq t < t_B$ is

$$\varphi(z,t) = \begin{cases} \varphi_{L1} & \text{for } z_2(t) \leq z \leq L, \\ \varphi_{L0} & \text{for } z_1(t) \leq z < z_2(t), \\ \varphi_\infty & \text{for } 0 \leq z < z_1(t), \end{cases}$$

and for $t_B \leq t < t_2$ is

$$\varphi(z,t) = \begin{cases} \varphi_{L1} & \text{for } z_3(t) \leq z \leq L, \\ \varphi_\infty & \text{for } 0 \leq z < z_3(t). \end{cases}$$

Lemma 8.5 *If*

$$t_2 \geq t_1 - \frac{L-A}{\sigma_1(\varphi_{L1}, \varphi_{L0})} - \frac{A}{\sigma(\varphi_{L1}, \varphi_\infty)}, \tag{8.39}$$

the ICT will empty.

Note that this lemma sets a restriction on the maximum value of t_2 since we require that the ICT not empty.

Proof. The time t_x of intersection of the shock $z_3(t)$ with $z = 0$ is

$$t_x = t_B - \frac{A}{\sigma_1(\varphi_{L1}, \varphi_\infty)} = t_1 - \frac{L-A}{\sigma_1(\varphi_{L1}, \varphi_{L0})} - \frac{A}{\sigma(\varphi_{L1}, \varphi_\infty)}, \tag{8.40}$$

which is greater than t_B. If t_2 is larger than t_x, the ICT will empty. This completes the proof. ∎

Solution for $t \geq t_2$

At $t = t_2$, the control action is taken. It consists in changing the volume average velocity from q_1 to q_2. This implies that the solid flux density function has changed from f_1 to f_2. Therefore the slope of the characteristics $f_1'(\varphi_{L1})$ (as we remarked before) changes to $f_2'(\varphi_{L2})$ and that of $f_1'(\varphi_\infty)$ to $f_2'(\varphi_\infty)$. Also the slope of the shock $z_3(t)$ has changed to $\sigma_2(\varphi_{L1}, \varphi_\infty)$, generating a new shock $z_4(t)$. At $z = L$, the concentration is now φ_{L2}.

Lemma 8.6 $\varphi_{L1} < \varphi_{L2}$.

Proof. We know by construction that

$$f_{F2} = q_1\varphi_{L1} + f_b(\varphi_{L1}) = q_2\varphi_{L2} + f_b(\varphi_{L2}).$$

Consider the function $g(\varphi) = q\varphi + f_b(\varphi) = f_{F2}$ and take its derivative with respect to φ to obtain

$$\frac{d\varphi}{dq} = \frac{-\varphi}{q + f_b'(\varphi)},$$

which positive for $\varphi \leq \varphi_{si}$, $i = 1, 2$. Then φ is an increasing function of q and since $q_1 < q_2$ then $\varphi_{L1} < \varphi_{L2}$. This completes the proof. ∎

Let us denote by C the height of the sediment at $t = t_2$,

$$C = z_3(t_2). \tag{8.41}$$

Then at $t = t_2$ we have a jump from φ_{L2} to φ_{L1} that is resolved through a rarefaction wave and the jump from φ_{L1} to φ_∞ through the shock $z_4(t) = C + \sigma_2(\varphi_{L1}, \varphi_\infty)(t - t_2)$. The rarefaction wave meets the shock $z_4(t)$ at the point

(D, t_D). The line of continuity $z_6(t) = L + f_2'(\varphi_{L1})(t - t_2)$ and the shock $z_4(t)$ intersect at the point

$$D = C + \frac{\sigma_2(\varphi_{L1}, \varphi_\infty)(L - C)}{\sigma_2(\varphi_{L1}, \varphi_\infty) - f_2'(\varphi_{L1})}, \quad t_D = t_2 + \frac{L - C}{\sigma_2(\varphi_{L1}, \varphi_\infty) - f_2'(\varphi_{L1})}. \quad (8.42)$$

Lemma 8.7 *The time t_D is greater than t_2.*

Proof. Since $\varphi_{L1} < \varphi_{\infty 2}^{**}$ then $\varphi_\infty < \varphi_{L1}^*$, which implies that $\varphi_\infty \in (\varphi_{L1}, \varphi_{L1}^*)$; then from Lemma 5.1 of Chapter 5 we get that $f_2'(\varphi_{L1}) < \sigma_2(\varphi_{L1}, \varphi_\infty)$ in (8.42), then $t_D > t_2$. This completes the proof. ∎

The line $z_7(t) = L + f_2'(\varphi_{L2})(t - t_2)$ is a line of continuity. Now we consider the interaction of the shock $z_4(t)$ with the rarefaction wave: the concentration φ_∞ encounters consecutively the values of φ between φ_{L1} and φ_{L2} generating the curved shock $z_5(t)$ that satisfies

$$z_5'(t) = \frac{dz_5(t)}{dt} = \sigma_2\left(h\left(\frac{z_5(t) - L}{t - t_2}\right), \varphi_\infty\right),$$

where h is the inverse of f_2' restricted to values of φ less than a and $z_4(t_D) = z_5(t_D) = z_6(t_D)$. Let $z_8(t) = \sigma_2(\varphi_{L2}, \varphi_\infty)t$ be the horizontal shock produced by the jump from φ_{L2} to φ_∞, we have $z_5(t_E) = z_7(t_E) = z_8(t_E)$ where (E, t_E) is the point of intersection of the shocks $z_5(t)$ and $z_8(t)$. Now we prove that such a point exists. The value E is the height of the new steady state.

Lemma 8.8 *There exists a time t_E when the curve $z_5(t)$ intersects $z_8(t)$.*

Proof. First we prove that z_5'' is positive. In fact,

$$z_5''(t) = \frac{d}{dt} \frac{f_2\left(h\left(\frac{z_5(t) - L}{t - t_2}\right)\right) - f_2(\varphi_\infty)}{h\left(\frac{z_5(t) - L}{t - t_2}\right) - \varphi_\infty}$$

$$= \frac{-\left(z_5'(t) - \frac{z_5(t) - L}{t - t_2}\right)^2}{(t - t_2)f_2''\left(h\left(\frac{z_5(t) - L}{t - t_2}\right)\right)\left(h\left(\frac{z_5(t) - L}{t - t_2}\right) - \varphi_\infty\right)},$$

which is positive since h is restricted to $\varphi < a$, i.e.

$$f_2''\left(h\left(\frac{z_5(t) - L}{t - t_2}\right)\right) > 0.$$

Furthermore,

$$\varphi_{L1} < h\left(\frac{z_5(t) - L}{t - t_2}\right) < \varphi_{L2}$$

by construction. Since $\sigma_2(\varphi, \varphi_\infty)$ is an increasing function of φ for $\varphi \leq \varphi_\infty^{**}$ as proved in Lemma 8.1, we have

$$z_5'(t) = \sigma_2\left(h\left(\frac{z_5(t) - L}{t - t_2}\tilde{z}_5(t)\right), \varphi_\infty\right) < \sigma_2(\varphi_{L2}, \varphi_\infty) = 0 = z_8'(t).$$

Since $z_5'(t)$ is increasing with time, there exists a time t_E when $z_5(t)$ meets $z_8(t)$. This completes the proof. ∎

To find the value of E, we use equation (2.25) and the fact that for $t_2 \leq t$ the feed solid flux density f_{F2} is equal to f_{D2}, obtaining

$$\frac{d}{dt}\int_0^L \varphi(z, t)dz = 0. \tag{8.43}$$

Evaluating this last integral, which is independent of time, at $t = t_2$ and at $t = t_E$, we obtain $\varphi_\infty C + \varphi_{L1}(L - C) = \varphi_\infty E + \varphi_{L2}(L - E)$, from which we get

$$E = \frac{C(\varphi_\infty - \varphi_{L1}) + L(\varphi_{L1} - \varphi_{L2})}{\varphi_\infty - \varphi_{L2}}. \tag{8.44}$$

The time t_E, at which the new steady state is attained, is obtained from the intersection of the line $z_7(t)$ with the line $z_8(t) = E$ giving

$$t_E = t_2 + \frac{E - L}{f_2'(\varphi_{L2})}. \tag{8.45}$$

This time t_E is greater than t_2 and sets a smaller bound for t_2 as we show in the following lemma.

Lemma 8.9 *The bound for t_2 given in (8.39) must be further decreased so that the thickener does not empty at $t = t_E$.*

Proof. The condition for the thickener to have sediment at $t = t_E$ is that $E > 0$. Requiring that E be positive in (8.45), we get $\sup_{E>0} t_E = t_2 - L/f_2'(\varphi_{L2})$, which is greater than t_2. Thus t_2 must be reduced. The new bound is

$$t_2 < t_1 - \frac{L - A}{\sigma_1(\varphi_{L1}, \varphi_{L0})} - \frac{A}{\sigma_1(\varphi_{L1}, \varphi_\infty)} + \frac{L}{f_2'(\varphi_{L2})}. \tag{8.46}$$

Since the fourth term on the right-hand side of (8.46) is negative, this new bound for t_2 is smaller that the one given in (8.39). ∎

The solution for $t_2 \leq t < t_D$ is

$$\varphi(z, t) = \begin{cases} \varphi_{L2} & \text{for } z_7(t) \leq z \leq L, \\ h((z - L)/(t - t_2)) & \text{for } z_6(t) \leq z < z_7(t), \\ \varphi_{L1} & \text{for } z_4(t) \leq z < z_6(t), \\ \varphi_\infty & \text{for } 0 \leq z < z_4(t). \end{cases}$$

For $t_D \leq t < t_E$, it is

$$\varphi(z,t) = \begin{cases} \varphi_{L2} & \text{for } z_7(t) \leq z \leq L, \\ h((z-L)/(t-t_2)) & \text{for } z_6(t) \leq z < z_7(t), \\ \varphi_\infty & \text{for } 0 \leq z < z_5(t), \end{cases}$$

and for $t_E \leq t$, it is

$$\varphi(z,t) = \begin{cases} \varphi_{L2} & \text{for } z_8(t) \leq z \leq L, \\ \varphi_\infty & \text{for } 0 \leq z < z_8(t). \end{cases}$$

Boundary conditions

At $z = L$, $\epsilon_L(\varphi_L) = \{\varphi_{L0}\} \cup \{\varphi_{L1}\} \cup \{\varphi_{L2}\} \cup \{\varphi_\infty\}$ and the boundary condition is satisfied classically. At $z = 0$, $\epsilon_0(\varphi_\infty) = [0, \varphi_{L0}] \cup \{\varphi_\infty\}$ or $\epsilon_0(\varphi_\infty) = [0, \varphi_{L2}] \cup \{\varphi_\infty\}$ if $\varphi_{L2} > \varphi_{L0}$; then the boundary condition is also satisfied classically.

Chapter 9

Mathematical theory for sedimentation with compression

In this chapter, we present recent results of the mathematical analysis of the governing equation (3.65) of the sedimentation of flocculated suspensions (or *sedimentation with compression*) with appropriate initial and boundary conditions. Detailed proofs of the theorems stated here may be found in Bürger's thesis (1996), in the articles by Bürger and Wendland (1998a,b) and in the paper by Bürger *et al.* (1999d).

9.1 The initial-boundary value problem

9.1.1 Initial and boundary conditions

We assume that a piecewise continuous initial concentration distribution

$$\varphi(z,0) = \phi_0(z), \ z \in \overline{\Omega}, \ \Omega = (0, L) \tag{9.1}$$

satisfying

$$0 \le \phi_0(z) \le \varphi_\infty \ \text{for} \ z \in \overline{\Omega} \tag{9.2}$$

is given, where $0 < \varphi_\infty \le 1$ is the maximum solid concentration. Furthermore we suppose that the control of the feed flux is equivalent to prescribing explicit concentration values in the thickener at feed height $z = L$ for $t \in \overline{\mathfrak{T}}$, $\mathfrak{T} = (0, T)$; thus at $z = L$ we require that

$$\varphi(L,t) = \varphi_L(t), \ t \in \overline{\mathfrak{T}} \tag{9.3}$$

where φ_L is a piecewise continuous function with

$$\varphi_L(t) \in [0, \varphi_\infty] \text{ for } t \in \overline{\mathcal{T}}. \tag{9.4}$$

Moreover, the total variations of ϕ_0 and φ_L with respect to z and t, defined by

$$\mathrm{TV}_{\overline{\Omega}}(\phi_0) = \sup_{\Delta z > 0} \frac{1}{\Delta z} \int_0^{L - \Delta z} |\phi_0(\xi + \Delta z) - \phi_0(\xi)| \, d\xi, \tag{9.5}$$

$$\mathrm{TV}_{\overline{\mathcal{T}}}(\varphi_L) = \sup_{\Delta t > 0} \frac{1}{\Delta t} \int_0^{T - \Delta t} |\varphi_L(\tau + \Delta t) - \varphi_L(\tau)| \, d\tau, \tag{9.6}$$

respectively, are assumed to be finite, and we assume that

$$\varphi_L(t) = \varphi_c \text{ at most at a finite number of times } t = t_1, \dots, t_n \in \overline{\mathcal{T}} \tag{9.7}$$

where φ_c is the critical concentration at which the solid particles get into contact (see Chapter 3) and that

$$\mathrm{TV}_{\overline{\Omega}}\left(\frac{\partial}{\partial z} A(\phi_0)\right) < \infty, \tag{9.8}$$

where

$$A(\varphi) = \int_0^\varphi a(\tau) d\tau, \quad a(\varphi) = -\frac{f_{\mathrm{bk}}(\varphi)\sigma_e'(\varphi)}{\Delta \rho g \varphi}. \tag{9.9}$$

The significance of the technical assumptions (9.7) and (9.8) is explained in detail elsewhere (Bürger *et al.* 1999d), and is satisfied for the numerical examples to be presented in Chapter 10.

It will turn out that prescribed concentration values at $z = L$ are not always assumed in a pointwise sense by the generalized solution, which occurs, for example, if the thickener overflows (see Chapter 8). Next, we assume that at $z = 0$ the solid volume flux per thickener unit area, φv_s, reduces to the fraction that corresponds to the control of the average flow velocity of the suspension by regulating the discharge, i.e. $\varphi v_s|_{z=0} = q(t)\varphi(0, t)$, which includes batch sedimentation processes for $q \equiv 0$. This leads to the boundary condition at $z = 0$:

$$f_{\mathrm{bk}}(\varphi)\left(1 + \frac{\sigma_e'(\varphi)}{\Delta \rho \varphi g} \frac{\partial \varphi}{\partial z}\right)\bigg|_{z=0} = 0. \tag{9.10}$$

9.1.2 Type degeneracy and smoothness assumptions

Defining, as in Chapter 3, $f_k(\varphi, t) = q(t)\varphi + f_{\mathrm{bk}}(\varphi)$, equation (3.65) can be written as

$$\frac{\partial \varphi}{\partial t} + \frac{\partial}{\partial z} f_k(\varphi, t) = \frac{\partial}{\partial z}\left(a(\varphi)\frac{\partial \varphi}{\partial z}\right). \tag{9.11}$$

Taking into account the assumptions (7.4)–(7.6) stated for $f_b(\varphi) = f_{bk}(\varphi)$ and the assumption (3.43) on σ_e, we see that

$$a(\varphi) \begin{cases} = 0 \text{ for } \varphi \leq \varphi_c : & \text{equation (9.11) is hyperbolic,} \\ > 0 \text{ for } \varphi_c < \varphi < \varphi_\infty : & \text{equation (9.11) is parabolic,} \\ = 0 \text{ for } \varphi \geq \varphi_\infty : & \text{equation (9.11) is hyperbolic.} \end{cases}$$

In what follows, we assume that $a(\varphi)$ is a differentiable function. Since common Kynch batch flux density functions $f_{bk}(\varphi)$ are smooth, this requirement implies that σ_e is given as a twice differentiable function of φ. In particular, σ_e' should be differentiable at $\varphi = \varphi_c$. This means that the transition from zero effective stress to its full value takes place smoothly within a range of concentration values, starting at φ_c, as supposed by Pane and Schiffman (1985), rather than abruptly at a single critical concentration value. However, existence of a generalized solution for discontinuous $\sigma_e'(\varphi)$, as used, for example, by Becker (1982) and Landman and White (1994), can also be shown (see Section 9.7). For $\varphi \geq \varphi_\infty$, equation (9.11) degenerates into the linear advection equation

$$\frac{\partial \varphi}{\partial t} + q(t) \frac{\partial \varphi}{\partial z} = 0.$$

However, it can be shown (Bürger 1996) that conditions (9.2) and (9.4) imply that the solution of the initial-boundary value problem assumes values from the interval $[0, \varphi_\infty]$ almost everywhere, such that it is sufficient to consider the degeneracy only for $0 \leq \varphi \leq \varphi_c$ and for $\varphi = \varphi_\infty$.

Similar boundary value problems, formulated more generally for quasilinear parabolic degenerate equations with homogeneous boundary data, were studied thoroughly by Wu (1982, 1983) and Wu and Wang (1982). Those results have been applied and transferred to the problem under consideration here.

9.2 Definition of generalized solutions

9.2.1 The space $BV(Q_T)$

As in the case of a purely hyperbolic equation, which might have a discontinuous generalized solution despite smooth initial data, the solvability of the initial-boundary value problem (9.1), (9.3), (9.10) and (9.11) must be discussed in a space of functions that are in general discontinuous. This is due both to the nonlinearity of f_k as a function of φ and to the degeneracy of $a(\varphi)$. Following Wu (1982), we consider discontinuous generalized solutions in the function space $BV(Q_T)$. The following criterion for membership of a function u in $BV(Q_T)$ is very useful for practical applications:

Lemma 9.1 *A bounded function u defined on Q_T belongs to the space $BV(Q_T)$ if and only if there exist positive constants K_1, K_2 such that*

$$\int_0^T \int_0^{L-\Delta z} |u(z + \Delta z, t) - u(z, t)| \, dz dt \leq K_1 \Delta z, \tag{9.12}$$

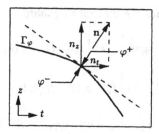

Figure 9.1: Notations

$$\int_0^{T-\Delta t} \int_0^L |u(z, t+\Delta t) - u(z,t)|\, dzdt \leq K_2 \Delta t. \tag{9.13}$$

Consider the set $\Gamma_\varphi \subset Q_T$ of jump points of φ and introduce the following notation (see Figure 9.1): if $(z_0, t_0) \in \Gamma_\varphi$, let $\mathbf{n} = (n_z, n_t)$ be the normal to Γ_φ, $\varphi^\pm = l_{\pm}\varphi$ the approximate limits of φ with respect to the respective half planes $\{(z, t) \in Q_T \,|\, ((z,t) - (z_0, t_0)) \cdot (\pm \mathbf{n}) > 0\}$, and

$$\overline{\varphi} = \frac{1}{2}\left(\varphi^+ + \varphi^-\right). \tag{9.14}$$

It is possible to derive differentiation formulas for measurable functions in BV which generalize the common formulas for differentiable functions (Vol'pert and Hudjaev 1969). In order to keep the formula for the differentiation of $(f \circ u)(z)$ valid in BV, the composition has to be replaced by the functional superposition

$$f(\widehat{u(z,t)}) = \int_0^1 f(\tau u^+(z,t) + (1-\tau)u^-(z,t))d\tau. \tag{9.15}$$

9.2.2 Definition of generalized solutions

The definition of generalized solutions of the initial-boundary value problem is based on an integral inequality with Kružkov's (1970) entropy pair.

Definition 9.1 *A function $\varphi \in L^\infty(Q_T) \cap BV(Q_T)$ is a generalized solution of the initial-boundary value problem which consists of the equation (9.11) with the initial data (9.1) and the boundary conditions (9.3) and (9.10) if the following is valid:*

1. *There exists a function $g \in L^2(Q_T)$ such that for $r(\tau) = \sqrt{a(\tau)}$ there holds*

$$\iint_{Q_T} \phi g\, dzdt = \iint_{Q_T} \phi \widehat{r(\varphi)} \frac{\partial \varphi}{\partial z} \text{ for all test functions } \phi \in C_0^\infty(Q_T).$$

2. *The function u satisfies the integral inequality*

$$
\iint_{Q_T} \left\{ |u-k|\frac{\partial\phi}{\partial t} + \mathrm{sgn}(\varphi - k)\left[f_k(\varphi,t) - f_k(k,t) - \widehat{a(\varphi)}\frac{\partial\varphi}{\partial z} \right] \frac{\partial\phi}{\partial z} \right\} dzdt
$$
$$
+ \int_0^T \left\{ -\mathrm{sgn}(\varphi_L(t) - k)\left[f_k(\gamma\varphi,t) - f_k(k,t) + \gamma\left(\widehat{a(\varphi)}\frac{\partial\varphi}{\partial z} \right) \right] \phi(L,t) \right.
$$
$$
\left. + [\mathrm{sgn}(\gamma\varphi - k) - \mathrm{sgn}(\varphi_L(t) - k)](A(\gamma\varphi) - A(k))\frac{\partial\phi}{\partial z}(L,t) \right\} dt \geq 0 \quad (9.16)
$$

for all $\phi \in C^\infty((0,L] \times \overline{\mathcal{J}})$ where $\phi \geq 0$ and $\mathrm{supp}\phi \subset (0,L] \times \mathcal{J}$, for all $k \in \mathbb{R}$.

3. *For almost all $t \in \overline{\mathcal{J}}$,*

$$
\gamma\left(-\widehat{a(\varphi)}\frac{\partial\varphi}{\partial z} + f_{\mathrm{bk}}(\varphi) \right)\bigg|_{z=0} = 0.
$$

4. *For almost all $z \in \overline{\Omega}$, $(\gamma\varphi)(z,0) = \phi_0(z)$.*

In this definition we use:

Lemma 9.2 (Wu and Wang 1982) *If $\partial v/\partial t$ and $\partial v/\partial z$ are measures (of bounded variation, which will not be mentioned explicitly in the sequel) for $v(z,t) \in L^1(Q_T)$, then for almost all $z \in \overline{\Omega}$ and for almost all $t \in \overline{\mathcal{J}}$, the limits*

$$
(\gamma v)(z,0) = \lim_{t \to 0^+} \tilde{v}(z,t), \tag{9.17}
$$

$$
(\gamma v)(0,t) = \lim_{z \to 0^+} \tilde{v}(z,t), \tag{9.18}
$$

$$
(\gamma v)(L,t) = \lim_{z \to L^-} \tilde{v}(z,t) \tag{9.19}
$$

exist, and $(\gamma v)(z,0) \in L^1(\Omega)$ and $(\gamma v)(0,t), (\gamma v)(L,t) \in L^1(\mathcal{J})$, respectively, where $\tilde{v}(z,t)$ is a certain function equivalent to $v(z,t)$ on Q_T. Here γv is the trace of v and obviously independent of the chosen function $\tilde{v}$.

9.3   Jump condition

The derivation of a jump condition valid on Q_T for generalized solutions follows from the existing results for Cauchy problems. The jump condition was derived by Wu and Yin (1989), who corrected a previous result by Vol'pert and Hudjaev (1969) stating that the jump condition for equation (9.11) coincides with the jump conditions for a purely hyperbolic equation. In the following we state the correct jump condition, based on a definition of generalized solutions of equation (9.11) without initial or boundary conditions:

Definition 9.2 *A function $u \in L^\infty(Q_T) \cap BV(Q_T)$ is said to be a generalized solution of equation (9.11) if the following conditions are satisfied:*

1. There exists a function $g \in L^1(Q_T)$ such that

$$\iint_{Q_T} \phi(z,t)g(z,t)dxdt = \iint_{Q_T} \phi(z,t)\widehat{a(\varphi)}\frac{\partial\varphi}{\partial z}dzdt \text{ for all } \phi \in C_0^\infty(Q_T).$$

2. For all $\phi \in C_0^\infty(Q_T)$, $\phi > 0$, $k \in \mathbb{R}$, the following inequality holds:

$$\iint_{Q_T} \left\{ |\varphi - k|\frac{\partial\phi}{\partial t} + \text{sgn}(\varphi - k)\left[f_k(\varphi,t) - f_k(k,t) - \widehat{a(\varphi)}\frac{\partial\varphi}{\partial z} \right]\frac{\partial\phi}{\partial z} \right\}dzdt \geq 0.$$

$$(9.20)$$

If we denote by $\varphi^l(z,t)$ and $\varphi^r(z,t)$ the left and right approximate limits of $\varphi(.,t)$ as a function of z and denote

$$\tilde{\varphi}(z,t) = \frac{1}{2}\left(u^l(z,t) + u^r(z,t) \right), \tag{9.21}$$

then the result will be:

Theorem 9.1 (Wu and Yin 1989) *If φ is a generalized solution of equation (3.65) on Q_T, then the following conditions hold H-almost everywhere on Γ_φ:*

1. $(\varphi^+ - \varphi^-)n_t + (f_k(\varphi^+,t) - f_k(\varphi^-,t))n_z$

$$- \left[\left(\widehat{a(\varphi)}\frac{\partial\varphi}{\partial z} \right)^r - \left(\widehat{a(\varphi)}\frac{\partial\varphi}{\partial z} \right)^l \right] |n_z| = 0,$$

2. *For all $\varphi \in I(\varphi^-, \varphi^+)$: $a(\varphi) = 0$.*

3. *For all $k \in \mathbb{R}$, the following inequality is valid:*

$$[\text{sgn}(\varphi^+ - k) - \text{sgn}(\varphi^- - k)] \times$$

$$\times \left[(\overline{\varphi} - k)n_t + (\overline{f_k(\varphi,t)} - f_k(k,t))n_z - \left(\widehat{a(\varphi)}\frac{\partial\varphi}{\partial z} \right)^\sim n_z \right] \leq 0. \quad (9.22)$$

Here, H denotes the one-dimensional Hausdorff measure (see, e.g., Halmos (1950)) and the theorem states that the conditions hold everywhere on Γ_φ except on a small set whose one-dimensional Hausdorff measure is zero. Note that in this formulation we merely assume $a(\varphi) > 0$. Using more detailed knowledge of the degeneracy of $a(\varphi)$, as implied by its functional representation given by (9.9), and assuming that the generalized solution of the initial-boundary problem assumes only values between zero and φ_∞ (they prove elsewhere that this is actually valid, see Bürger and Wendland 1998a), Bürger and Wendland (1998b) formulated the jump condition seen in the following corollary.

Corollary 9.1 *If φ is a generalized solution of the initial-boundary value problem (9.1), (9.3), (9.10) and (9.11), then a jump between two values of the solution φ^- and φ^+ can occur only for $0 \leq \varphi^-$, $\varphi^+ \leq \varphi_c$. This jump satisfies one of the following conditions:*

1. *For $0 \le \varphi^-$, $\varphi^+ < \varphi_c$ the usual Rankine-Hugoniot condition for hyperbolic conservation laws is valid. The propagation velocity σ of the jump between φ^- and φ^+ is given by*

$$\sigma = \frac{f_k(\varphi^+, t) - f_k(\varphi^-, t)}{\varphi^+ - \varphi^-} = q(t) + \frac{f_{bk}(\varphi^+) - f_{bk}(\varphi^-)}{\varphi^+ - \varphi^-}.$$

Inequality (9.22) then reads for all $k \in \mathbb{R}$:

$$[\operatorname{sgn}(\varphi^+ - k) - \operatorname{sgn}(\varphi^- - k)][(\overline{\varphi} - k)n_t + (\widehat{f_k(\varphi, t)} - f_k(k, t))n_z] \le 0$$

and is equivalent to Oleinik's condition E (4.7).

2. *For $0 \le \varphi^+ < \varphi^- = \varphi_c$, the velocity of a jump at $(z_0, t_0) \in Q_T$ is given by*

$$\sigma = \frac{1}{\varphi^+ - \varphi_c} \left[f_k(\varphi^+, t_0) - f_k(\varphi_c, t_0) + \lim_{z \to z_0^-} \widehat{a(\varphi)} \frac{\partial \varphi}{\partial z} \right],$$

and inequality (9.22) reads for all $k \in \mathbb{R}$

$$[\operatorname{sgn}(\varphi^+ - k) - \operatorname{sgn}(\varphi_c - k)] \times$$

$$\times \left[(\overline{\varphi} - k)n_t + (\widehat{f_k(\varphi, t_0)} - f_k(k, t_0))n_z - \left(\frac{1}{2} \lim_{z \to z_0^-} \widehat{a(\varphi)} \frac{\partial \varphi}{\partial z} \right) n_z \right] \le 0.$$

$$(9.23)$$

Inequality (9.23) is satisfied if and only if for all $k \in [\varphi^+, \varphi_c]$:

$$\frac{f_k(\varphi^+, t_0) - f_k(k, t_0)}{\varphi^+ - k} \le \sigma \le \frac{1}{\varphi_c - k} \left[f_k(\varphi_c, t_0) - f_k(k, t_0) - \lim_{z \to z_0^-} \widehat{a(\varphi)} \frac{\partial \varphi}{\partial z} \right].$$

3. *Analogously, for $0 \le \varphi^- < \varphi^+ = \varphi_c$, we obtain*

$$\sigma = \frac{1}{\varphi^- - \varphi_c} \left[f_k(\varphi^-, t_0) - f_k(\varphi_c, t_0) + \lim_{z \to z_0^+} \widehat{a(\varphi)} \frac{\partial \varphi}{\partial z} \right]$$

and for all $k \in \mathbb{R}$:

$$[\operatorname{sgn}(\varphi_c - k) - \operatorname{sgn}(\varphi^- - k)] \left[(\overline{\varphi} - k)n_t \right.$$

$$\left. + (\widehat{f_k(\varphi, t_0)} - f_k(k, t_0))n_z - \left(\frac{1}{2} \lim_{z \to z_0^+} \widehat{a(\varphi)} \frac{\partial \varphi}{\partial z} \right) n_z \right] \le 0. \quad (9.24)$$

Inequality (9.24) is satisfied if and only if for all $k \in [\varphi^-, \varphi_c]$

$$\frac{1}{\varphi_c - k} \left[f_k(\varphi_c, t_0) - f_k(k, t_0) - \lim_{z \to z_0^-} \widehat{a(\varphi)} \frac{\partial \varphi}{\partial z} \right] \le \sigma \le \frac{f_k(\varphi^-, t_0) - f_k(k, t_0)}{\varphi^- - k}.$$

Note that Corollary 9.1 implies that the propagation speed of the suspension-sediment interface during a sedimentation process, where $\varphi^+ < \varphi_c$ and $\varphi^- = \varphi_c$ is valid (corresponding to the second case), cannot be predicted a priori, since $\lim_{z \to z_0^-} \widehat{a(\varphi)} \partial\varphi / \partial z$ is in general unknown. Therefore an extension of the control model by Bustos *et al.* (1990) for ideal suspensions (see Chapter 8) to continuous sedimentation of flocculated suspensions is not feasible at present.

9.4 Entropy boundary condition

Lemma 9.3 (Bürger and Wendland 1998b) *Condition (9.16) in the definition of generalized solution is satisfied if and only if the following conditions are fulfilled:*

1. *For all $\phi \in C_0^\infty(Q_T)$, $\phi \geq 0$, inequality (9.20) is valid.*

2. *For almost all $k \in \mathbb{R}$, there holds almost everywhere on $\overline{\mathcal{T}}$:*

$$[\operatorname{sgn}(\gamma\varphi - k) - \operatorname{sgn}(\varphi_L(t) - k)] \times$$
$$\times \left[f_k(\gamma\varphi, t) - f_k(k, t) - \gamma \left(\widehat{a(\varphi)} \frac{\partial\varphi}{\partial z} \right) \right] \Bigg|_{z=L} \geq 0. \quad (9.25)$$

3. *For all τ between $\varphi_L(t)$ and $(\gamma\varphi)(L, t)$: $a(\tau) = 0$.*

Consequently, the boundary datum for $z = L$, that is the function $\varphi_L(t)$, is assumed exactly in any case only for $\varphi_\infty > \varphi_L(t) > \varphi_c$, i.e. for $a(\varphi_L(t)) > 0$. This is valid also for $\varphi_L(t) = \varphi_\infty$ if we anticipate that the generalized solution assumes only values $\varphi \in [0, \varphi_\infty]$. Otherwise, relation (9.25) is valid. We see later that under additional assumptions on $a(\varphi)$ we even have

$$\gamma \left(\widehat{a(\varphi)} \frac{\partial\varphi}{\partial z} \right) = 0 \quad \text{for almost all } t \text{ where } \varphi_L(t) \leq \varphi_c,$$

such that the entropy boundary inequality (9.25) reduces almost everywhere to the entropy boundary inequality derived in Bustos *et al.* (1996) for the hyperbolic initial-boundary value problem of continuous sedimentation of an ideal suspension; that is, for all $k \in \mathbb{R}$ the following inequality holds almost everywhere on $\overline{\mathcal{T}}$:

$$[\operatorname{sgn}(\gamma\varphi - k) - \operatorname{sgn}(\varphi_L(t) - k)][f_k(\gamma\varphi, t) - f_k(k, t)]|_{z=L} \geq 0.$$

9.5 Existence, uniqueness and stability of generalized solutions

9.5.1 The regularized problem

To prove existence of generalized solutions to problem (9.1), (9.3), (9.10) and (9.11), we consider the regularized quasilinear parabolic initial-boundary value problem

$$\frac{\partial \varphi_\epsilon}{\partial t} + \frac{\partial}{\partial z} f_k(\varphi_\epsilon, t) = \frac{\partial}{\partial z}\left((a(\varphi_\epsilon) + \epsilon)\frac{\partial \varphi_\epsilon}{\partial z}\right), \quad \epsilon > 0, \quad (z,t) \in Q_T, \qquad (9.26a)$$

$$\varphi_\epsilon(z,0) = \phi_0^\epsilon(z), \quad z \in \overline{\Omega}, \qquad (9.26b)$$

$$f_{bk}(\varphi_\epsilon) - (a(\varphi_\epsilon) + \epsilon)\frac{\partial \varphi_\epsilon}{\partial z}\bigg|_{z=0} = 0, \quad t \in \overline{\mathfrak{I}}, \qquad (9.26c)$$

$$\varphi_\epsilon(L,t) = \varphi_L^\epsilon(t), \quad t \in \overline{\mathfrak{I}}, \qquad (9.26d)$$

and show that the limit of its solutions for $\epsilon \to 0$ exists. Here, $\varphi_L^\epsilon(t)$ and $\phi_0^\epsilon(z)$ are the following regularizations of the initial and boundary data: set first

$$\widetilde{\phi_0}(z) = \begin{cases} \phi_0(L) & \text{for } z \geq L - 2\epsilon, \\ \phi_0\left((z - 2\epsilon)L/(L - 4\epsilon)\right) & \text{for } 2\epsilon < z < L - 2\epsilon, \\ \varphi_\infty & \text{for } z \leq 2\epsilon, \end{cases} \qquad (9.27)$$

$$\widetilde{\varphi_L}(t) = \begin{cases} \phi_0(L) & \text{for } t \leq 2\epsilon, \\ \varphi_L((t - 2\epsilon)T/(T - 4\epsilon)) & \text{for } 2\epsilon < t < T - 2\epsilon, \\ \varphi_L(T) & \text{for } t \geq T - 2\epsilon. \end{cases} \qquad (9.28)$$

Now let ω be a function satisfying

$$\omega \in C_0^\infty(\mathbb{R}), \quad \omega(x) \geq 0, \quad \omega(x) = 0 \text{ for } |x| \geq 1, \quad \int_{-\infty}^\infty \omega(\zeta)d\zeta = 1, \qquad (9.29)$$

and define a standard mollifier ω_ϵ (Kufner *et al.* 1977, Málek *et al.* 1996) with support in $(-\epsilon, \epsilon)$ by setting

$$\omega_\epsilon(x) = \frac{1}{\epsilon}\omega\left(\frac{x}{\epsilon}\right). \qquad (9.30)$$

The approximations $\phi_0^\epsilon(z)$ and $\varphi_L^\epsilon(t)$ are then defined by the convolutions

$$\phi_0^\epsilon(z) = \int_{-\epsilon}^\epsilon \widetilde{\phi_0}(z - \zeta)\omega_\epsilon(\zeta)d\zeta, \quad z \in [0,L]; \qquad (9.31a)$$

$$\varphi_L^\epsilon(t) = \int_{-\epsilon}^\epsilon \widetilde{\varphi_L}(t - \tau)\omega_\epsilon(\tau)d\tau, \quad t \in \overline{\mathfrak{I}}. \qquad (9.31b)$$

Lemma 9.4 (Bürger *et al.* 1999d) *The functions* $\phi_0^\epsilon(z)$ *and* $\varphi_L^\epsilon(t)$ *given by (9.27)–(9.31) satisfy the regularity assumptions (to be made precise in Theorem 9.2) necessary to establish the existence of a smooth solution of the regularized initial-boundary value problem (9.26) and the first order compatibility conditions*

$$\phi_0^\epsilon(L) = \varphi_L^\epsilon(0), \qquad (9.32)$$

$$-q(0)(\phi_0^\epsilon)'(L) - f_{bk}(\phi_0^\epsilon(L)) - a'(\phi_0^\epsilon(L))\left((\phi_0^\epsilon)'(L)\right)^2$$
$$-(a(\phi_0^\epsilon(L)) + \epsilon)(\phi_0^\epsilon)''(L) = (\varphi_L^\epsilon)'(0), \qquad (9.33)$$

$$(a(\phi_0^\epsilon(0)) + \epsilon)(\phi_0^\epsilon)'(0) - f_{bk}(\phi_0^\epsilon(0)) = 0. \qquad (9.34)$$

Moreover, the functions $\phi_0^\epsilon(z)$ *and* $\varphi_L^\epsilon(t)$ *assume only values from the interval* $[0, \varphi_\infty]$ *and satisfy*

$$TV_{\overline{\Omega}}(\phi_0^\epsilon) \leq TV_{\overline{\Omega}}(\phi_0) + \varphi_\infty - \phi_0(0), \qquad (9.35)$$

$$TV_{\overline{\mathcal{J}}}(\varphi_L^\epsilon) \leq TV_{\overline{\mathcal{J}}}(\varphi_L) + |\phi_0(L) - \varphi_L(0)|. \qquad (9.36)$$

9.5.2 Existence of the solution of the regularized problem

To discuss the solvability of the regularized parabolic initial-boundary value problems, we consider the Hölder spaces $H^l(\overline{\Omega})$ and $H^{l,l/2}(\overline{Q_T})$, where l is always a non-integer positive number. Here, $H^l(\overline{\Omega})$ is the Banach space whose elements are continuous functions on Ω that possess continuous derivatives on Ω up to order $[l]$ and a particular finite norm, while $H^{l,l/2}(\overline{Q_T})$ is the Banach space of functions defined on Q_T which are continuous together with all their derivatives of the form $\partial/\partial t^r \partial z^s$ for $2r + s < l$ on $\overline{Q_T}$, and which analogously possess a particular finite norm. These norms are given by Ladyženskaja *et al.* (1968).

Theorem 9.2 (Bürger 1996) *If the initial and boundary data satisfy the regularity assumptions*

$$\phi_0^\epsilon \in H^{2+\beta}(\overline{\Omega}) \qquad (9.37)$$

and

$$\varphi_L^\epsilon \in H^{1+\beta/2}(\overline{\mathcal{J}}) \qquad (9.38)$$

and the compatibility conditions (9.32)–(9.34), then the regularized problem (9.26) has a unique smooth solution $\varphi_\epsilon \in H^{2+\beta,1+\beta/2}(\overline{Q_T}) \subset C^{2,1}(\overline{Q_T})$.

Note that any functions $\phi_0^\epsilon \in C^3(\overline{\Omega})$ and $\varphi_L^\epsilon \in C^2(\overline{\mathcal{J}})$ satisfy (9.37) and (9.38), respectively, so the functions ϕ_0^ϵ and φ_L^ϵ constructed by (9.27)–(9.31) satisfy (9.37) and (9.38) as well since they even belong to the respective spaces $C^\infty(\overline{\Omega})$ and $C^\infty(\overline{\mathcal{J}})$.

9.5.3 Existence of a generalized solution

Theorem 9.3 *The family* $\{\varphi_\epsilon\}_{\epsilon>0}$ *of solutions of the regularized problems (9.26) is compact in* $L^1(Q_T)$; *that is, there exists a sequence* $\epsilon = \epsilon_n \to 0$ *such that* $\{\varphi_{\epsilon_n}\}$ *converges in* $L^1(Q_T)$ *to a bounded function* $u \in BV(Q_T)$. *The limit* φ *satisfies the definition of generalized solutions of the initial-boundary value problem.*

The crucial step of the proof of Theorem 9.3 consists in deriving the following estimates (see Wu and Wang (1982), Bürger and Wendland (1998a), and Bürger *et al.* (1999d)):

Lemma 9.5 *Let $\{\varphi^\epsilon\}_{\epsilon>0}$ be the family of solutions of the regularized problem (9.26). Then there exist positive constants $M_1,\ldots,M_4$ such that the following estimates hold uniformly for $\epsilon > 0$:*

$$|\varphi^\epsilon(z,t)| \le M_1 \text{ for } (z,t) \in Q_T, \quad \int_0^L \left|\frac{\partial\varphi^\epsilon}{\partial z}\right| dz \le M_2 \text{ for } t \in [0,T],$$

$$\iint_{Q_T} \left|\frac{\partial\varphi^\epsilon}{\partial t}\right| dzdt \le M_3, \quad \iint_{Q_T} (a(\varphi^\epsilon) + \epsilon)\left(\frac{\partial\varphi^\epsilon}{\partial z}\right)^2 dzdt \le M_4.$$

9.5.4 Stability and uniqueness of generalized solutions

Using a similar result by Wu and Wang (1982) and the inequality

$$\iint_{Q_T} \left\{|\varphi - \psi|\frac{\partial\phi}{\partial t} + \text{sgn}(\varphi - \psi)\left[f_k(\varphi,t) - f_k(\psi,t)\right.\right.$$

$$\left.\left. - \left(\widehat{a(\varphi)}\frac{\partial\varphi}{\partial z} - \widehat{a(\psi)}\frac{\partial\psi}{\partial z}\right)\right]\frac{\partial\phi}{\partial z}\right\} \ge 0 \quad (9.39)$$

shown by Wu and Yin (1989) to be valid for all $\phi \in C_0^\infty(Q_T \cup \{L\} \times \{t \in \mathcal{T} : \varphi_L(t) > \varphi_c\})$ for any two generalized solutions of equation (9.11), Bürger and Wendland (1998a) obtain the following stability theorem:

Theorem 9.4 *Let $a^\alpha(u)$ be locally Lipschitz continuous and condition (9.7) be satisfied. If ϕ and ψ are generalized solutions of (9.11) with the same boundary conditions (9.3) and (9.10) and the initial conditions $(\gamma\phi)(z,0) = \phi_0(z)$ and $(\gamma\psi)(z,0) = \psi_0(z)$ for almost all $z \in \overline{\Omega}$, then*

$$\int_0^L |\phi(z,t) - \psi(z,t)|dx \le \int_0^L |\phi_0(z) - \psi_0(z)|dz$$

holds almost everywhere on $\overline{\mathcal{T}}$.

Setting $\phi_0 \equiv \psi_0$, we immediately obtain uniqueness of generalized solutions:

Corollary 9.2 *Under the conditions of Theorem 9.4, the initial-boundary value problem (9.1), (9.3), (9.10) and (9.11) has at most one generalized solution.*

9.6 Properties of generalized solutions

9.6.1 Range of generalized solutions

As a consequence of the proof of Lemma 9.5, it can be shown (Bürger and Wendland 1998a), as mentioned before, that the generalized solution of the initial-boundary value problem (9.1), (9.3), (9.10) and (9.11) satisfies $0 \le u(z,t) \le \varphi_\infty$ almost everywhere on Q_T, which means that the generalized solution assumes almost everywhere values which are physically relevant as volumetric solid concentrations.

9.6.2   Construction of the boundary value at $z = L$

For the mathematical analysis, it was assumed that the control of the feed flux corresponds to prescribing a volumetric solid concentration in the thickener. However, the quantities given in practical applications are the volume flux of the feed suspension $Q_F(t)$ and its volumetric concentration $\varphi_F(t)$. It is assumed that the feed suspension mixes immediately with the suspension in the thickener at $z = L$ such that the concentration is constant across the whole cross section of the thickener at height $z = L$. If we denote by S the cross-sectional area of the thickener, the quantity that is actually prescribed is the solid volume flux per unit area, which amounts to $f_F(t) = -Q_F(t)\varphi_F(t)/S$ for $t \in \bar{\mathcal{I}}$. The concentration value $\varphi_L(t)$ which has to be prescribed at $z = L$ can be calculated for the purely hyperbolic case, $\sigma_e \equiv 0$, from

$$f_k(\varphi_L(t), t) = f_F(t). \tag{9.40}$$

If we assume that $f_F(t) \geq \min_{0 \leq \varphi \leq \varphi_\infty} f_k(\varphi, t)$ is valid, then (9.40) has, for a Kynch batch flux density function f_{bk} with exactly one inflection point, up to three distinct solutions $\varphi_L^{(1)}(t)$, $\varphi_L^{(2)}(t)$ and $\varphi_L^{(3)}(t)$. The relevant value is selected by the physical argument that the suspension is immediately diluted upon entering the thickener, that is, $\varphi_L(t) \leq \varphi_F(t)$ holds. Thus $\varphi_L(t)$ is selected as

$$\varphi_L(t) = \begin{cases} \varphi_L^{(1)}(t) & \text{if } \varphi_L^{(1)}(t) \leq \varphi_F(t) < \varphi_L^{(2)}(t), \\ \varphi_L^{(2)}(t) & \text{if } \varphi_L^{(2)}(t) \leq \varphi_F(t) < \varphi_L^{(3)}(t), \\ \varphi_L^{(3)}(t) & \text{if } \varphi_L^{(3)}(t) \leq \varphi_F(t) < \varphi_\infty \end{cases}$$

and as

$$\varphi_L(t) = \begin{cases} \varphi_L^{(1)}(t) & \text{if } \varphi_L^{(1)}(t) \leq \varphi_F(t) < \varphi_L^{(2)}(t), \\ \varphi_L^{(2)}(t) & \text{if } \varphi_L^{(2)}(t) \leq \varphi_F(t) < \varphi_\infty \end{cases}$$

if (9.40) has only two solutions. This shows that, by a corresponding choice of $\varphi_F(t)$ and $Q_F(t)$, all concentration values $\varphi \in [0, \varphi_\infty]$ are physically relevant as boundary values.

9.6.3   Entropy boundary condition at $z = L$

Under the assumptions of this chapter, which means in particular that $a(\varphi)$ is Lipschitz continuous, the generalized solution satisfies

$$\gamma\left(\widehat{a(\varphi)\frac{\partial\varphi}{\partial z}}\right)(L, t) = 0 \text{ for almost all } t \text{ satisfying } \varphi_L(t) \leq \varphi_c$$

and hence the following inequalities are equivalent:

1. For almost all $k \in \mathbb{R}$ almost everywhere on $\bar{\mathcal{I}}$,

$$[\mathrm{sgn}(\gamma\varphi - k) - \mathrm{sgn}(\varphi_L(t) - k)][f_k(\gamma\varphi, t) - f_k(k, t)]|_{z=L} \geq 0.$$

2. For all $k \in I((\gamma\varphi)(L,t), \varphi_L(t))$,

$$\text{sgn}((\gamma\varphi)(L,t) - \varphi_L(t))[f_k(\gamma\varphi, t) - f_k(k,t)]|_{z=L} \geq 0.$$

3. For all $k \in I(\gamma\varphi)(L,t), \varphi_L(t))$,

$$\frac{f_k((\gamma\varphi)(L,t), t) - f_k(k,t)}{(\gamma\varphi)(L,t) - k} \geq 0.$$

This means that for those $t \in \bar{J}$ for which $\varphi_L \leq \varphi_c$, the same entropy boundary condition holds as for the hyperbolic initial-boundary value problem studied by Bustos and Concha (1992) and Bustos *et al.* (1996) (see Chapter 8), and that then $(\gamma\varphi)(L,t) = \varphi_L(t)$ does not hold in general. Consequently, the boundary condition at $z = L$ should be reformulated in a set-valued manner, which means that to every value $\varphi_L(t) \leq \varphi_c$, the set of admissible boundary values

$$\epsilon(t) := \left\{ \varphi \in [0, \varphi_\infty] \Big| \frac{f_k(\varphi, t) - f_k(k, t)}{\varphi - k} \geq 0 \text{ for all } k \in I(\varphi, \varphi_L(t)) \right\}$$

is assigned. Hence the initial-boundary value problem is not properly posed unless we replace condition (9.3) by

$$\varphi(L,t) \begin{cases} = \varphi_L(t) & \text{if } \varphi_L(t) > \varphi_c, \\ \in \epsilon(t) & \text{if } \varphi_L(t) \leq \varphi_c. \end{cases}$$

The set $\epsilon(t)$ can be visualized for a fixed time t using the graph of $f_k(\varphi, t)$; see Figure 9.2. This figure shows the flux density function $f_k(\varphi, t) = f_k(\varphi)$ with $0 > q = const.$ which possesses a local minimum at $\varphi_{\min}$ and a local maximum at $\varphi_M < \varphi_\infty$ satisfying $f_k'(\varphi_\infty) < 0$ and $f_k(\varphi_{\min}) < f_k(\varphi_\infty) < f_k(\varphi_M) < 0$. We then obtain:

(a) If $0 \leq \varphi_M^* < \varphi_M$ with $f_k(\varphi_M^*) = f_k(\varphi_M)$, then $\epsilon(t) = \{\varphi_L(t)\}$ if $\varphi_L(t) \in [0, \varphi_M^*)$; see Figure 9.2 a).

(b) If

$$f_k(\overline{\varphi_L(t)}) = f_k(\varphi_L(t)) \text{ with } \varphi_{\min} < \overline{\varphi_L(t)} < \varphi_M, \tag{9.41}$$

then $\varphi_L(t) \in [\varphi_M^*, \varphi_{\min}]$ implies $\epsilon(t) = [\overline{\varphi_L(t)}, \varphi_M] \cup \{\varphi_L(t)\}$; see Figure 9.2 b).

(c) For $\varphi_L(t) \in [\varphi_{\min}, \varphi_M]$ we have $\epsilon(t) = [\varphi_{\min}, \varphi_M]$; see Figure 9.2 c).

(d) If $\varphi_M < \varphi_L(t) \leq \varphi_\infty$ and

$$f_k(\underline{\varphi_L(t)}) = f_k(\varphi_L(t)) \text{ with } \varphi_{\min} < \underline{\varphi_L(t)} < \varphi_M, \tag{9.42}$$

then $\varphi_L(t) \in (\varphi_M, \varphi_\infty]$ implies $\epsilon(t) = [\varphi_{\min}, \underline{\varphi_L(t)}] \cup \{\varphi_L(t)\}$; see Figure 9.2 d).

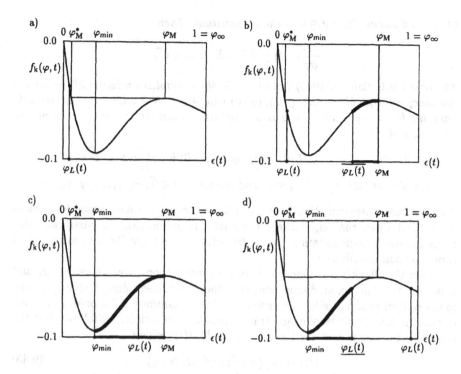

Figure 9.2: The set $\epsilon(t)$ of admissible boundary values, $f_k(\varphi) = -0.06\varphi - \varphi(1-\varphi)^4$.

9.6.4   Boundary condition at $z = 0$

The boundary condition at $z = 0$ given by (9.10) is assumed exactly by the generalized solution for almost all $t \in \mathcal{T}$, such that the discharge can in fact be controlled by the choice of $q(t)$. While the entropy boundary condition at $z = L$ is valid for smooth flux density functions in general, the existence and stability proof by Bürger and Wendland (1998a) makes explicit use of the assumption $q(t) \leq 0$, which corresponds to an outflow condition at $z = 0$.

9.6.5   Monotonicity of concentration profiles

The solution is monotonically decreasing with z, which means that the solid concentration increases downwards in the thickener, as expected, if the initial concentration $\phi_0(z)$ and the boundary function $\varphi_L(t)$ are also decreasing. This is first stated for the solution of the regularized problem.

Theorem 9.5 (Bürger 1996) *Let $\varphi_\epsilon(z, t) \in C^{2,1}(Q_T)$ be the solution of the initial-boundary value problem (9.26), and assume that the initial and boundary data satisfy the additional conditions $(\phi_0^\epsilon)'(z) \leq 0$ for all $z \in \overline{\Omega}$ and $(\varphi_L^\epsilon)'(t) \leq 0$ for all*

$t \in \overline{\mathcal{J}}$, and assume that $q \leq 0$ is chosen constant. Then

$$\frac{\partial \varphi_\epsilon}{\partial z}(z, t) \leq 0 \text{ for all } (z, t) \in \overline{Q_T}.$$

Theorem 9.6 (Bürger 1996) *Let $\varphi(z, t)$ be the generalized solution of the initial-boundary value problem (9.1), (9.3), (9.10) and (9.11). If in addition to the smoothness and boundary conditions, q is a nonpositive constant, ϕ_0 and φ_L are nonincreasing and*

$$(\gamma\varphi)(L, t_0) - (\gamma\varphi)(L, 0) \leq \varphi_L(t_0) - \varphi_L(0) \text{ for almost all } t_0 \in \overline{\mathcal{J}},$$

then for almost all $0 \leq z_1 < z_2 \leq L$ and almost all $t \in \overline{\mathcal{J}}$, $\varphi(z_1, t) \geq \varphi(z_2, t)$.

This theorem states conditions under which the concentration increases downwards for a fixed time, that is, thickening actually is performed. In particular, this holds for batch sedimentation ($\varphi_L \equiv 0$) with a monotonically decreasing initial concentration distribution.

From Corollary 9.1 it is clear that in the compression zone, where $\varphi > \varphi_c$ and hence $a(\varphi) > 0$ are valid, discontinuities of the generalized solution, that is, jumps in the concentration profiles, are excluded. If the conditions of Theorem 9.6, which ensure the monotonicity of concentration profiles, are satisfied, it follows that the compression zone is contained in an interval $[0, \lambda(t))$ where

$$\lambda(t) = \sup\{z \in \overline{\Omega} : \varphi(z, t) > \varphi_c\} \tag{9.43}$$

is the uniquely defined height of the interface or sediment level, and $[\lambda(t), L]$ is the hyperbolic or settling zone where $0 \leq \varphi(z, t) \leq \varphi_c$ is valid and jumps might occur, which can be denoted *shocks* due to the entropy jump condition they satisfy.

9.7 Sedimentation of flocculated suspensions with a discontinuous solid effective stress function

The study of degenerate parabolic equations is a relatively new topic of mathematical analysis and considerably less elaborate than the theory of hyperbolic conservation laws, on which Kynch's theory of sedimentation is based. In particular, at the writing of this book, it seems that not all results presented in this chapter do also apply to equation (9.11) under the assumption that $\sigma_e'(\varphi)$ and hence $a(\varphi)$ are discontinuous at $\varphi = \varphi_c$, which is the case, for example, for solid effective stress functions of the types

$$\sigma_e(\varphi) = \begin{cases} 0 & \text{for } \varphi \leq \varphi_c, \\ \alpha \exp(\beta\varphi) & \text{for } \varphi > \varphi_c; \alpha, \beta > 0 \end{cases} \tag{9.44}$$

used by Becker (1982) or

$$\sigma_e(\varphi) = \begin{cases} 0 & \text{for } \varphi \leq \varphi_c, \\ k\left[(\varphi/\varphi_c)^n - 1\right] & \text{for } \varphi > \varphi_c; k, n > 0 \end{cases} \tag{9.45}$$

as implied by Green *et al.* (1996); see Figure 12.1 in Chapter 12. For discontinuous $a(\varphi)$, existence of a generalized solution of the initial-boundary value problem considered in this chapter can be established if not only the initial and boundary data ϕ_0 and φ_L of the degenerate problem but also the diffusion coefficient $a(\varphi)$ are mollified. The uniform estimates given in Lemma 9.5 will be still valid since the constants M_1 to M_4 of that lemma depend on the L^1 norms (and not on pointwise values, except for the 'corner points' of the computational domain Q_T) of the mollified initial and boundary data, which approximate the respective L^1 norms of the data of the degenerate problem (Bürger and Wendland 1998a; an improved analysis is given in Bürger *et al.* 1999d). The numerical examples presented in the following chapter were all calculated with a discontinuous diffusion coefficient and solutions are similar to previous calculations perfomed with a diffusion coefficient whose jump had been smoothed out (Bürger and Concha 1998). The numerical results in Chapter 10 illustrate that it is possible to extend the analysis to a discontinuous diffusion coefficient and thereby to complete the mathematical theory of sedimentation.

Chapter 10

Numerical simulation of sedimentation with compression

In this chapter, we present a numerical algorithm to calculate approximate solutions of the initial-boundary value problem (9.1), (9.3), (9.10) and (9.11) of sedimentation of flocculated suspensions or sedimentation with compression. By a series of examples, the behavior of suspension and sediment which is predicted for batch and continuous sedimentation processes is visualized. The principal objective of the presentation of these examples (see also Bürger *et al.* 1997, 1999f, Bürger and Concha 1997a,b, 1998, Bürger and Wendland 1997, 1998c and Concha and Bürger 1998) is to illustrate that the phenomenological theory developed in Chapter 3 predicts correctly the transient behavior of a flocculated suspension in most applications. The numerical results also give rise to new questions and indicate that, although the phenomenological theory is generally valid, the constitutive equations selected are not appropriate for certain cases and should be modified. In this chapter, we limit ourselves to one pair of model functions $f_{bk}(\varphi)$ and $\sigma_e(\phi)$ so that all calculations demonstrate the behaviour of the same (hypothetical) suspension under different initial and boundary conditions. However, the numerical algorithm presented here has also been employed for the simulation of the behaviour of different flocculated suspensions using parameters obtained from published experiments. This made comparison of the predictions of the phenomenological theory with the experimentally observed behaviour possible, see Bürger *et al.* (1999c).

10.1 Numerical algorithm

The following viscous splitting algorithm (Bürger and Wendland 1998c), an operator splitting technique, is used to simulate batch and continuous sedimentation of
flocculated suspensions by calculating numerical solutions of the IBVP (9.1), (9.3),
(9.10) and (9.11). The splitting methods presented here are similar to the splitting
methods that have been used over the years to simulate multiphase flows in oil
reservoirs, see Karlsen *et al.* (1998, 1999), Karlsen and Lie (1999) and Karlsen
and Risebro (1997, 1999). An overview of this application of splitting methods is
provided by Espedal and Karlsen (1999).

We set

$$\Delta t = T/N, \quad N \in \mathbb{N}; \quad t_n = n\Delta t \text{ for } n = 0, 1, 2, \dots, N. \tag{10.1}$$

Let $\mathcal{C}_t$ denote the solution operator which is defined by $\varphi(z, t) = \mathcal{C}_t \phi_0(z)$ if $\varphi(z, t)$
is the solution of the parabolic equation

$$\frac{\partial \varphi}{\partial t} = \frac{\partial^2}{\partial z^2} A(\varphi) \tag{10.2a}$$

with the initial condition

$$\varphi(z, 0) = \phi_0(z). \tag{10.2b}$$

As in Chapter 9, we define

$$A(\varphi) = \int_0^{\varphi} a(s)\, ds,$$

where $a(\varphi)$ is given in equation (9.9). Similarly, let $\mathcal{Q}_t$ denote the solution operator
associated with the linear advection equation

$$\frac{\partial \varphi}{\partial t} + q(t)\frac{\partial \varphi}{\partial z} = 0 \tag{10.3a}$$

with the initial condition

$$\varphi(z, 0) = \phi_0, \tag{10.3b}$$

and $\mathcal{F}_t$ the solution operator for the nonlinear hyperbolic equation

$$\frac{\partial \varphi}{\partial t} + \frac{\partial}{\partial z} f_{bk}(\varphi) = 0 \tag{10.4a}$$

with initial and boundary conditions

$$\varphi(z, 0) = \phi_0, \quad \varphi(L, t) = \varphi_L(t), \quad f_{bk}(\varphi) - \frac{\partial}{\partial z} A(\varphi)\bigg|_{z=0} = 0. \tag{10.4b}$$

The viscous splitting algorithm is then based on the semi-discrete approximation

$$\varphi(z, n\Delta t) \approx [\mathcal{F}_{\Delta t} \circ \mathcal{Q}_{\Delta t} \circ \mathcal{C}_{\Delta t}]^n \, \phi_0.$$

Furthermore, set

$$\Delta z = L/J, \quad J \in \mathbb{N}; \quad z_j = j\Delta z \quad \text{for } j = -2, \dots, J + 2 \tag{10.5}$$

and consider cell-average approximations

$$v_j^n \approx A(j)(\varphi(\cdot, t_n)) = \frac{1}{\Delta z} \int_{z_{j-1/2}}^{z_{j+1/2}} \varphi(\zeta, t_n) \, d\zeta; \quad v_j^0 = A(j)\phi_0, \tag{10.6}$$

where $A(j)$ is the averaging operator of the 'cell' $[z_{j-1/2}, z_{j+1/2}]$. If

$$V^n = \left(v_{-2}^n, \dots, v_{J+2}^n \right) \tag{10.7}$$

is given, we calculate V^{n+1} by solving for each time step

$$V^{n+1/3} = \mathcal{C}_{\Delta t}^{\Delta z} V^n, \quad V^{n+2/3} = \mathcal{Q}_{\Delta t}^{\Delta z} V^{n+1/3} \quad \text{and } V^{n+1} = \mathcal{F}_{\Delta t}^{\Delta z} V^{n+2/3}$$

succesively, or, in short, by solving

$$V^{n+1} = \left[\mathcal{F}_{\Delta t}^{\Delta z} \circ \mathcal{Q}_{\Delta t}^{\Delta z} \circ \mathcal{C}_{\Delta t}^{\Delta z} \right] V^n,$$

where $\mathcal{C}_{\Delta t}^{\Delta z}$, $\mathcal{Q}_{\Delta t}^{\Delta z}$ and $\mathcal{F}_{\Delta t}^{\Delta z}$ correspond to the following spatial discretizations of $\mathcal{C}_{\Delta t}$, $\mathcal{Q}_{\Delta t}$ and $\mathcal{F}_{\Delta t}$, respectively:

1. The time step $\tau = \Delta t/N_\tau$, $N_\tau \in \mathbb{N}$, is chosen small enough such that

$$2\frac{\tau}{\Delta z^2} \max_u |A'(u)| \leq 1; \quad \text{that is, } N_\tau \geq 2\frac{\Delta t}{\Delta z^2} \max_u |A'(u)|$$

is satisfied, and we set $v_{-1}^1 = v_{-1}^0$ and $v_{-2}^1 = v_{-2}^0$ and calculate

$$v_j^0 = A(j)\phi_0, \quad j = -1, \dots, J + 1,$$
$$v_j^{m+1} = v_j^m + \frac{\tau}{\Delta z^2} \left[A\left(v_{j-1}^m\right) - 2A\left(v_j^m\right) + A\left(v_{j+1}^m\right) \right],$$
$$j = 0, \dots, J, \quad m = 0, \dots, N_\tau - 1.$$

2. Since $q(t) \leq 0$, the advection step can be solved by first-order upwind method if the CFL condition $\lambda \max_t |q(t)| \leq 1$, $\lambda = \Delta t/\Delta z$ is satisfied:

$$v_j^0 = A(j)\phi_0, \quad j = -2, \dots, J + 2,$$
$$v_j^1 = v_j^0 - \lambda q(t_n) \left[v_{j+1}^0 - v_j^0 \right], \quad j = -2, \dots, J + 1.$$

3. For the nonlinear hyperbolic equation, we use a variant of the non-oscillatory central difference method introduced by Nessyahu and Tadmor (1990), a finite difference scheme in conservation form. From

$$v_j^0 = A(j)u_0, \quad j = -2, \ldots, J + 2,$$

calculate first

$$v_j^{1/2} = v_j^0 - \lambda f_j'/2 \text{ for } j = -1, \ldots, J$$

and then

$$v_j^1 = \begin{cases} \varphi_L(t_n + \Delta t) & \text{for } j = J, \ldots, J + 2, \\ \frac{1}{2}\left[v_{j-1}^0 + v_{j+1}^0\right] + \frac{1}{4}\left[v_{j-1}' + v_{j+1}'\right] & \\ \quad -\frac{\lambda}{2}\left[f_{bk}(v_{j+1}^{1/2}) - f_{bk}(v_{j-1}^{1/2})\right] & \text{for } j = 0, \ldots J - 1. \end{cases}$$

Using the minmod limiter mm defined by

$$\text{mm}(x, y, z) = \begin{cases} \text{sgn}(x)\min\{|x|, |y|, |z|\} & \text{if } \text{sgn}(x) = \text{sgn}(y) = \text{sgn}(z) \\ 0 & \text{otherwise,} \end{cases}$$

$$(10.8)$$

the numerical derivatives v_j' and f_j' are given by

$$v_j' = \text{mm}\left\{\gamma\left(v_j^0 - v_{j-1}^0\right), \left(v_{j+1}^0 - v_{j-1}^0\right)/2, \gamma\left(v_{j+1}^0 - v_j^0\right)\right\}, \quad (10.9)$$
$$f_j' = \text{mm}\left\{\gamma\left(f_{bk}\left(v_j^0\right) - f_{bk}\left(v_{j-1}^0\right)\right), \right. \quad (10.10)$$
$$\left(f_{bk}\left(v_{j+1}^0\right) - f_{bk}\left(v_{j-1}^0\right)\right)/2, \gamma\left(f_{bk}\left(v_{j+1}^0\right) - f_{bk}\left(v_j^0\right)\right)\right\},$$

where $\gamma \in [1, 4)$ is a parameter. This method is second-order accurate and total variation diminishing for Cauchy problems if

$$\lambda \max_u |f_{bk}'(u)| \leq \left(\sqrt{1 + 2\gamma - \gamma^2} - 1\right)/\gamma$$

is satisfied. The boundary condition at $z = 0$ is approximated in the following way: if $v_0^1 > \varphi_c$, define for a small parameter $\varepsilon > 0$

$$\theta = \left(A\left(v_{j+1}^1 + \varepsilon/2\right) - A\left(v_{j+1}^1 - \varepsilon/2\right)\right)/\varepsilon, \quad (10.11)$$

and set for $j = -1$ and $j = -2$

$$v_j^1 = \begin{cases} v_0^1 + j\left[v_1^1 - v_0^1\right] & \text{if } v_0^1 \leq \varphi_c \text{ and } q(t_n + \Delta t) = 0, \\ v_j^1 & \text{if } v_0^1 \leq \varphi_c \text{ and } q(t_n + \Delta t) < 0, \\ \varphi_\infty & \text{if } v_0^1 > \varphi_c \text{ and } \theta = 0, \\ v_{j+1}^1 - \Delta z f_{bk}(v_{j+1}^1)/\theta & \text{otherwise.} \end{cases}$$

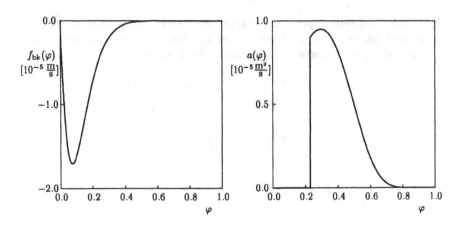

Figure 10.1: The Kynch batch flux density function $f_{bk}(\varphi)$ defined in (10.12) and the diffusion coefficient $a(\varphi)$ given by (9.9) and (10.13).

The numerical algorithm is constructed in a straightforward manner, but its convergence still has to be shown. Evje and Karlsen (1997a,b) apply a similar operator splitting procedure to an initial value problem of a degenerate quasilinear parabolic equation. The extension of their arguments to the initial-boundary value problem (9.1), (9.3), (9.10) and (9.11) considered here is currently being studied (Bürger et al. 1999e), which should eventually lead to the proof of convergence of this algorithm. Recently, Evje and Karlsen (1999a,b,c) have developed and analysed finite difference methods for degenerate parabolic equations. For details on numerical methods for hyperbolic conservation laws, we refer to the books by Le Veque (1992), Godlewski and Raviart (1991, 1996), Serre (1996), Kröner (1997) and Toro (1997).

10.2   Experimental data and numerical parameters

Consider a Kynch batch flux density function of the well-known Richardson and Zaki (1954) type with parameters obtained by experimental measurements (Becker 1982, Concha and Bustos 1992, Concha et al. 1996),

$$f_{bk}(\varphi) = -6.05 \times 10^{-4}\varphi(1 - \varphi)^{12.59} \ [m/s], \tag{10.12}$$

and choose for σ_e the function

$$\sigma_e(\varphi) = \begin{cases} 0 & \text{for } \varphi \leq \varphi_c = 0.23, \\ 5.35 \exp(17.9\varphi) \ [N/m^2] & \text{for } \varphi > \varphi_c \end{cases} \tag{10.13}$$

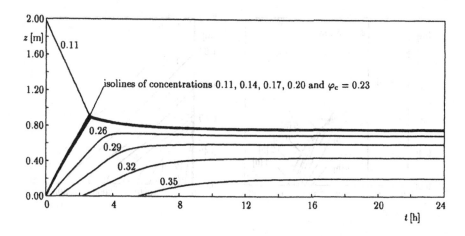

Figure 10.2: Settling plot for batch sedimentation showing sedimentation with compression. The isolines correspond to the annotated concentration values.

determined by Becker (1982). Note that this choice of σ_e leads to a discontinuous diffusion coefficient $a(\varphi)$ defined by equation (9.9) which does not satisfy the smoothness assumptions stated in Chapter 9. We use the parameters $\Delta \rho = \rho_s - \rho_f = 1500 \, [\text{kg/m}^3]$ and $g = 9.81 \, [\text{m/s}^2]$. Figure 10.1 shows the Kynch batch flux density function $f_{bk}(\varphi)$ and the resulting diffusion coefficient $a(\varphi)$.

The numerical parameters used in all calculations are $\Delta z = L/500$, $\gamma = 1.3$ and $\lambda = 400 \, [\text{s/m}]$.

10.3 Simulation of batch sedimentation

10.3.1 Batch settling of a uniform suspension

Consider a settling column closed at the bottom ($q \equiv 0$) without feeding ($\varphi_L \equiv 0$), and a suspension of initial concentration $\phi_0 = 0.123$. Figure 10.2 shows the settling plot consisting of isolines in the z-t-plane, and Figure 10.3 displays the corresponding concentration and excess pore pressure profiles at selected times. The total simulation time is $T = 24 \, [\text{h}]$. The numerical results agree qualitatively with the experimental measurements by Been and Sills (1981) and by Tiller (1991). A direct recalculation of Been and Sills' data is given by Bürger and Concha (1998), while Tiller's measurements have been simulated by Bürger and Wendland (1998c).

10.3.2 Repeated batch sedimentation

An interesting modification of the previous example referring to a certain industrial application is obtained if we suppose that after some time, the pure liquid above the nearly compressed sediment is replaced by a suspension of the homogeneous initial

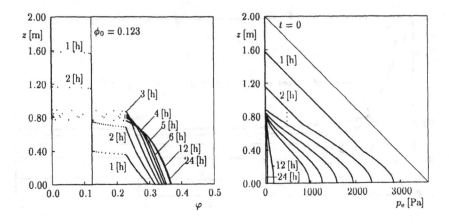

Figure 10.3: Concentration and excess pore pressure profiles at $t = 1, 2, 3, 4, 5,$ 6, 12 and 24 [h] for the batch sedimentation process of Figure 10.2.

concentration ϕ_0. Here, the same data are used as in the previous example, but the settling column is now of height $L = 6$ [m], and after 96 hours, all concentration values less than φ_c are replaced by the value $\phi_0 = 0.123$. Figure 10.4 shows the corresponding numerical simulation. Clearly, the addition of new suspension to be separated above a given sediment causes additional compression.

10.3.3   A membrane problem

Next, the transient settling behaviour of a hypothetical 'membrane' problem is studied (Bürger and Concha 1997a). We assume that for $t < 0$, the settling column contains pure liquid in its lower half and suspension of the concentration $\varphi = 0.18$ above:

$$\phi_0(z) = \begin{cases} 0 & \text{for } 0 \leq z \leq 3\,[\text{m}], \\ 0.18 & \text{for } 3\,[\text{m}] < z \leq 6\,[\text{m}]. \end{cases}$$

Both sections are assumed to be separated by a 'membrane' which is destroyed instantaneously at $t = 0$. Figure 10.5 shows the numerical solution of this problem given as a perspective view of the sequence of concentration profiles calculated for $t = 0, 1, 2, \ldots, 39, 40$ [h]. The result clearly shows that at $z = 3$ [m], the solid flocs immediately start to flow downwards, in fact the solution includes a rarefaction wave which soon interferes with the sediment layer building up from the bottom of the settling column. After about 12 hours, the settling of solid flocs due to the specific choice of the initial state is terminated, and the whole solid component is contained in the sediment layer which is further compressed with time.

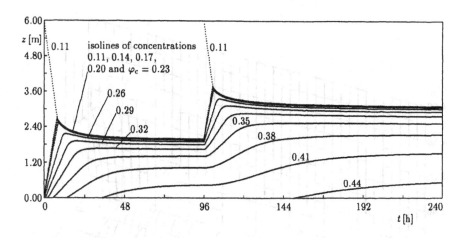

Figure 10.4: Settling plot for repeated batch sedimentation of a homogeneous suspension. The isolines correspond to the annotated concentration values.

10.3.4 Expansion of overcompressed sediment

In this example (Figure 10.6), a settling column of height 6 [m] is assumed to contain at $t = 0$ in its lower 25% overcompressed sediment of concentration 0.6:

$$\phi_0(z) = \begin{cases} 0.6 & \text{for } 0 \leq z \leq 1.5\,[\text{m}], \\ 0 & \text{for } 1.2\,[\text{m}] < z \leq 6\,[\text{m}]. \end{cases}$$

The numerical solution predicts that at $t = 0$ the sediment immediately starts to expand and that it tends to an equilibrium state with the sediment-liquid interface at a height of about 2.4 [m]. This expansion after entering in contact with water is due to the description of the solid as an elastic fluid, and is not observed in nature. This indicates that, although the phenomenological model presented in Chapter 3 is still valid, better constitutive equations of an elasto-plastic type should be formulated, permitting the deformation of the sediment.

10.3.5 Simultaneous expansion of overcompressed sediment and batch sedimentation

In the last example for batch sedimentation, we consider an overcompressed sediment of concentration 0.55 in the lowest 10% of the settling column and a homogeneous suspension of concentration 0.13 above as initial state in a settling column of height 2 [m]:

$$\phi_0(z) = \begin{cases} 0.55 & \text{for } 0 \leq z \leq 0.2\,[\text{m}], \\ 0.13 & \text{for } 0.2\,[\text{m}] < z \leq 2\,[\text{m}]. \end{cases}$$

The numerical result (Figure 10.7) shows the combined effects of expanding sediment and batch settling.

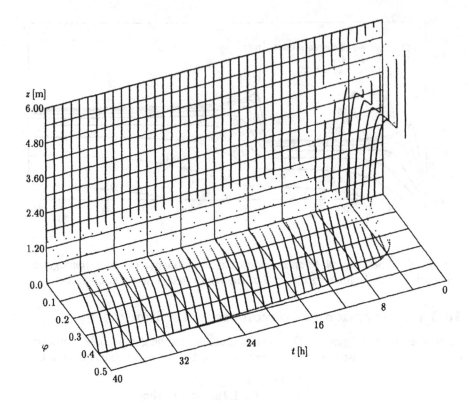

Figure 10.5: Batch settling: aspect of concentration profiles for a membrane problem of sedimentation with compression.

The numerical simulations of batch sedimentation presented here have in common that the sediment apparently tends to an equilibrium state for $t \to \infty$. It should be desirable to show that this equilibrium state is independent of the initial concentration distribution, and that, moreover, convergence to this state actually takes place within finite time.

10.4 Simulation of continuous thickening

10.4.1 Filling and emptying of a thickener

This example (Figure 10.8) shows the behavior of the sediment under usual thickener operations. We start with a thickener of feed height $L = 6\,[\mathrm{m}]$ which is assumed to be full of water initially ($\phi_0 \equiv 0$). It is kept closed till $t = 120\,[\mathrm{h}]$ and

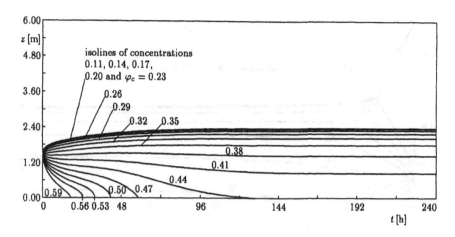

Figure 10.6: Concentration lines for the expansion of overcompressed sediment. The isolines correspond to the annotated concentration values.

then opened:

$$q(t) = \begin{cases} 0 & \text{for } 0 \le t \le 120\,[\text{h}], \\ -2.5 \times 10^{-5}\,[\text{m/s}] & \text{for } 120\,[\text{h}] < t \le T = 168\,[\text{h}]. \end{cases}$$

For $0 < t \le 33\,[\text{h}]$, it is fed with a feed flux of $f_F = -9.3826 \times 10^{-6}\,[\text{m/s}]$; since $f_F = f_{bk}(0.02)$, this is equivalent to prescribing

$$\varphi_L(t) = \begin{cases} 0.02 & \text{for } 0 < t \le 33\,[\text{h}], \\ 0 & \text{for } 33\,[\text{h}] < t \le T = 168\,[\text{h}]. \end{cases}$$

We observe that the sediment rises at constant speed. At $t = 33\,[\text{h}]$, a discontinuity between the concentration values 0.02 and zero starts to propagate from $z = L$ into the thickener at speed

$$\sigma = \frac{f_{bk}(0.02)}{0.02} = -4.6913 \times 10^{-4}\,[\text{m/s}],$$

and arrives at the sediment level of approximate height 3.3 [m] at

$$t = 118800\,[\text{s}] - \frac{2.7\,[\text{m}]}{\sigma} \approx 124555\,[\text{s}] = 34.6\,[\text{h}].$$

As the thickener remains closed, the sediment compresses further by its own weight, until the equipment is opened at $t = 120\,[\text{h}]$. It has emptied entirely by $t \approx 151\,[\text{h}]$.

10.4.2 Transition between three steady states

A steady state $\Phi = \Phi(z)$ is the concentration profile of the stationary mode in which a continuous thickener should operate normally in industrial applications.

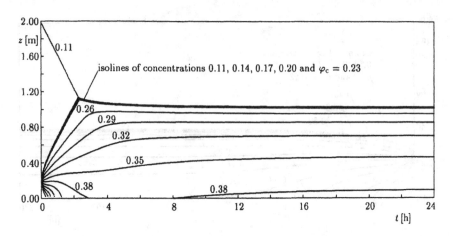

Figure 10.7: Concentration lines for simultaneous expansion of overcompressed sediment and batch sedimentation. The marked isolines correspond to the annotated concentration values, while the isolines ending at $z = 0$ before $t = 2$ [h] correspond to the concentration values 0.41, 0.44, 0.47, 0.50 and 0.53 (from top to bottom).

It is fed with a suspension of constant concentration such that $\Phi(L) = \Phi_L$ or, equivalently, the feed solid flux density f_F amounts to

$$f_F = f_k(\Phi_L) = q\Phi_L + f_{bk}(\Phi_L),$$

and its discharge flux density $f_D = q\Phi(0)$ must equal the feed volume flux density, $f_F = f_D$. The conditions for the existence of a steady state can be obtained by considering time-independent solutions of the field equation (9.11), that is, of the ordinary differential equation

$$q\Phi' + f'_{bk}(\Phi)\Phi'(z) = \left(-\frac{f_{bk}(\Phi)\sigma'_e(\Phi)}{\Delta\rho g\Phi}\Phi'(z)\right)'$$

which yields

$$q\Phi + f_{bk}(\Phi) + \frac{f_{bk}(\Phi)\sigma'_e(\Phi)}{\Delta\rho g\Phi}\Phi'(z) = C.$$

A desired discharge concentration Φ_D is assumed to be given. By the boundary condition at $z = 0$, the integration constant is $C = q\Phi_D$. Thus the concentration profile in the compression zone is the solution of the ordinary differential equation

$$\Phi'(z) = -\frac{\Delta\rho g\Phi}{\sigma'_e(\Phi)f_{bk}(\Phi)}(q\Phi + f_{bk}(\Phi) - q\Phi_D), \quad z > 0 \qquad (10.14)$$

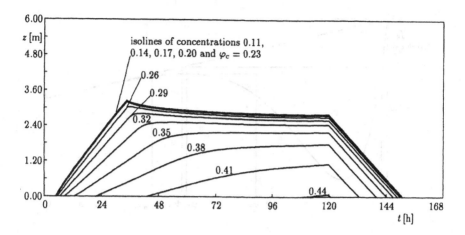

Figure 10.8: Continuous thickening: numerical solution for filling up and emptying of a thickener. The isolines correspond to the annotated concentration values.

with the boundary condition

$$\Phi(0) = \Phi_D. \tag{10.15}$$

(Mathematically, this is an initial condition for the starting the integration of (10.14)). Equation (10.14) can be solved, for example, by a Runge-Kutta method, and is integrated until the solution assumes the value $\Phi = \varphi_c$ at a sediment height $z = \lambda$. For sedimentation to take place, the concentration must increase downwards, that is, $\Phi'(z) \le 0$; hence, the condition

$$q\varphi + f_{\mathrm{bk}}(\varphi) \le q\Phi_D \text{ for all } \varphi \in [\varphi_c, \Phi_D] \tag{10.16}$$

must be satisfied so that the right-hand side of equation (10.14) is nonpositive. In fact, condition (10.16) is a restriction on the choice of the maximal value $\Phi_{D\max}$ of the discharge concentration $\Phi_D > \varphi_c$. If the flux density function $f_k(\varphi) = q\varphi + f_{\mathrm{bk}}(\varphi)$ possesses a local maximum at $\varphi = \varphi_M$, then $\Phi_{D\max}$ is obtained from solving $f_k(\varphi_M) = q\Phi_{D\max}$. In the case of a monotonically decreasing flux density function, $\Phi_{D\max}$ is obtained from $f_k(\varphi_c) = q\Phi_{D\max}$. The boundary value at $z = L$ is obtained by solving the equation

$$f_k(\Phi_L) = q\Phi_D, \quad \Phi_L < \varphi_M. \tag{10.17}$$

Figure 10.9 shows an example of the construction of $\Phi_{D\max}$ and of Φ_L for a given value of Φ_D.

Now a numerical simulation of transitions between approximate steady states is presented. The following steady states $\Phi^i(z)$, $i = 1, 2, 3$ were calculated in advance

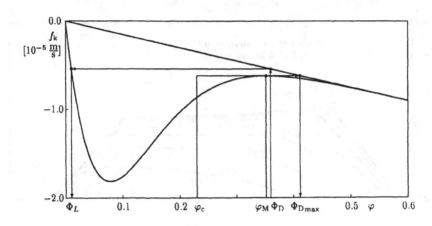

Figure 10.9: Construction of the maximum steady state discharge concentration $\Phi_{D\,\text{max}}$ and the construction of the boundary datum Φ_L for $q = -1.5 \times 10^{-5}$ [m/s], $\varphi_c = 0.23$ and $\Phi_D = 0.36$.

for a feeding level height of $L = 6$ [m]:

i	q^i [10^{-5}m/s]	Φ_D^i	$f_F^i = f_D^i$ [10^{-5}m/s]	Φ_L^i	Sediment height λ_i [m]
1	-1.5	0.36	-0.54	0.00983189	1.00 [m]
2	-0.5	0.42	-0.21	0.00369112	2.49 [m]
3	-3.0	0.35	-1.05	0.02139377	1.54 [m]

The steady state Φ^1 is selected as the initial state of the continuous thickener, and the boundary datum is chosen correspondingly: $\varphi_L(t) = \Phi_L^1$. At $t = 14$ [h], we wish to change to the next steady state. Therefore q is changed from q^1 to q^2, causing a rise of the sediment level. As the feeding flux remains constant during this change, $\varphi_L(t)$ has to be changed from Φ_L^1 to the value Φ_L^{1*}, which is calculated from

$$f_F^1 = q^2 \Phi_L^{1*} + f_{bk}\left(\Phi_L^{1*}\right). \tag{10.18}$$

We obtain $\Phi_L^{1*} = 0.010040585$. The sediment rises at an apparently constant velocity of 0.97×10^{-5} [m/s] and reaches the sediment level λ_2 at $t = 57.1$ [h]. Taking into account the propagation velocity of the discontinuity between Φ_L^{1*} and Φ_L^2, $\varphi_L(t)$ is changed at $t = 55.2$ [h]. We observe that the concentration profile tends to the calculated second steady state, and that the discharge concentration approximates the corresponding value $\Phi_D^2 = 0.42$.

At $t = 144$ [h], we start proceeding to the third steady state by changing $q(t)$ from q^2 to q^3, and by the continuity of the feed flux, $\varphi_L(t)$ from Φ_L^2 to $\Phi_L^{2*} = 0.003532479$. This value is obtained by solving

$$f_F^2 = q^3 \Phi_L^{2*} + f_{bk}\left(\Phi_L^{2*}\right). \tag{10.19}$$

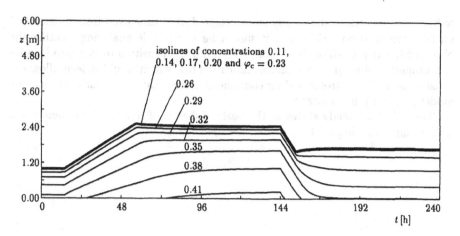

Figure 10.10: Settling plot for continuous sedimentation with transition between three approximate steady states. The dotted isolines correspond to the annotated concentration values.

The sediment will have fallen to height λ_3 by $t = 154.7\,[\mathrm{h}]$. The change from Φ_L^{2*} to Φ_L^3 has to be carried out before such that $\varphi(z, t) = \Phi_L^3$ is valid above the sediment level for $t \geq 154.7\,[\mathrm{h}]$. The change from Φ_L^{2*} to Φ_L^3 causes a rarefaction wave propagating into the thickener. Assuming that the speed at which the value $\varphi = \Phi_L^3$ propagates is the relevant one, we obtain that $\varphi_L(t)$ has to be changed at $t = 150.9\,[\mathrm{h}]$. The desired third steady state $\Phi_3(z)$ is approximated well by the concentration profile at $t = T = 240\,[\mathrm{h}]$, when the simulation ends. Summing up, we consider the following initial, boundary and operating conditions:

$$\phi_0(z) = \begin{cases} \text{the solution of (10.14) and (10.15) with} & \\ q = -1.5 \times 10^{-5}\,[\mathrm{m/s}] \text{ and } \Phi_D = 0.36 & \text{for } 0 \leq z \leq \lambda_1, \\ \Phi_L^1 \text{ calculated from (10.17)} & \text{for } \lambda_1 < z \leq L, \end{cases}$$

$$\varphi_L(t) = \begin{cases} \Phi_L^1 = 0.00983189 & \text{for } 0 \leq t \leq 14\,[\mathrm{h}], \\ \Phi_L^{1*} = 0.010040585 & \text{for } 14\,[\mathrm{h}] < t \leq 55.2\,[\mathrm{h}], \\ \Phi_L^2 = 0.00369112 & \text{for } 55.2\,[\mathrm{h}] < t \leq 144\,[\mathrm{h}], \\ \Phi_L^{2*} = 0.003532479 & \text{for } 144\,[\mathrm{h}] < t \leq 150.9\,[\mathrm{h}], \\ \Phi_L^3 = 0.02139377 & \text{for } 150.9\,[\mathrm{h}] < t \leq 240\,[\mathrm{h}], \end{cases}$$

$$q(t) = \begin{cases} q^1 = -1.5 \times 10^{-5}\,[\mathrm{m/s}] & \text{for } 0 \leq t \leq 14\,[\mathrm{h}], \\ q^2 = -0.5 \times 10^{-5}\,[\mathrm{m/s}] & \text{for } 14\,[\mathrm{h}] < t \leq 144\,[\mathrm{h}], \\ q^3 = -3.0 \times 10^{-5}\,[\mathrm{m/s}] & \text{for } 144\,[\mathrm{h}] < t \leq 240\,[\mathrm{h}]. \end{cases}$$

Figure 10.10 shows the settling plot of this simulation. Here, a stabilization effect is observed, that is, if the feed and discharge conditions belong to a steady state, the sediment converges to the steady state profile during time even from a disturbed initial concentration. Moreover, it should be worthwhile analysing whether the observed linear growth of the sediment level during transition from a steady state is a property inherent to the mathematical model, which would indeed allow the construction of a control model for continuous sedimentation similar to the control model discussed in Chapter 8.

The study of steady states as the basis of design of industrial thickeners will be continued in Chapter 11.

Chapter 11

Critical review of thickener design methods

11.1 Introduction: definition, equipment and operation

In this chapter we analyze different methods of thickener design that have been proposed in the literature, in the light of their physical foundations. We distinguish three types of methods: (1) those based on the macroscopic balances, (2) those based on kinematical models and (3) those based on dynamical models. This classification permits the analysis of thickener design procedures with a clear perspective of their applicability and limitations.

Thickener design includes the fundamental choice of type of thickener (standard, deep (Tiller and Tarng 1995), high capacity (Concha *et al.* 1995) etc.) and the specification of the feed well, pumps, and rake mechanism. Here, we assume that a standard thickener has been chosen and that its size must now be specified. The objective is to use the smallest thickener which has the required capacity. That is, it should discharge a clear fluid at the top and an underflow with the required solids content.

As discussed by Coe and Clevenger (1916), thickeners will overflow if their solid handling capacity is surpassed. Dixon (1979) explains that a thickener can overflow in three ways: when the feed has very fine particles that cannot settle, when the feed rate exceeds the settling capacity of the settling zone II, and when the feed rate is higher than the discharge rate. In this case solid particles accumulate in the thickener and will eventually be transported to the clear water zone and thence to the overflow. Overflow and underflow of ideal slurries were discussed extensively in Chapter 8. Better control is possible when the flocs are compressible.

When, due to a change in the solid concentration of the feed or when the solid feed rate changes, the discharge concentration diminishes, it is possible to get it back to its original value by controlling the discharge volume flow rate. This, in

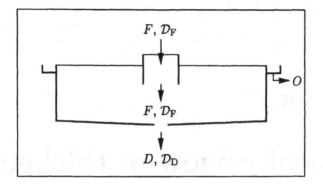

Figure 11.1: Macroscopic balance in a continuous thickener.

turn, is obtained by manipulating a variable speed pump or, when the discharge is by gravity, by changing the size of the outlet aperture of the thickener. The result is that, by slowing down the discharge flow rate, more solids accumulate in the compression zone, increasing its depth and therefore producing a more concentrated discharge. The opposite happens when the discharge rate is increased. These effects are usually used for thickener control.

11.2   Macroscopic mass balance in a continuous thickener at steady state: classical methods of thickener design

11.2.1   Mishler's equation

The first equation to predict the capacity of a thickener was developed by Mishler in 1912 and corresponds to a simple macroscopic mass balance in the equipment. Consider a thickener working at steady state, as shown in Figure 11.1. Using the same variables as Mishler, we write the respective solid and water mass balances as

$$F = D, \tag{11.1}$$

$$F\mathcal{D}_F = D\mathcal{D}_D + O, \tag{11.2}$$

where F and D are the solid mass flow rates in the feed and the discharge respectively, O is the water mass flow rate in the overflow and $\mathcal{D}_F$ and $\mathcal{D}_D$ are the *dilutions* of the feed and discharge. Dilution is a measure of concentration defined as the ratio of the mass of water to the mass of solid, i.e.

$$\mathcal{D} = \frac{\rho_f(1 - \varphi)}{\rho_s \varphi}. \tag{11.3}$$

The volume flow rate of water at the overflow is then

$$Q_O = F\left(\mathcal{D}_F - \mathcal{D}_D\right)/\rho_f, \tag{11.4}$$

where ρ_f is the water density. According to Mishler (1912), the flow rate of water per unit thickener area Q_O/S, must be equal to the rate of water formed in a batch test with the same pulp, at the concentration of the feed. Since this rate is equal to the rate of descent of the water-suspension interface in a batch settling test, which we denote by $\sigma_I(\mathcal{D}_F)$, we can write

$$\sigma_I(\mathcal{D}_F) = \frac{F(\mathcal{D}_F - \mathcal{D}_D)}{\rho_f S},$$

and the settling area required to treat a feed rate of F is then

$$S = \frac{F(\mathcal{D}_F - \mathcal{D}_D)}{\rho_f \sigma_I(\mathcal{D}_F)}, \tag{11.5}$$

where S is the thickener area. Mishler used the following units: F in [short tons], σ_I in [ft/min] and ρ_f in [lb/ft^3] and obtained

$$S = 0.0222 \frac{F(\mathcal{D}_F - \mathcal{D}_D)}{\rho_f \sigma_I(\mathcal{D}_F)} \text{ [ft}^2\text{]}. \tag{11.6}$$

11.2.2   Coe and Clevenger's method

As we have already discussed, Coe and Clevenger (1916) distinguished four zones in a continuous thickener at steady state, each with different concentration. Therefore they proposed to apply Mishler's equation for every dilution $\mathcal{D}_k$ present in the thickener (they did not actually mention Mishler as the proponent of the equation) and find the minimum solid handling capacity F/S in the settling column. Then, from equation (11.5),

$$\min\left(\frac{F}{S}\right) = \min_{\mathcal{D}_k}\left(\frac{\rho_f \sigma_I(\mathcal{D}_k)}{F(\mathcal{D}_k - \mathcal{D}_D)}\right). \tag{11.7}$$

Coe and Clevenger used the following units: F in [lb], ρ_f in [lb/ft^3], S in [ft^2] and σ_I in [ft/h], to give

$$\min\left(\frac{F}{S}\right) = \min_{\mathcal{D}_k}\left(62.35 \frac{\sigma_I(\mathcal{D}_k)}{F(\mathcal{D}_k - \mathcal{D}_D)}\right) \text{ [lb/(h ft}^2\text{)]}. \tag{11.8}$$

The design method consists in measuring, in the laboratory, the initial settling rate $\sigma_I(\mathcal{D}_k)$ of a suspension at a range of concentrations between that of the feed and that of the discharge, applying equation (11.8) to find the minimum value of the solid handling capacity F/S. Defining the *basic unit area* UA_0 as the reciprocal of the minimum solid handling capacity, we can write according to equation (11.7):

$$UA_0 = \max_{\mathcal{D}_k}\left(\frac{\mathcal{D}_k - \mathcal{D}_D}{\rho_f \sigma_I(\mathcal{D}_k)}\right), \qquad \mathcal{D}_D < \mathcal{D}_k < \mathcal{D}_F. \tag{11.9}$$

Taggart (1927) used the following units: $\rho_f = 62.5\,[\text{lb/ft}^3]$ and σ_I in $[\text{ft/h}]$ giving UA_0 in $[\text{ft}^2/(\text{short ton} \cdot \text{day})]$; then

$$UA_0 = \max_{\mathcal{D}_k} \left(1.33 \frac{\mathcal{D}_k - \mathcal{D}_D}{\rho_f \, |\sigma_I(\mathcal{D}_k)|} \right) \qquad [\text{ft}^2/(\text{short ton} \cdot \text{day})]. \qquad (11.10)$$

For future reference we will express equation (11.9) in terms of the solid volume fraction φ. Since the dilution is given by equation (11.3), the unit area becomes

$$UA_0 = \max_{\varphi_k} \left(\frac{1}{\rho_s \sigma_I(\varphi)} \left(\frac{1}{\varphi_k} - \frac{1}{\varphi_D} \right) \right), \qquad \varphi_F \leq \varphi_k < \varphi_D. \qquad (11.11)$$

We call equation (11.11), which gives the unit area of a thickener based on laboratory initial settling tests, the *Mishler equation*, and the design method we call the *Coe and Clevenger method of thickener design*. If the following units are selected: ρ_s in $[\text{kg/m}^3]$, σ_I in $[\text{m/s}]$ and UA_0 in $[\text{m}^2/\text{TPD}]$, we have

$$UA_0 = \max_{\varphi_k} \left(1.1574 \times 10^{-2} \frac{1}{\rho_s \, |\sigma_I(\varphi_k)|} \left(\frac{1}{\varphi_k} - \frac{1}{\varphi_D} \right) \right) \qquad [\text{m}^2/\text{TPD}]. \qquad (11.12)$$

According to Coe and Clevenger (1916), when the discharge concentration of a thickener is still in the range of hindered settling, the depth of the tank is of no consequence, except to permit ample depth of clear liquid to accomodate fluctuations of the feed. On the other hand, when the consistency of the pulp at the discharge is in the range where it is necessary to expel fluid by compression, sufficient capacity must be given to the tank so that the pulp in compression is retained in the thickener the necessary period of time to reach the required density. To calculate the height of the compression zone, the time t^* to reach the desired concentration φ_D is measured in a batch test. According to the nomenclature of Figure 11.2, t^* is the time necessary for $z_I = z_D = z_0/\varphi_D$. Then, t^* is divided in n equal (or almost equal) intervals of length $\Delta t_i = t^*/n$, that is, the following intervals are considered:

$$\Delta t_1 = [0, t_1], \quad \Delta t_2 = [t_1, t_2], \ldots, \Delta t_n = [t_{n-1}, t_n] \qquad (11.13)$$

(in their original paper, Coe and Clevenger considered four intervals of this kind as an example of calculations, even though they did not consider the Δt_i strictly constant, only for numerical convenience). Then, the necessary height in the thickener is calculated so that the pulp goes from a concentration φ_{i-1} to φ_i, giving to the pulp the necessary volume V_i so that

$$V_i = Sh_i. \qquad (11.14)$$

One considers that the concentration during this period is equal to the average concentration

$$\overline{\varphi}_i = (\varphi_{i-1} + \varphi_i)/2. \qquad (11.15)$$

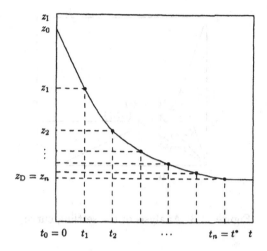

Figure 11.2: Coe and Clevenger's method to determine the height of a thickener.

The value V_i of pulp of concentration between φ_{i-1} and φ_i is given by $V_i = F\Delta t_i/\bar{\rho}_i$, where $\bar{\rho}_i$ is the average density of a pulp of concentration $\bar{\varphi}_i$. In this way, the height interval φ_{i-1} to φ_i is given by

$$h_i = \frac{V_i}{S} = \frac{F\Delta t_i}{\bar{\rho}_i S} = \frac{1}{UA_0} \frac{\Delta t_i}{[\Delta\rho\bar{\varphi}_i + \rho_f]}. \tag{11.16}$$

Then the total height necessary is given by

$$h = \sum_{i-1}^{n} h_i = \frac{1}{UA_0} \sum_{i-1}^{n} \frac{\Delta t_i}{\Delta\rho\bar{\varphi}_i + \rho_f}. \tag{11.17}$$

To the height obtained from this calculation, about 0.5 to 1 additional meters, depending on the size of the thickener, must be added to accommodate the feed and the clear liquid region.

11.3 Design methods based on kinematical sedimentation precesses

The development of Kynch's theory of sedimentation in 1952 immediately opened a new field of research: the search to put thickener design on a theoretical basis and, in this way, find faster and more accurate methods of thickener design. Several research workers were involved in this search, leaving their names associated with thickener design procedures. We will restrict our review to some of them, namely W.P. Talmage, B. Fitch, J.H. Wilhelm, Y. Naide, H. Oltmann, N.J. Hassett and N. Yoshioka. In this section we will review Kynch's theory of sedimentation and those design methods based on it.

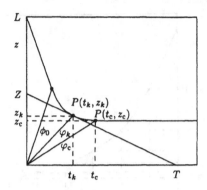

Figure 11.3: Analysis of the settling curve.

11.3.1   Analysis of the batch sedimentation curve

Consider a batch Kynch sedimentation process and draw a settling plot and a characteristic line for the concentration φ_k, such as that shown in Figure 11.3. The line ZT is tangent to the curve at the point (t_k, z_k). As we have seen, for all the regions of the settling plot where the variables are continuous it is possible to obtain the settling parameters φ, $\sigma_{\rm I}(\varphi)$, $f_{\rm bk}(\varphi)$ and $f'_{\rm bk}(\varphi)$ graphically.

Settling rate

From the solution of the equation

$$\frac{\partial \varphi}{\partial t} + \frac{\partial f_{\rm bk}}{\partial z} = 0, \tag{11.18}$$

we know that the rate of fall of the water-suspension interface $\sigma_{\rm I}(\varphi_k)$ is given by

$$-\sigma_{\rm I}(\varphi_k) = \sigma(0, \varphi_k) = f_{\rm bk}(\varphi_k)/\varphi_k; \tag{11.19}$$

therefore, $\sigma_{\rm I}(\varphi_k)$ will be equal to the slope of the settling curve at point (t_k, z_k):
$\sigma_{\rm I} = Z/T$.

Concentration

Let W_0 be the total volume of solid present in the settling column per unit cross sectional area. Then, the flux of solid crossing the iso-concentration wave φ_k, as it travels from $z = 0$ to $z = z_k$, is related to W_0 by

$$W_0 = \int_0^{t_k} \varphi_k(-v_s(\varphi_k) + f'_{\rm bk}) \, dt, \tag{11.20}$$

where $v_s(\varphi_k) = -\sigma_{\rm I}(\varphi_k)$ is the settling velocity of the suspension of concentration φ_k. Since the slope $f'_{\rm bk}(\varphi_k)$ of the characteristic of concentration φ_k is constant and the velocity $v_{\rm s}(\varphi_k)$ is also constant, we can integrate equation (11.20) directly:

$$W_0 = \varphi_k \left(-v_{\rm s}(\varphi_k) + \frac{z_k}{t_k} \right) t_k. \qquad (11.21)$$

From Figure 11.3 we can see that

$$\frac{Z}{T} = \frac{Z - z_k}{t_k}. \qquad (11.22)$$

On the other hand, since at $t = 0$ the suspension was homogeneous and had a concentration of ϕ_0, the volume of solid per unit cross-sectional area present in the column is

$$W_0 = L\phi_0. \qquad (11.23)$$

Substituting the last two equations into equation (11.21) yields

$$\varphi_k = \phi_0 L/Z. \qquad (11.24)$$

Knowing the settling curve of a batch sedimentation process (KSP) for a given suspension with initial concentration ϕ_0 and initial height L, we can obtain (graphically from the curve) the parameters for any concentration φ_k which appears in the KSP (see Chapter 7). Summarizing, we can write

$$\varphi_k = \phi_0 L/Z, \qquad \sigma_{\rm I}(\varphi_k) = -v_{\rm s}(\varphi_k) = Z/T, \qquad (11.25)$$
$$-f_{\rm bk}(\varphi_k) = \phi_0 L/T, \qquad f'_{\rm bk}(\varphi_k) = z_k/t_k. \qquad (11.26)$$

11.3.2 Design of continuous thickeners based on a batch Kynch sedimentation process

As we have seen, Coe and Clevenger's method of thickener design uses Mishler's equation to calculate the basic unit area,

$$\text{UA}_0 = \max_{\varphi_k} \left(\frac{1}{\rho_{\rm s}\sigma_{\rm I}(\varphi_k)} \left(\frac{1}{\varphi_k} - \frac{1}{\varphi_{\rm D}} \right) \right), \qquad (11.27)$$

where $\rho_{\rm s}$ is the density of the solid, $\sigma_{\rm I}(\varphi_k)$ is the initial settling rate of a suspension of concentration φ_k and $\varphi_{\rm D}$ is the discharge concentration. Coe and Clevenger suggested making many laboratory tests with suspensions of concentration ranging from that of the feed to that of the discharge to find $\sigma_{\rm I}(\varphi_k)$ satisfying equation (11.27). If the suspension to be thickened can be considered an ideal suspension, that is, if $(\varphi_k, f_{\rm bk}(\varphi_k))$ constitutes a Kynch sedimentation process (KSP), *one properly selected sedimentation test* will give all the information necessary to calculate UA_0, see equations (11.25) and (11.26). To calculate φ_k and $\sigma_{\rm I}(\varphi_k)$, we draw a tangent at any point in the settling curve and calculate φ_k and $\sigma_{\rm I}(\varphi_k)$ from

$$\varphi_k = \phi_0 L/Z \text{ and } \sigma_{\rm I}(\varphi_k) = Z/T, \qquad (11.28)$$

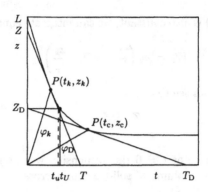

Figure 11.4: Thickener design method based on batch Kynch theory

where Z and T are the coordinates of the intersection point of the tangent with the coordinate axis and where L is the initial height of the suspension. See Figure 11.2. A completely graphical procedure can be designed by realizing that equations (11.28) should also hold for φ_D (remember that the pulp follows Kynch's theory); that is (see Figure 11.4)

$$\varphi_D = \phi_0 L / Z_D. \tag{11.29}$$

Substituting equations (11.28) and (11.29) into equation (11.27) yields

$$UA_0 = \frac{1}{\rho_s \phi_0 L} \max_{\varphi_k} \left(\frac{T(Z - Z_D)}{Z} \right). \tag{11.30}$$

By a properly selected concentration we imply an initial condition for the KSP that will give a continuous settling curve. The best concentration would be that of the inflection point in the flux density function, because it would give a Mode of Sedimentation 3 (see Chapter 7). Obviously, we do not know the concentration of the inflection point and must make our best guess. Too low a value for the initial concentration will lead to a Mode of Sedimentation 1, which would give no possibility of drawing tangents.

Talmage and Fitch's method

Talmage and Fitch (1955) observed that the settling velocity for a concentration φ_k could be expressed, in relation to Figure 11.2, by

$$\sigma_I(\varphi_k) = \frac{Z}{T} = \frac{Z - Z_D}{t_u}. \tag{11.31}$$

Substituting this equation into (11.30) gives

$$UA_0 = \frac{1}{\rho_s \phi_0 L} \max(t_u). \tag{11.32}$$

From Figure 11.2 we can see that the maximum value of t_u is obtained when $z_k = Z_D$ and t_u coincides with t_k. We will call this time t_U:

$$\text{UA}_0 = \frac{t_U}{\rho_s \phi_0 L}. \tag{11.33}$$

Talmage and Fitch's method of thickener design may be summarized in the following steps:

1. Perform one settling test at an intermediate concentration and obtain all initial settling rates $\sigma_I(\varphi)$ by drawing tangents to the settling curve, according to Kynch's theory.

2. Calculate the height Z_D by $Z_D = \phi_0 L/\varphi_D$.

3. Draw a horizontal line in the settling plot and determine the intersection with the settling curve. This point defines the time t_U.

4. Calculate the unit area using equation (11.33).

Since this method has been developed to design industrial thickeners, very often the horizontal line, drawn through $Z_D = \phi_0 L/\varphi_D$, does not intersect the settling curve (settling curves of compressible suspensions). In this case, the limiting concentration is the critical concentration and a tangent should be drawn at this point to the settling curve. The intersection of this tangent with the horizontal line through z_D defines the time t_U. See Figure 11.5.

Oltman's design method

The Oltman method (Fitch and Stevenson 1976) is an empirical variant of Talmage and Fitch's method. Both rely upon identifying the compression point. Oltman proposes drawing a straight line from the point $(0, L)$ to the critical point (t_c, z_c) extending it beyond this point. The intersection of this line and the horizontal line drawn through the point $(0, Z_D)$ gives the value of t_U, see Figure 11.5. There is no theoretical justification for this method.

Steady state capacity of an ICT for ideal suspensions

The analysis of the continuous Kynch sedimentation process for ideal suspensions having a flux density function with one inflection point shows that the only possible steady state is an MCS-2 with discharge concentration $\varphi_D = \varphi_M$ and conjugate concentration $\varphi_L = \varphi_M^{**}$. See Figure 8.5 of Chapter 8. Then the following equations are valid:

$$f_F = q\varphi_M^{**} + f_{bk}(\varphi_M^{**}), \tag{11.34}$$
$$f_D = q\varphi_M + f_{bk}(\varphi_M). \tag{11.35}$$

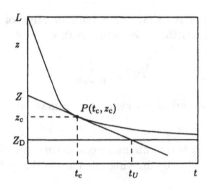

Figure 11.5: Talmage and Fitch's thickener design method for compressible pulps using the alternative method with the compression point.

Since at steady state $f_F = f_D$, obtaining q from equation (11.34) and substituting it into equation (11.35) and rearranging yields

$$f_F \left(\frac{1}{\varphi_M^{**}} - \frac{1}{\varphi_M} \right) = \frac{f_{bk}(\varphi_M^{**})}{\varphi_M^{**}} - \frac{f_{bk}(\varphi_M)}{\varphi_M}.$$

The term $f_{bk}(\varphi)/\varphi$ can be identified as the falling velocity of the clear liquid-suspension interface $-\sigma_I(\varphi)$ in batch settling of a suspension of concentration φ. Then the previous equation may be written in the form

$$f_F = -\frac{\sigma_I(\varphi_M^{**}) - \sigma_I(\varphi_M)}{\dfrac{1}{\varphi_M^{**}} - \dfrac{1}{\varphi_M}}. \tag{11.36}$$

The steady-state solution of a continuous Kynch sedimentation process, giving the capacity of the thickener, is represented graphically in Figure 11.6. The capacity of an ICT in terms of mass flow rate per unit area F/S is $F/S = -\rho_s f_F$. Then, from equation (11.36), the steady state capacity of an ICT is given by

$$F/S = \rho_s \frac{\sigma_I(\varphi_M) - \sigma_I(\varphi_M^{**})}{\dfrac{1}{\varphi_M^{**}} - \dfrac{1}{\varphi_M}}, \tag{11.37}$$

where $\varphi_M = \varphi_D$, $\varphi_M^{**} = \varphi_L$. According to the definition of Unit Area, $UA_0 = S/F$, for an ICT treating an ideal suspension with discharge concentration $\varphi_D = \varphi_M$ we have

$$UA_0 = \frac{1}{\rho_s(\sigma_I(\varphi_M^{**}) - \sigma_I(\varphi_M))} \left(\frac{1}{\varphi_M^{**}} - \frac{1}{\varphi_M} \right). \tag{11.38}$$

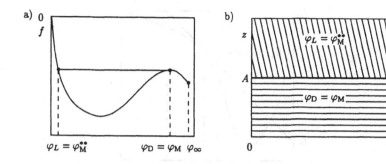

Figure 11.6: Solution for an MCS-3. a) Flux density function with $f'(\varphi_\infty) < 0$ and $\varphi_D = \varphi_M$. b) Global solution showing a steady state of the ICT with $\varphi_L = \varphi_M^{**}$ and $\varphi_D = \varphi_M$.

11.3.3 Thickener design methods based on a continuous Kynch sedimentation process

We analyze in this section those thickener design methods in which the continuous flux density function has been mentioned explicity. The research workers involved are N. Yoshioka, N.J. Hassett, J.H. Wilhelm and Y. Naide.

Yoshioka-Hassett method

Yoshioka *et al.* (1957) developed a graphical thickener design method based on the total solid flux density function. From the previous section we know that

$$f_k(\varphi) = q\varphi + f_{bk}(\varphi), \tag{11.39}$$

and at steady state $f_k(\varphi) = f_F$, so that

$$f_F = q\varphi + f_{bk}(\varphi). \tag{11.40}$$

Solving equation (11.40) for $f_{bk}(\varphi)$ with $q = q_D$ leads to

$$f_{bk}(\varphi) = f_F - q_D\varphi, \tag{11.41}$$

where q_D is the volume average velocity at the discharge. Equation (11.41) represents a straight line passing through φ_D with q_D as the slope ($q_D = -f'_{bk}(\varphi_M)$) at the point $\varphi_M = \varphi_D^*$ and with f_F as the intercept of the ordinate in a plot of $f_{bk}(\varphi)$ versus φ. See Figure 11.7. Therefore, the intercept of the straight line with the vertical axis in Figure 11.7 gives the continuous solid flux density at steady state. The unit area, of course, is proportional to the reciprocal of the feed flux density: $UA_0 = 1/(-\rho_s f_F)$. Yoshioka *et al.* (1957) and Hassett (1958) independently interpreted the result of Figure 11.7 still in another way. If the continuous flux density function $f_k(\varphi)$ is plotted instead of $f_{bk}(\varphi)$ against φ, Figure 11.8 is obtained. Here

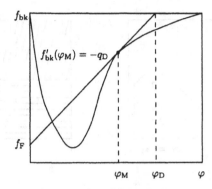

Figure 11.7: Plot of the batch flux density function and the Yoshioka construction.

the solid flux density at steady state is the horizontal line tangent to the continuous flux density curve at its maximum with concentration φ_M. Then, the design method consists in choosing a discharge concentration φ_D and drawing a tangent to the $f_k(\varphi)$ curve from $(\varphi_D, 0)$. The intercept f_F is the solid feed density wanted.

Hassett (1968) realized that there was a problem of interpretation in this approach, because Figure 11.8 shows that only two concentrations are possible in the thickener: the limiting concentration φ_M and its conjugate concentration φ_M^{**} (see our Figure 11.6 and equations (11.36) and (11.38)). Hassett says: *"Thus the theory predicted the absence of the feed and discharge concentration within the thickener, and shows that there must be an abrupt increase (of the concentration from φ_M) to the discharge concentration at the moment of discharge...".* It is obvious that this conclusion is absurd, because it would mean that the passage of a suspension through a series of restrictions would increase its concentration making the thickener unnecessary. The principal objection to these graphical methods of thickener design is that they use the Kynch flux density function for values of the concentration that are outside its range of validity. Remember that the Kynch batch flux density functions, is obtained with *initial settling experiments* and therefore they are valid up to the critical concentration only. Obviously, the definition of a flux density is valid beyond this concentration, but in this range it is not a function of concentration only. We will discuss this fact further in a later section and will give an explanation to Hassett's problem.

Wilhelm and Naide's method

Wilhelm and Naide (1981) also use the continuous flux density function at steady state in the form

$$f_F = q_D \varphi + f_{bk}(\varphi). \tag{11.42}$$

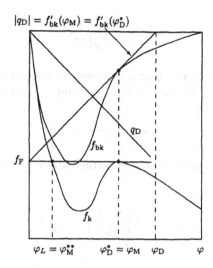

Figure 11.8: Plot of the continuous Kynch flux density function and the Yoshioka-Hassett construction.

At the limiting concentration, φ_M, the feed flux which maintains steady state will be a minimum. Thus we differentiate this equation with respect to φ and note that $f'_F = 0$ at the limiting concentration $\varphi = \varphi_M$. This yields

$$q_D = -f'_{bk}(\varphi)|_{\varphi_M}. \tag{11.43}$$

Since $f_F = q_D \varphi_D$, the unit area $UA_0 = -1/(\rho_s f_F)$ is given by

$$UA_0 = \frac{1}{\rho_s \varphi_D f'_{bk}(\varphi_M)}. \tag{11.44}$$

This expression is equivalent to Talmage and Fitch's method of thickener design. From Figure 11.5, we see that

$$\phi_0 L = \varphi_D Z_D \quad \text{and} \quad t_U = Z_D/f'_{bk}(\varphi_M). \tag{11.45}$$

Substitution of the expressions in equation (11.33) yields equation (11.44). This also demonstrate the equivalence of the Yoshioka-Hassett and Talmage and Fitch methods.

Wilhelm and Naide suggest expressing the settling velocity of a suspension of uniform concentration, within a given range of concentrations, by the equation

$$v_s(\varphi) = \sigma_I(\varphi) = -a\varphi^{-b}, \tag{11.46}$$

where the constant a has dimensions of LT^{-1} and the constant b is dimensionless. Since $f_{bk}(\varphi) = \varphi v_s(\varphi)$, we have

$$f_{bk}(\varphi) = -a\varphi^{1-b}, \tag{11.47}$$

$$q_D = a(1-b)\varphi_M^{-b}, \tag{11.48}$$

$$f_F = -ab\varphi_M^{1-b}. \tag{11.49}$$

On the other hand, since $f_F = q_D\varphi_D$, substituting into equation (11.49) yields a relationship between the discharge and the limiting concentration:

$$\varphi_M = \frac{b-1}{b}\varphi_D. \tag{11.50}$$

Then, in terms of the discharge concentration, the steady state flux density is given by

$$f_F = -ab\left(\frac{b-1}{b}\right)^{1-b}\varphi_D^{1-b} \tag{11.51}$$

and

$$UA_0 = \frac{1}{\rho_s ab}\left(\frac{b-1}{b}\right)^{b-1}\varphi_D^{b-1}. \tag{11.52}$$

According to Wilhelm and Naide, when the effect of compressive forces is negligible, for example for thickeners with shallow beds, the method described give unique results, otherwise different results are obtained for each bed height. The recommendation they give in this case is to perform the batch experiments at bed heights similar to those expected in the continuous thickener. Wilhelm and Naide's method of thickener design may be summarized in the following steps:

1. Perform batch settling experiments with suspensions at initial concentrations between that of the feed and that of the discharge of the thickener to be designed and record the initial settling velocity $\sigma_I(\varphi)$.

2. Alternatively, perform one settling test at an intermediate concentration and obtain all initial settling velocities by drawing tangents to the settling curve, according to Kynch theory.

3. Plot $\log(\sigma_I(\varphi))$ versus $\log(\varphi)$, as in Figure 11.9, and approximate the curve with one or more straight lines.

4. From each straight line in Figure 11.9 calculate the parameters a and b graphically or by linear regression.

5. Using the values of a and b, determined for each section, calculate the unit area with equation (11.52). The value of UA will be the maximum value thus obtained.

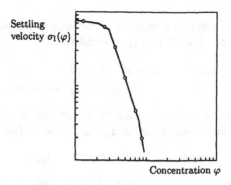

Figure 11.9: Settling velocity versus concentration for coal refuse, according to Wilhelm and Naide (1981).

11.4 Design method based on a dynamical sedimentation process

11.4.1 Sedimentation of a compressible suspension at steady state in an ICT

The continuous sedimentation of a flocculated suspension may be described, as was shown in Chapter 3, by the following set of variables: the solid concentration $\varphi(z,t)$, the solid flux density function $f(\varphi(z,t),t)$, the volume average velocity $q(z,t)$ and the excess pore pressure $p_e(z,t)$. In those regions where the variables are continuous they must satisfy the following equations (see Chapter 3):

$$\frac{\partial \varphi}{\partial t} + \frac{\partial}{\partial z} f_k(\varphi,t) = \frac{\partial}{\partial z}\left(a(\varphi)\frac{\partial \varphi}{\partial z}\right), \tag{11.53}$$

$$\frac{\partial q}{\partial z} = 0, \tag{11.54}$$

$$f_k(\varphi,t) = q(t)\varphi + f_{bk}(\varphi), \tag{11.55}$$

$$\frac{\partial p_e}{\partial z} = -\Delta\rho\varphi g - \sigma'_e(\varphi)\frac{\partial \varphi}{\partial z} \tag{11.56}$$

with the diffusion coefficient

$$a(\varphi) = -\frac{f_{bk}(\varphi)\sigma'_e(\varphi)}{\Delta\rho\varphi g}. \tag{11.57}$$

At discontinuities they obey the jump condition

$$\sigma = \frac{f^+ - f^-}{\varphi^+ - \varphi^-} \tag{11.58}$$

(see Corollary 9.1 for a precise formulation of the jump condition). At steady state
the solution of equations (11.53) and (11.54) satisfies

$$f_k(\varphi(z)) - a(\varphi(z))\varphi'(z) = f_F \quad \text{for all } z, \tag{11.59}$$

$$q(z) = q_D \quad \text{for all } z. \tag{11.60}$$

Since $D(\varphi)$ vanishes for $\varphi < \varphi_c$, to obtain the concentration profile it is necessary
to divide the domain for values smaller and greater than z_c (see Section 3.9). Then

$$\varphi(z) = \varphi_L \quad \text{for } \varphi < \varphi_c, \; z > z_c, \tag{11.61}$$

while for $z \le z_c$, $\varphi(z)$ is the solution of the ordinary differential equation

$$\varphi'(z) = \frac{f_F - q_D\varphi(z) - f_{bk}(\varphi(z))}{a(\varphi(z))}, \quad z > 0 \tag{11.62}$$

with the boundary condition $\varphi(0) = \varphi_D$. In fact, the height z_c is usually determined by solving equation (11.62) numerically, e.g. by a Runge-Kutta method,
until $\varphi(z) = \varphi_c$ is reached. Once the concentration profile is obtained, the excess pore pressure is calculated by integrating equation (11.56) numerically with
boundary condition $p_e(L) = 0$. Then

$$p_e = \begin{cases} \Delta\rho\varphi_L g(L - z) & \text{for } z_c < z \le L, \\ p_e(z_c) + \Delta\rho g \displaystyle\int_z^{z_c} \varphi(\xi)\frac{f_F - q_D\varphi(\xi)}{f_{bk}(\varphi(\xi))}d\xi & \text{for } z \le z_c. \end{cases} \tag{11.63}$$

11.4.2   Capacity of an ICT treating a flocculated suspension

As we have seen, a thickener at steady state must obey equations (11.59), (11.60)
and

$$f_F \ge q_D\varphi_D + f_{bk}(\varphi_D). \tag{11.64}$$

In a solid flux density versus concentration plot these conditions are expressed by
the fact that the horizontal straight line $f(\varphi) = f_F$ lies above the maximum of the
function $f(\varphi) = f_k(\varphi)$. This result shows that the necessary and sufficient condition for the existence of a steady state in an ICT treating flocculated suspensions
is that, in a flux density plot, the extended continuous Kynch flux density function
lies below the line $f = f_F$.

Consider a suspension with concentration $\varphi_F = 0.1$ as the feed of a continuous
thickener. Figure 11.10 shows three possible steady states and one condition where
a steady state is impossible. The extended Kynch flux density function is

$$f_k = -2.4 \times 10^{-5} - 6.05 \times 10^{-4}\varphi(1 - \varphi)^{12.59}[\text{m/s}].$$

By changing the value of f_F, a different intersection of f_F and $q_D\varphi$ is obtained giving pairs of values of φ_L and φ_D. For $f_F \times 10^5 = -0.70, -0.84$ and $-0.90\,[\text{m/s}]$, the

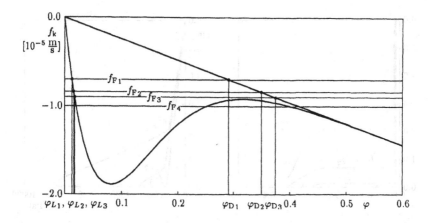

Figure 11.10: Extended Kynch flux density function versus concentration plot showing possible steady states for a thickener with feed concentration $\varphi_F = 0.1$, feed flux densities $f_{F_1} = -0.7 \times 10^{-5}$ [m/s], $f_{F_2} = -0.84 \times 10^{-5}$ [m/s], $f_{F_3} = -0.9 \times 10^{-5}$ [m/s], $f_{F_4} = -1.0 \times 10^{-5}$ [m/s], and $f_k(\varphi) = -2.4 \times 10^{-5}\varphi - 6.05 \times 10^{-4}\varphi(1 - \varphi)^{12.59}$ [m/s].

following steady states are observed: $(\varphi_L, \varphi_D) = (0.0131, 0.293)$, $(0.0164, 0.352)$ and $(0.0177, 0.374)$ (to three significant digits). The last value of $f = -1.00 \times 10^{-5}$ [m/s] gives no steady state. If, on the other hand, we consider a thickener with a discharge concentration $\varphi_D = 0.3$ and feed concentration $\varphi_F = 0.1$, several steady states can be obtained with different values of q_D, as is shown in Figures 11.12 and 11.13. For a given value of φ_D, any value of q_D would fix f_F. Note that the value φ_F is not used in these computations. However, the feed suspension of concentration φ_F should be diluted to the concentration value φ_L, therefore we assume for a given flux density function f_k that φ_F is chosen large enough such that dilution to all possible values of φ_L is actually possible.

The concentration and excess pore pressure profiles in Figures 11.11 and 11.13 were calculated using the solid effective stress function

$$\sigma_e(\varphi) = \begin{cases} 0 & \text{for} \quad \varphi \leq \varphi_c = 0.23, \\ 5.35 \exp(17.9\varphi) \, [\text{N/m}^2] & \text{for} \quad \varphi > \varphi_c \end{cases} \qquad (11.65)$$

determined by Becker (1982) for copper ore tailings.

Maximum capacity of an ICT

The maximum value of q_D is that for which f_F becomes tangent to $f_k(\varphi)$ for a given value of φ_D. Figure 11.10 illustrates that a steady state given by q_D and φ_D (or by q_D and $f_F = q_D\varphi_D$) is admissible if

$$f_F \geq f_k(\varphi) = q_D\varphi + f_{bk}(\varphi) \quad \text{for all} \quad \varphi \in [\varphi_L, \varphi_D]. \qquad (11.66)$$

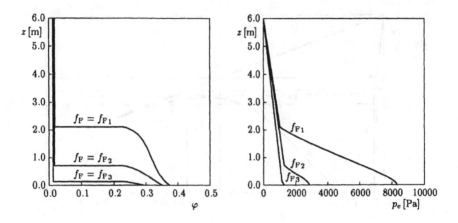

Figure 11.11: Concentration and excess pore pressure profiles for the conditions of Figure 11.10.

Note that, at steady state, $f_k(\varphi)$ can be written as

$$f_k(\varphi) = \frac{f_F}{\varphi_D}\varphi + f_{bk}(\varphi).$$

Substituting in equation (11.66) yields

$$f_F \geq f_F \frac{\varphi}{\varphi_D} + f_{bk}(\varphi) \quad \text{for all } \varphi \in [\varphi_L, \varphi_D].$$

Dividing by φ we obtain

$$f_F \geq \frac{\dfrac{f_{bk}(\varphi)}{\varphi}}{\dfrac{1}{\varphi} - \dfrac{1}{\varphi_D}} \quad \text{for all } \varphi \in [\varphi_L, \varphi_D], \tag{11.67}$$

hence, using the definition of capacity $F/S = -\rho_s f_F$ and of the initial settling rate $\sigma_I(\varphi) = -f_{bk}(\varphi)/\varphi$, we can write

$$\frac{F}{S} \leq \frac{-\dfrac{\rho_s f_{bk}(\varphi)}{\varphi}}{\dfrac{1}{\varphi} - \dfrac{1}{\varphi_D}} = \frac{\rho_s \sigma_I(\varphi)}{\dfrac{1}{\varphi} - \dfrac{1}{\varphi_D}} \quad \text{for all } \varphi \in [\varphi_L, \varphi_D] \tag{11.68}$$

and therefore, using $UA = S/F$:

$$UA \geq \frac{1}{\rho_s \sigma_I(\varphi)} \left(\frac{1}{\varphi} - \frac{1}{\varphi_D} \right) \quad \text{for all } \varphi \in [\varphi_L, \varphi_D], \tag{11.69}$$

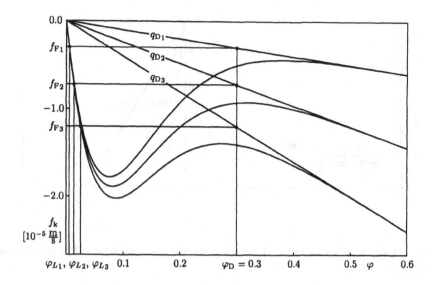

Figure 11.12: Extended Kynch flux density functions $f_k = q_{D_i}\varphi - 6.05 \times 10^{-4}\varphi(1-\varphi)^{12.59}$ [m/s] with $q_{D_1} = -1.0 \times 10^{-5}$ [m/s], $q_{D_2} = -2.4 \times 10^{-5}$ [m/s] and $q_{D_3} = -4.0 \times 10^{-5}$ [m/s] with possible steady states for the discharge concentration $\varphi_D = 0.3$. The corresponding concentration values at $z = L$ are $\varphi_{L_1} = 0.00520$, $\varphi_{L_2} = 0.0135$ and $\varphi_{L_3} = 0.0250$.

which implies

$$UA \geq \max_{\varphi_L \leq \varphi \leq \varphi_D} \left\{ \frac{1}{\rho_s \sigma_I(\varphi)} \left(\frac{1}{\varphi} - \frac{1}{\varphi_D} \right) \right\} = \max_{\varphi_L \leq \varphi \leq \varphi_D} \left\{ \frac{1}{\rho_s f_{bk}(\varphi)} \left(\frac{\varphi}{\varphi_D} - 1 \right) \right\}. \tag{11.70}$$

This is Mishler's equation in Coe and Clevenger's method of thickener design. This result indicates that *the minimum area* of an ICT, with a flocculated suspension, depends on the hindered settling rate $\sigma_I(\varphi)$ of the suspension and *does not depend on the compressibility of the sediment*. Note that (11.70) can also be expressed as

$$UA \geq \max_{\varphi_L \leq \varphi \leq \varphi_D} G(\varphi; \varphi_D), \quad G(\varphi; \varphi_D) = \frac{1}{\rho_s f_{bk}(\varphi)} \left(\frac{\varphi}{\varphi_D} - 1 \right). \tag{11.71}$$

As an example, the minimum unit area function $G(\varphi; 0.3)$ is shown in Figure 11.14. Assume that the ICT is to be operated at $q = -2.4 \times 10^{-5}$ [m/s], for which we obtain $\varphi_L = 0.0135$. Taking into account that $G(0.0135; 0.3) = 55.5$ [m²s/kg] and that $G(\varphi; 0.3)$ has a local maximum $G(0.2107; 0.3) = 18.37$ [m²s/kg], we see that the unit area should satisfy

$$UA \geq 55.5 \, [\text{m}^2\text{s/kg}] = 0.642 \, [\text{m}^2/\text{TPD}]$$

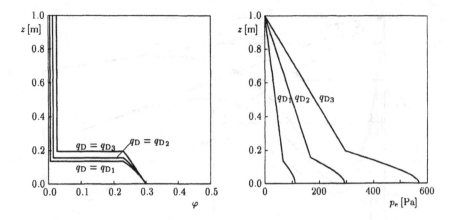

Figure 11.13: Concentration and excess pore pressure profiles for the conditions of Figure 11.12.

for these operating conditions. Comparing these results with Coe and Clevenger's method of design leads to the conclusion that the unit area of a thickener is always greater than the basic unit area UA_0, $UA \geq UA_0$.

11.4.3   Adorján's method of thickener design

Adorján's (1975, 1976) was the first work to express in a clear and consistent way the sizing of an industrial thickener based on mass and momentum balances. His results are equivalent to those deduced in the previous section, especially to equations (11.62) and (11.68).

Thickener area

According to Adorján, a thickener operating under limiting conditions requires a considerable pulp depth and, therefore, it should be operated at only a fraction λ, called *loading factor*, of the limiting feed rate. Then

$$\lambda = \frac{F}{F_0} = \frac{f_F}{f_k} \quad \text{with } 0 < \lambda < 0.1, \tag{11.72}$$

where F and F_0 are the actual and the limiting mass feed rate. For the unit area we have $UA = UA_0/\lambda$. For Adorján, λ represented a safety factor, and it could be chosen arbitrarily.

Thickener height

The height of the thickener has three terms: c is the depth of clear water zone, h is the depth of the hindered settling zone and z_c is the thickness of the sediment

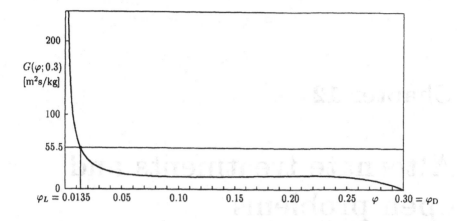

Figure 11.14: The function $G(\varphi, \varphi_D)$ defined in (11.71) for $\varphi_D = 0.3$.

layer or compression zone. The total height H is

$$H = c + h + z_c. \tag{11.73}$$

Arbitrary criteria can be used to establish the size of c and h, for example $c = h = 0.5\,[\text{m}]$. The depth z_c is obtained from (11.62). Then

$$H = c + h + \int_{\varphi_D}^{\varphi_c} \frac{-a(\varphi)}{f_F - q_D\varphi - f_{bk}(\varphi)}\,d\varphi. \tag{11.74}$$

From (11.72) we obtain

$$H = c + h + \int_{\varphi_D}^{\varphi_c} \frac{-a(\varphi)}{\lambda f_k(\varphi) - f_{bk}(\varphi)}\,d\varphi = c + h + \int_{\varphi_a(\varphi)}^{\varphi_c} \frac{D}{f_k(1 - \lambda)}\,d\varphi. \tag{11.75}$$

Chapter 12

Alternate treatments and open problems

12.1 Introduction

The topic on which Kynch's theory is based, quasilinear hyperbolic differential equations, is now mathematically mature, so this is an appropriate time to write a comprehensive analysis of the subject. In presenting what we hope will be the definitive work on Kynch's theory for ideal thickeners, we have incorporated these post-Kynch developments. We have also carefully noted the underlying assumptions of the theory and placed it in its proper context as a special case of the theory of mixtures. It is clear that these assumptions are very restrictive and that many other possibilities can be exploited within this theory. We illustrated one of these possibilities in Chapters 3, 9 and 10.

Kynch's theory postulates that all particles at a given concentration settle with the same velocity. This is never true for rigid spheres and probably for flocs as well. Identical particles have different velocities (Tory *et al.* 1992, Tory and Hesse 1996, Davis 1996) and particles or flocs in real slurries will always differ somewhat in size, shape and/or density. Kynch's theory assumes that velocities change instantaneously when particles enter a region of different concentration and, in particular, when they hit the packed bed. This assumption cannot be strictly true.

In this chapter, we look at alternate treatments of sedimentation, especially those that use less restrictive assumptions. In particular, we look at cases where Kynch's theory is a useful approximation even though his assumptions do not hold. We also examine open problems and try to assess which can be resolved within the theory of mixtures and which may require other treatments.

12.2   Inertial and end effects

We have already shown (in Chapter 3) that unsteady and inertial terms are negligible at low Reynolds numbers. Nevertheless, concerns have been raised about this feature (Hassett 1965, Dixon 1977, Fitch 1993). We know that the finite forces on a particle will require a finite time to change its velocity. We also know that a smooth particle slows down as it approaches a smooth surface such as the bottom of the container or a sphere in the packed bed (Happel and Brenner 1983, Kim and Karrila 1991, Jeffrey 1996). Each sphere in a dilute dispersion of spheres above an infinite flat plate is slowed down as it approaches the plate (Beenakker *et al.* 1984). This is a long-range effect which would apply to arbitrary particles. In at least some cases, it would be augmented by lubrication terms not considered by them. Sedimenting spheres near the bottom must approach the plate and those above must approach each other. As separation decreases, resistance increases rapidly because fluid must be squeezed out from between them. Resistance is inversely proportional to gap width for the very small gaps which characterize concentrated dispersions (Ladd 1993). Part of the downward force on an upper sphere is transmitted through the fluid to the spheres below, pushing them closer together (on the average) than those above. Thus a concentration gradient forms and propagates upward (Tory and Pickard 1982). However, this downward force accelerates the compaction at the bottom and the gradients terminate in those cases where Kynch's theory predicts a shock rising from the bottom. A particle moving from a dilute region through a very steep concentration gradient (approximated in Kynch's theory by a discontinuity) will experience an extra drag as the transient term quickly reduces its velocity to the value appropriate to its new environment. Only during this short time is the transient term important. Contrary to the arguments of Dixon (1977) and others, both the change in concentration and the change in velocity occur over such short intervals that their approximation as a shock is entirely appropriate.

Many experimental studies indicate that the final concentration of a randomly packed bed is roughly 0.64 (Jodrey and Tory 1985). However, it is possible to obtain beds with lower concentrations which can then be collapsed to 0.64 by gentle vibration (Scott and Kilgour 1969). Simulations based on very slow sedimentation into a packed bed (Tory *et al.* 1973) yield densities of about 0.58 and densities from 0.565 to 0.58 are obtained experimentally in the absence of vibration (Verhoeven 1963). Thus, the problem of solids settling slowly into a packed bed is not a simple one. Should the flux curve be represented by one based on a truncated Richardson-Zaki (1954) equation which drops suddenly to zero at 0.64 or should the curve reflect the lubrication forces just mentioned? Perhaps the best solution is to consider the former as the extended Kynch batch flux density function (Chapter 3) and the end effects as a form of compression even though they are purely hydrodynamic. Assuming that we could obtain a uniformly mixed dispersion with $\phi_0 = 0.60$, we could determine $f_{bk}(0.60)$ by using a very tall column and vibrating the packed bed gently (Scott and Kilgour 1969, Shannon *et al.* 1964). The velocity of the spheres at the top should be determined almost entirely by the hydrodynamic resistance. Thus, the resistance there should be essentially the same as the resistance in fluidization.

We shall consider both this and the inertial problem in more detail when we discuss the Markov model.

12.3   Heterogeneity problems

12.3.1   Spatial heterogeneity of homogeneous components

The existence of a distribution of velocities does not necessarily invalidate conclusions based on Kynch's theory. By using averages or weighted averages, we can apply the theory of mixtures in situations in which there is considerable random variation. Average quantities are then assumed to be representative. The averaging volume should be large enough to produce smooth values of the averaged variables and small enough that the size of that volume is negligible compared to the total (Quintard and Whitaker 1993).

Pickard and Tory (1977) proposed a Markov model which explicitly takes into account both the variability and continuity of particle velocities. They and their colleagues, notably Kamel and Hesse, have developed and refined the model over many years. Tory and Pickard (1982) summarize the mathematical approach, Tory et al. (1992) provide a comprehensive review of experimental results and derivations from fluid dynamics while Hesse and Tory (1996) and Tory and Hesse (1996) discuss the current status of the model. All of these papers provide many references to earlier work by them and many others. Rather than interrupt the discussion by a multitude of references, we refer the reader to these papers and simply state results.

Three parameters are required to describe a one-dimensional sedimentation process. These are the mean, variance and autocorrelation of particle velocities. Generalizing Kynch, these researchers assume that these parameters are functions of the solids concentration. However, their solids concentration parameter is expressed as a weighted average which is a simple convolution of a univariate kernel with the (one-dimensional) local solids concentration. Hence, the concentration in a simulation can be updated efficiently by fast Fourier-transform methods. Also, this parameter allows the particles to 'see' regions of higher concentration (including the packed bed) below and reduce the magnitude of their velocities. Initially by a theoretical analysis and later by simulations, they showed that using Kynch's theory with the mean velocity (in place of a uniform velocity) closely approximates the Markov model in slurries with intermediate to high solids concentrations. Despite the variability of particle velocities, the simulated interface was fairly sharp and the rise of the packed bed was clearly apparent. Discontinuities were smoothed to rapid (but continuous) changes in concentration. However, the simulation was fairly small and the parameters were only rough estimates. Thus no more than qualitative agreement can be claimed. Though little information about the parameters is available, numerical computations (Ladd 1993, 1996) and experiments are now yielding some values. Also, new statistical methods have been developed to extract the variance and the autocorrelation from data that do not give them directly. As the data base grows, it will be possible to assess more closely the extent of agreement. In particular, the diffusive effect, in concentration gradients, of the variability of particle velocities should be studied (Gösele and Wambsganss 1983).

Experimental results for very dilute dispersions of closely sized spheres indicate that the magnitude of the mean velocity of spheres in the interior greatly exceeds that of the fall of the interface and even the Stokes velocity. The first result is easily understood because the region near the interface is populated mainly by spheres which are fairly evenly spaced, pairs and triplets having settled faster. Thus, the distribution approaches that of a cell model and it is well known (Batchelor 1972) that such models predict lower magnitudes of velocities than do random models. The second result is surprising because theoretical treatments based on a random distribution predict a monotonic decrease in the magnitude of the mean velocity with increasing φ (Batchelor 1972, Tory and Kamel 1997). This experimental result has been attributed to cluster settling and it is clear that clusters can grow. More recently, a fluid-dynamics analysis showed that regions of a dispersion move rapidly en masse. This might lead to streaming of dense and light regions, leading to these large increases. As a consequence of the difference between the magnitude of the mean velocity and that of the interface, the solids fraction in the upper levels of the dispersion decreases as settling proceeds, establishing a concentration gradient at the top of the dispersion. This gradient spreads out with time and thus becomes less sharp. This is in accordance with the Markov model, but in sharp contrast with previous analyses using Kynch's theory. Using the new results (Bürger and Tory 1999) in Chapter 7 and recognizing that the mean velocity of spheres in the interior is the appropriate velocity for calculating $f_{bk}(\phi_0)$, it might be possible to interpret this result as an MS-6, thereby extending the KSP approximation to dilute dispersions. Previous analyses of sedimentation in dilute dispersions were incapable of finding an MS-6 because they assumed that the rate of fall of the interface gave the velocity of particles in the interior.

Experimental evidence indicates that the mean particle velocity in very dilute dispersions increases with increasing container size. Experimental and fluid-dynamic evidence indicates that the variance does likewise. Using periodic boundaries and an efficient computational method, Ladd (1996) found no increase in the mean velocity with size at intermediate solids concentrations, but extended the result for the variance to slurries with $\varphi = 0.10$. However, Nicolai and Guazzelli (1995) carried out experiments which did not show an increase at intermediate concentrations. The difference may be due to the different boundary conditions, especially the presence of an interface in the experiment, but not in the simulation. More studies are necessary to see if the results for very dilute dispersions apply at higher concentrations. In any case, the results refute the idea that sedimentation is always the same in large and small containers except for small wall effects.

These results predict that differential settling will be substantial in clarification where solids concentrations are often very small. For example, depletion occurs in dispersions of very closely sized spheres when $\phi_0 < 0.15$, but not when $\phi_0 > 0.15$.

For particles other than spheres, the settling velocity may depend on the orientation. For example, the heterogeneity of orientations of an ellipsoid will lead to a distribution of velocities. The dyadics (second-order tensors) which govern the Stokes resistance of various types of particles, such as cylinders, provide the basis for considering their sedimentation as a Markov process. The three parameters of that process for small cylinders have been determined.

It is often claimed (e.g. Fitch, 1993) that flocculated slurries form a loose network which forces all particles to settle with the same velocity, but Gösele and Wambsganss (1983) observed differential settling of flocs and advanced the hypothesis that diffusion from higher to lower concentrations in gradients would reduce the magnitude of velocities there compared to those in regions of uniform concentration. Been and Sills (1981) found that natural flocculation of a soil sample with a broad size distribution greatly reduced the extent of differential settling, but did not eliminate it, except at very high concentrations.

The most significant spatial variability in flocculated slurries is the presence of channels at certain concentrations and under certain conditions. As the nature and extent of channeling depend on the constitutive equations which apply to such slurries, we will consider this phenomenon after those equations have been discussed.

12.3.2   Heterogeneity of solid particles

A complicating factor in many industrial slurries is the heterogeneity of solid particles. The Markov model extends to slurries of multifarious particles and particles with appropriately small differences can be treated as a single class of particles. The experimental results of Shannon et al. (1963, 1964) showed that Kynch's theory holds closely for polydisperse suspensions of spheres even when the Stokes velocities of the spheres vary considerably. Aside from a small tail of fines, their size distribution was approximately normal with a mean diameter of 66.5 [μm] and a standard deviation of 6.7 [μm]. Nevertheless, the rise of the packed bed was linear and the settling curves predicted from the flux curve (based on initial rates) were very close to the actual values at intermediate and high solids concentrations. Also, the interface appeared sharp and the fines were not left behind at the higher concentrations. For $\phi_0 = 0.15$, the interface was fuzzy (but clearly visible) and the concentration gradient predicted for an MS-4 was apparent. Furthermore, its dimensions and position were in complete accord with predictions. The ratio of Stokes velocities of spheres with diameters of 73.2 [μm] ($\mu + \sigma$) and 59.8 [μm] ($\mu - \sigma$) is roughly 1.5, so this is a remarkable result. For $\varphi_0 < 0.10$, no interface was visible and the rise of the packed bed was not linear. For $\varphi_0 = 0.10$, however, the nonlinearity was confined to the very last stage of sedimentation and probably reflects the loss of many of the smallest spheres, including most of the fines which made up only 1.5% of the weight but 14% of the number. This would account for the fact that the interface could not be discerned. The nonlinearity increased markedly as φ_0 was decreased to 0.075 and then to

0.05. Since the value at which depletion occurred in this work was the same as that found by Koglin (1976) for much more closely sized spheres, it is possible that much of the differential settling of the normally distributed spheres (not the fines) was due to the spatial distribution of velocities. Without an analysis of the size distribution of different levels of the packed bed, it is impossible to separate these effects. Been and Sills (1981) settled flocculated dispersions of a natural soil and measured the size distribution at different heights in the compacted sediment. Despite a very wide size distribution, there was essentially no differential settling at very high concentrations. Differential settling did occur at lower concentrations. Ralston (1916) notes: *"With more concentrated suspensions it is possible to reach a consistency in which the distance between the larger particles is so small that the particles of smaller size are trapped and carried down. This suspension of the larger particles might be regarded as a filter for the suspension of the smaller particles"*.

The equations derived in Chapter 1 allow for the possibility of many different species, so slurries with either discrete or continuous distributions can (in principle) be handled, the latter by approximation as the former. It is then necessary to take the interaction of species into account and find the appropriate constitutive equations. A considerable amount of theoretical and experimental work on polydisperse suspensions of spheres (summarized by Davis and Gecol (1994) and Davis (1996)) has been done, but there is no general agreement on the best equations. Concha et al. (1992) have studied such systems as a Kynch sedimentation process, and they have been solved numerically by Bürger et al. (1999b). At present, however, it is not possible to describe analytically the different modes of sedimentation for such systems.

Turney et al. (1995) studied the hindered settling of rod-like particles. The behavior of these suspensions resembled that of spheres in many respects. Depletion of the region near the interface continued throughout, resulting in a nonlinear rise of the packed bed. This behavior is very similar to that observed by Shannon et al. (1963, 1964). The principal difference is that the rods settled initially to $\varphi = 0.155$ and then very slowly to $\varphi = 0.225$. Turney et al. noted that the value of 0.155 suggests that the loosely packed structure is isotropic. It seems probable, therefore, that this is the value of φ_c, and that the intermediate concentrations arise from differential sedimentation. The rods varied somewhat in diameter, length and aspect ratio, so it is impossible to separate differential settling due to these properties from that which also occurs in dilute dispersions of equal spheres.

The heterogeneity of size, shape and density of crushed or ground materials is clearly a very difficult problem. Surprisingly, a good deal is known about the Stokes resistance of an arbitrary particle or a dispersion of such particles (Happel and Brenner 1983). Also, efficient computational techniques are evolving (Kim and Karrila 1991), but are still limited to relatively small numbers. In any case, the physical characteristics of the particles are never known exactly, so the problem reduces to finding constitutive equations based on a few parameters which can be measured fairly easily. Concha and Barrientos (1986) have developed expressions for a modified drag coefficient and a modified settling velocity for single isometric particles. Their Figures 4 and 5 show that particles with very different sphericities are remarkably well approximated by a single curve over a wide range of (modified) Reynolds numbers. Using the concept of an effective hydrodynamic sphericity, Concha and Christiansen (1986) obtained a similar but even better correlation for suspensions of particles of arbitrary shape. It appears that the particle-particle interaction smoothes the random fluctuations.

The concentration dependence of closely sized spheres settling at high Reynolds numbers is weaker than for Stokes flow (Richardson and Zaki 1954). Tory (1965) has shown that this implies that concentration gradients do not form. It appears that suspensions of larger particles do go directly from dilute suspension to compression (Fitch 1993).

12.4 Constitutive equations

Much of the apparent disagreement in the literature reflects a difference in what is meant by settling in compression. For many years, the prevailing view was that settling in compression was a function of time and thus sufficient depth must be provided to allow enough time. Roberts' equation (Roberts 1949), which we discuss later, appeared to support this view. Thus, a slurry was thought to be in compression when it followed Roberts' equation. It is now known that this phase is the last of several (Tory and Shannon 1965, Fitch 1990) and certainly does not represent the point at which the slurry enters compression. A subsequent view was that the point of compression was the concentration at which the settling rate decreased sharply. Flux plots of Mishler's slimes (Mishler 1912) showed that #4 fell abruptly between 15.1 and 15.5 [lb/ft^3] (based on the laboratory test shown as a dashed line in his Figure 3). According to his Table 1, therefore, this drop occurred between $\varphi_0 = 0.092$ and $\varphi_0 = 0.094$. Using the Michaels and Bolger modification of the Richardson-Zaki equation, Scott (1968b) showed that the experimental values of Tory (1961) and Hassett (1965) for calcium carbonate slurries with different properties fell on the same dimensionless flux plot. Hassett's data showed a clear discontinuity in the curve at $k\varphi = 1.15$. Here, k denotes the volume of solids, measured in kg/m^3, see Scott (1968b). Owing to the sudden and uneven collapse of the slurry-supernate interface in his tests at slightly lower values than this, Tory was unable to extend the upper portion to this value. He noted that channeling was extensive at the concentrations plotted on the upper portion and contributed to the high velocity. Of course, the sudden collapse indicates an extreme version of that phenomenon. (We will discuss channeling later.) Thus Tory's observations of the unstable range of concentrations indicate a fairly large but ill-determined velocity. Mallareddy (1963) also found a discontinuity in the flux plot. Indeed his values for slurries of praseodymium oxide show a small region of overlapping values, one high and the other low. This discontinuity might be taken as the point of compression (Scott 1968b). Since $k\varphi = 1$ indicates that flocs, of solids fraction φ, have deformed to occupy the entire volume, the slightly higher value of 1.15 indicates some loss of fluid from the original flocs according to the view of

Michaels and Bolger (1962). This may not be a true picture, but Scott's procedure based on this concept appears to unify the data, including the few values available from Comings *et al.* (1954). Furthermore, the value of the concentration which suddenly appeared in Comings' continuous thickener is quantitatively predicted by this flux plot (Scott 1968b).

The modern view, which we have adopted in this book, is to separate sedimentation into two regimes: one in which only hydrodynamic interactions occur and one in which the solids interact directly. Considerable evidence indicates that the resistance to compression is initially very slight. Thus, Tory and Shannon (1965) found that the rate of fall of the interface was (within experimental error) a unique function of the solids concentration just below. Talmage (1959) found that fluid pressure at any depth was approximately what would be predicted if all the solids were hydrodynamically supported. Data were erratic, but there was little difference which could be ascribed to squeeze. Gösele and Wambsganss (1983) also concluded that their sludge was supported by hydraulic forces rather than by a solids structure. Nevertheless, the distinction remains. Buscall and White (1987) explain that *"...most of the bed is in essentially a 'free fall' state. As such, particles at any level, although connected to all the others via the network, experience a greatly reduced effect of that part of the network above that level. It is therefore only in the vicinity of the container bottom where this 'free fall' state is not permitted by virtue of the zero-flux boundary condition that the compression or consolidation zone begins"*. Thus, the concept (defined in Chapter 3) of the extended Kynch batch flux density function, $f_{bk}(\phi)$, is very useful. Tory (1961) plotted interface velocity versus interface height for several thick slurries and found two strikingly linear portions (Tory and Shannon 1965). The lower is the well known Roberts' exponential decay of height (Roberts 1949). Fitch (1990) similarly plotted all of Tory's runs which were carried to completion. Aside from some small random and possibly systematic variation, Tory plots of interface velocity versus reduced height (height divided by the weight of solids per unit area) fell on the same straight line. Furthermore, the greater the weight of solids, the more this line extended into lower values of reduced height, i.e., higher values of φ. This suggests that this line represents the hydrodynamic resistance and Fitch identifies it as $f_{bk}(\varphi)$. Using Tory's data, Scott (1968b) plotted the weight of overlying solids versus the maximum concentration at the base of the settling vessel and found a smooth curve very similar to Figure 12.1. Using $k\phi_0 = 1.15$ as the value where the discontinuity occurs, he also plotted Comings' data. Though the curves are different, they have the same shape and both appear to originate at roughly 1.15. However, no data points are near this value, so this interpretation is uncertain. As we shall see, the value of φ_c in such slurries is almost certainly less than the value at which the discontinuity appears.

We believe that the proper setting for thickening of flocculent dispersions is the mathematical theory of degenerate parabolic equations (Bénilan and Touré 1995, Bénilan and Gariepy 1995, Bürger 1996, Bürger and Wendland 1998a,b). In this setting, these are second-order parabolic differential equations which degenerate into first-order hyperbolic equations for $\varphi < \varphi_c$. In Chapters 9 and 10, we developed this approach on the basis of particles with elastic properties. Most flocs, however, appear to be fragile and, once collapsed, do not spring back. Free (1916) states a remarkably modern view: *"In concentrated slimes all the particles may be thought of as in contact. Consolidation is due to their acceptance of a closer arrangement. This closer arrangement is induced by the gravity, including not only the weight of the individual particles that are slipping closer together, but also the weight of the superposed particles resting on them. Accordingly the compressive force and hence the consolidation will be greater the length of the column of slimes"*. The collapse of flocculated structures in settling in compression is closely related to that in filtration and many of the ideas developed for the latter are useful for the former.

According to Landman *et al.* (1995), once the average volume fraction of flocs is high enough, a network of connected particles will form throughout the suspension which will assume the properties of a solid structure. Provided that the solid effective stress, $\sigma_e(\varphi, \varphi_c)$ is sufficiently small, the network will remain solid-like and resist compression (their p_s is our σ_e). Ultimately, as the applied compressive force is increased, a point will be reached where the network is no longer able to resist elastically and will yield and irreversibly consolidate. The network is assumed to have a compressive yield stress, $P_y(\varphi)$, which increases with φ since the more particles per unit volume, the more interconnections between particles and, consequently, the stronger the network structure. $P_y(\varphi)$ vanishes for $\varphi < \varphi_c$. Experimental data are fitted by power-law expressions such as

$$P_y(\varphi) = k\left[\left(\frac{\varphi}{\varphi_c}\right)^n - 1\right] \text{ for } \varphi > \varphi_c \qquad (12.1)$$

$P_y(\varphi)$ is the stress at the elastic limit, k is a constant and n is an integer which is greater than 2. For example, Green *et al.* (1996) use $k = 100$ [Pa] and $n = 8$; from their Figure 7 we infer that $\varphi_c \approx 0.1805$ was used. Figure 12.1 shows the curve for these values. Thus P_y is continuous at φ_c. The simplest constitutive equation of this type is

$$\frac{D\varphi}{Dt} = \begin{cases} 0 & \text{if } \sigma_e \leq P_y, \\ \kappa(\varphi)\,(\sigma_e - P_y) & \text{if } \sigma_e > P_y, \end{cases} \qquad (12.2)$$

where $\kappa(\varphi)$ is the dynamic compressibility (Buscall and White 1987). However if hydrodynamic drainage of water from between the network structure is the rate-determining step, we can assume that collapse rapidly readjusts the network local volume fraction until $P_y(\varphi)$ at that volume fraction exactly matches σ_e. Then

$$\sigma_e(t) = P_y(\varphi(t)) \qquad (12.3)$$

at any point and time in the suspension (Landman *et al.* 1995). The viewpoint of these authors differs somewhat from that expressed in Chapter 3 in that they consider a network or gel structure rather than flocs resting on one another. In their model, it is the network that changes rather than the flocs which deform.

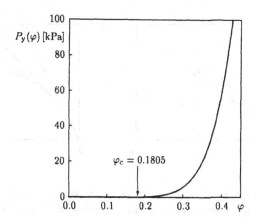

Figure 12.1: Power-law yield stress curve, after Green *et al.* (1996)

This need not affect the mathematical formulation. The dependence of σ_e on φ expressed in these equations explains a good deal of data, in particular that, in tall columns, the easy collapse of the structure below implies that the resistance to settling of the solids near the top is almost entirely hydrodynamic (Fitch 1966). It is also consistent with the prescient views of Free (1916) (some of which were later quoted and advocated by Tory and Shannon (1965)): *"With dilute slimes just beginning to settle, the amount of possible consolidation due to the breakdown of floccules is large. All of the floccules are nearly of their original size and looseness of structure. All will consolidate a little on touching or slight pressure. The total consolidation thus caused is relatively large in amount, easy and rapid. Accordingly the rate at which the top floccules in a dilute suspension can fall is not limited by the rate of consolidation below. ... When the slimes become more concentrated or if they start at a higher concentration, the rate at which total consolidation is permitted by breakdown of floccules is less. The floccules, being already partly consolidated, suffer further breakdown only slowly and with appreciable resistance. The total rate of consolidation becomes, therefore, less than the proper rate of fall of the upper floccules through the medium. These find themselves supported importantly from below and add their weight to the force urging further consolidation below. The rate of consolidation ceases to be constant, and the depth of the slime column begins to affect it"*.

It is well known that this consolidation produces a decreasing rate of rise of the sediment (though we show below that this result must be interpreted cautiously). Thus, the question arises as to what solids concentrations lie between the compression zone and the initial characteristic with the least slope. Fitch (1983) has argued that new characteristics arise tangentially from the top of this zone, but no rigorous proof has yet been given (Tiller *et al.* 1990). Some have not found Fitch's arguments convincing and the question remains controversial. We give a

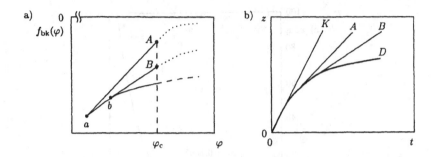

Figure 12.2: Generation of characteristics tangential to the slurry-sediment discontinuity. The long dashes represent the extension of $f_{bk}(\varphi)$ beyond φ_c.

short proof which is mathematically rigorous and physically transparent. Consider a flocculent suspension settling as an MS-4 or MS-5 from an initial height of L into a compressible sediment now at $h(t)$. The boundary conditions are

$$\varphi(L, t) = 0, \quad t > 0 \tag{12.4}$$

and

$$v_s(0, t) = 0. \tag{12.5}$$

A discontinuity develops between the concentration φ^+ and the concentration at $h(t)$ given by the Rankine-Hugoniot condition

$$h'(t) = \frac{\varphi_c v_s^-(h, t) - \varphi^+(h, t)v_s^+(h, t)}{\varphi_c - \varphi^+}, \tag{12.6}$$

which is the usual expression for $\sigma(\varphi^+, \varphi_c)$ representing the velocity of the discontinuity for solids settling from a Kynch regime into a sediment or packed bed. Figure 12.2 (a) shows a flux plot for values near φ_c. Here we have used a curve truncated at φ_c. This might be based on a Michaels and Bolger (1962) adaptation of the Richardson-Zaki equation or the Maude and Whitmore (1958) generalization of it:

$$v_s = u_\infty(1 - \varphi)^n, \quad \varphi < \varphi_c. \tag{12.7}$$

In choosing such a form, we are recognizing that fluid can flow fairly easily through the interstices as long as the solids below are not obstructed. As we show below, the rise of characteristics tangential to the discontinuity holds very broadly. However, to facilitate the discussion, we assume that the effective solids pressure is given by equations (12.3) and (12.1) because this behavior is supported by experimental evidence and explains many other features of sedimentation. According to these equations, the flocs on top crush those below and this allows those on top to move slowly downward. Thus, $v(h, t)$ is negative. Since $\varphi_c v_s^-$ is also negative, this implies that the sediment rises more slowly than it would if the flocs were incompressible. Furthermore every addition of flocs to the sediment pushes $\varphi(0, t)$ to still higher values. Thus the flocs at the top of the sediment move ever more rapidly downward. The rate of rise of the sediment at $t = t_1$, $h'(t_1)$, is given by the slope of the chord aA in Figure 12.2 (a) and a rarefaction wave (see Figures 7.7 and 7.8 of Chapter 7) has correspondingly expanded as shown in Figure 12.2 (b). According to equation (12.1), the flocs compress very easily when φ is only slightly greater than φ_c. Very little weight of solids is required for the initial compression, so the concentration at the bottom rises very quickly and attains a considerably higher value by the time the sediment has reached a very small height. When the velocity of the flocs at the top has increased to $v_s^-(h_2, t_2) > v_s^-(h_1, t_1)$, the velocity of the discontinuity is now given by the slope of the chord bB and the rarefaction wave has become correspondingly broader as shown in Figure 12.2 (b). The lines shown there are truly characteristics because they are all in the range of concentrations in which the equation is hyperbolic. It follows from equation (12.1) that the concentration at the bottom of the sediment will continue to increase, but slightly less quickly. Every increase in φ at the bottom increases the velocity of the flocs at the top and it is clear from equation (12.6) that this decreases the rate of rise of the sediment and broadens the rarefaction wave. This proves Fitch's contention. Note that all that is required is that additional weight of solids produce additional compression at the bottom. This is satisfied by many functions, including the elastic response discussed in Chapter 3.

Figure 3 of Been and Sills (1981) shows the curved boundaries, expected from the tangential generation of characteristics from the sediment, of a large density change. The curvature will be greater for very compressible slurries. For a slurry of a given ϕ_0, a larger value of L will continue the curved boundary with a reduced slope. Equation (12.6) and Figure 12.2 show that, for an infinitely long column,

$$\lim_{t \to \infty} \varphi^+(h, t) = \varphi_c \tag{12.8}$$

and

$$\lim_{t \to \infty} h'(t) = \lim_{\varphi \to \varphi_c^-} f'_{bk}(\varphi), \tag{12.9}$$

where it is clear from equation (12.7) that the one-sided derivative exists. Equation (12.8) implies that the magnitude of the discontinuity will become smaller and smaller until the discontinuity eventually becomes too small to be detected. This means that what is measured in the later stages of sedimentation is not the original discontinuity, but a steep (continuous) gradient in a region of the slurry which is entirely in compression. It follows that many of the measurements of "discontinuities" in the literature must be interpreted with caution. While it remains, the path of the discontinuity becomes asymptotic to a straight line with the same slope as the flux curve at φ_c. Thus, equation (12.9) provides a simple test: any iso-concentration line with a slope less than $f'_{bk}(\varphi_c)$ is not a characteristic.

Fitch (1990) has discussed the effect on settling curves of this tangential generation of characteristics. Noting the curved path of the discontinuity, he states
that "...*the more the compression zone builds up, the more Kynch characteristics
arise. When there are more solids per unit area present in a test, the compression
zone ultimately builds up further. Therefore free-settling characteristics of higher
concentration are formed and propagated to the surface. In reduced coordinates*
[values divided by the weight of solids per unit area], *the free-settling section of
the settling plot should then be longer, and the compression section should run
below that of the tests with less solids. The same should be true for reduced Tory
plots*". Figure 7 of Tory and Shannon (1965) shows the first effect for three tests
with the same ϕ_0, but different values of L. Fitch's Figure 6 shows both of these
effects for two tests with the same ϕ_0, but different values of L (Tory 1961). Both
the increasing concentration at the bottom and the decreasing rate of rise of the
sediment have been well documented by experiments. According to a strict interpretation of Kynch's theory all concentrations less than φ_c are formed immediately.
However, the particles near the bottom slow down before they reach it, so even
the first characteristics do not really emanate from $z = 0$. However, the first new
characteristics leave the discontinuity when it is still near the origin. Thus, the
initial phase is close to the behavior predicted by Kynch's theory and precisely
in accordance with his hope that the initial increase would be very rapid. Despite the increase in concentration near the bottom, most of the weight of solids is
supported hydrodynamically. Moreover, the decreased porosity causes the rate of
subsidence to decrease and this slows down the rate of compaction of the lowest
levels. Since a large force is required to reach the higher values of φ, these do not
appear at the bottom until much later. Therefore, aside from the extreme curvature near the origin, the discontinuity is only slightly curved. Lines of constant
concentration below the discontinuity are not characteristics since they represent
values of $\varphi > \varphi_c$. In general, these will be curved and must be separated by a
height representing the additional weight of unbuoyed solids to produce the higher
concentration. In comparing theoretical and experimental values, we may find
differences in the behavior in the first few seconds. As noted above, end effects
occur at all concentrations, including the very dilute. Thus, it will take a little
time before Kynch's theory could be a reasonable approximation. Nevertheless,
the theory applies to the propagation of characteristics; it is only the assumption of an instantaneous increase in concentration which does not hold. If some
characteristics are propagated rapidly to the surface, they might be detectable as
a decrease in the velocity there. In some cases, these effects may be masked or
overwhelmed by induction or depletion effects.

Many researchers have experimentally determined lines of constant concentration in the compression region. Figure 6 of Tiller *et al.* (1991) is slightly distorted
by what appears to be an induction period of about 400 seconds. After that, the
lines are approximately straight and parallel. Although their line OB divides the
regions, it may not be a Kynch characteristic. Kynch's theory assumes a sudden
increase in φ at the bottom and an interface velocity which is initially constant.
The concentration increase near the bottom in their experiments suggests that the

initial concentration is close to φ_c and may already exceed it. If the straight line representing the slurry-supernate interface is extrapolated to the initial height and the value of the time there is taken as the new origin, the results agree closely with those expected. Their Figure 4 shows some depletion of the upper region which, as discussed earlier, accounts for at least part of the strong curvature at the top of the lines. Their profiles are very similar to those of Been and Sills (1981), but the latter show a more rapid increase at the bottom and more depletion of the upper levels. Profiles constructed by Tory (1961) and Gaudin and Fuerstenau (1962) also show a fairly rapid increase at the bottom. Figure 4 (20 minutes elapsed time) of the latter shows more depletion near the top, but is otherwise very similar to Figure 4 of Tory and Shannon (1965). Both pairs of authors believed that their slurries were initially in compression. In all four papers, the concentration increased gradually at the bottom and the highest concentrations were attained only after a very long time. Scott (1968a) found that $\varphi = 0.20$ formed quickly and propagated in a straight line for almost 4000 minutes. It then curved sharply and remained about 7 cm below the surface (his Figure 6). Thus it is not a characteristic. It took more than 3000 minutes for $\varphi = 0.30$ to appear at the bottom and after 28 days, it had reached a height of only 3 or 4 cm. Even after 6 months, the solids profile was strongly curved. Gaudin and Fuerstenau (1962) also found that the highest concentrations took a long time to appear and were propagated very slowly. These papers and others show a pattern which is qualitatively consistent with the constitutive equations above. The differences that exist appear to arise from the somewhat different properties of the materials used.

As discussed in Chapters 9 and 10, a suitable method of handling the sedimentation of compressible slurries has been initiated, but additional development is required.

12.5 Channeling and collapse

Observations of channeling date back to the early part of this century. Coe and Clevenger (1916) stated that *"Free-settling pulp can be recognized by the evenly flocculated appearance of the pulp surface without channels coming through and by the uniform texture of the pulp as seen in a glass cylinder. The water liberated by compression finds its way out of zone D* [the compression zone] *through tubes or channels which form drainage systems upward through the zone. The trunk channel for any system has its outlet at the top of zone D"*. Their Figure 1 shows these channels schematically. Fitch (1972) states that *"Particularly with metallurgical pulps* [channeling] *visibly takes the form described by Coe and Clevenger, great drainage systems flow upward through the pulp. At their mouths these rivers of fluid may be of the order of a millimeter across, and they erupt through the surface as "volcanos". After eruption there is a steady and rapid streaming of fluid upward through them, evidenced occasionally by flutter of semi-detached floccules fluidized in channels which form next to the glass column wall"*. Mishler (1912) gives this description: *"The settling of excessively thick slime, on the other hand, appeared*

to be caused by the contraction of the spaces between the particles, due to cohesion.
A system of cracks was formed, which ultimately developed into channels, through
which the clear liquid reached the surface. The liquid coming from the lower part
of a deep vessel had a tendency to open up the channels in the upper part of the
vessel, thus facilitating settling near the surface. This was especially noticeable at
the critical dilution and explains why, at this point, an increase in depth caused a
large increase in the settling rate. Occasionally at the critical dilution the settling
rate in the deeper columns was found to be greater than it was for the same slime
when slightly more dilute". In a later paper, Mishler (1918) says: *"The critical*
dilution, or the boundary between dilute and thick pulp [D_c] *is the highest dilution*
[lowest concentration] *at which channels form in the pulp; clear liquid reaching the*
surface through the channels. Only a positive tensile force, like cohesion, could
form channels and keep them open afterward. It is in the [final] *stage of thickening*
that the channels collapse". In choosing the beginning rather than the end of
channeling as the critical dilution, Mishler and Coe and Clevenger were expressing
a very modern view.

Tory (1961) conjectured that the shape of the flux plot for calcium carbonate
(Figure 8 of Shannon and Tory (1965)) indicated that slurries of intermediate concentrations with no concentration gradient were fundamentally unstable (Tory and
Shannon 1965). These are regions which would produce an MS-1 in the absence of
direct particle-particle interaction (Tory 1961, Shannon and Tory 1965) (MS-1 is
also predicted for very dilute slurries and, as we have seen, their velocities are also
enhanced. Eventually these slurries are stabilized by differential sedimentation.).
Tory's settling curves show evidence of instability even before the substantial increase in velocity and, as described above, were completely unstable for several
concentrations in the range $0.10 < \varphi < 0.14$. Noting that the absolute value of the
slope of their settling curves (their Figure 8) increased with time, Michaels and
Bolger (1962) surmised that the initial flow paths were tortuous, but tended to
straighten out with time. This increasing speed has been noted many times and
raises the question of which rate to use. They used the maximum rate while Tory
(1961) used the initial rate on the grounds that it is least affected by channeling.
This was confirmed by Scott (1968b) who showed that slow stirring eliminated
any increase in channeling and yielded a constant rate which agreed fairly well
with the unstirred rate in the first few minutes. Scott notes that channel formation takes time and therefore would be expected to be favored in tall channels.
Though Tory reported initial velocities as similar in slurries with different initial
heights, his settling curves show larger maximum rates (which occurred later) for
the greater heights. Slurries of praseodymium oxide also showed a substantial increase in velocity as settling progressed (Mallareddy 1963). Several observers have
commented on the openings at the slurry-water interface. Harris *et al.* (1975) note
that *"These outlets have the appearance of miniature volcanoes both in shape and*
in the behaviour of fine particles entrained in the exiting liquid". All observers
agree with this description and the terminology is a common one. Glasrud *et al.*
(1993) show photographs of channels and volcanoes.

Reasoning that the return flow was almost exclusively through these channels,

Scott (1968a) showed that the Carman-Kozeny equation implied that a plot of $\sigma_I(\varphi)/\varphi$ against φ should yield a straight line. Using data from Coe and Clevenger, he showed that, despite some scatter, the points for $0.04 < \varphi < 0.08$ do appear to lie on such a line. Further support comes from the observation (Coe and Clevenger 1916) that the pulp was no longer free settling above the latter value. Scott then calculated that the mean diameter was about ten times the average distance between primary particles at $\varphi = 0.056$.

As noted above, stirring prevents the normal growth of these channels. In a later paper, Scott (1970) shows this dramatically. A flux plot based on the velocities of the interface of slurries settled without stirring had a 'second hump' so large that it contained a local maximum. At 1 rpm, the maximum had disappeared but the hump was still large. At 2 rpm, it had almost disappeared. At 3 rpm, the hump had completely disappeared and, at 4.5 rpm, the flux values were dramatically less than even those at 3 rpm. It is known (Harris *et al.* 1975, Glasrud *et al.* 1993) that air bubbles or large particles tend to create vertical tears which can become channels. The initial sizes, shapes and positions of the cracks appear to be random. It might be possible, therefore, to model the initial cracks as fractals and channeling as a stochastic process with a drift toward larger-diameter channels.

12.6   Roberts' equation

Roberts (1949) proposed the equation

$$\frac{d\mathcal{D}}{dt} = k_1(\mathcal{D} - \mathcal{D}_\infty) \tag{12.10}$$

where $\mathcal{D}$ is the dilution, see equation (11.3) in Chapter 11, and let $\mathcal{D}_c$ and $\mathcal{D}_\infty$ denote the dilutions corresponding to $\varphi = \varphi_c$ and $\varphi = \varphi_\infty$, respectively. Equation (12.10) integrates to

$$\mathcal{D} - \mathcal{D}_\infty = (\mathcal{D}_c - \mathcal{D}_\infty)\exp(-k_1 t). \tag{12.11}$$

Since $\mathcal{D} - \mathcal{D}_\infty$ is proportional to $z - z_\infty$ where z_∞ denotes the final sediment height, these equations may be expressed in terms of slurry height by merely substituting z for $\mathcal{D}$ (Tory and Shannon 1965). In this form, Roberts had been anticipated by Deerr (1920). Both developed it empirically. Behn and Liebman (1963) claim to have verified Roberts' equation by applying soil consolidation theory (Behn 1957), but we find his explanation unconvincing. Fitch, who worked with Roberts at Dorr-Oliver and has often used his equation, continues to refer to it as empirical (Fitch 1993). As an empirical equation, it is outstanding in the universality of its applicability. Of Tory's 42 settling tests, 36 were continued until completion. Initial concentrations ranged from $\varphi = 0.0037$ to 0.0708, but in all cases, a Roberts' plot produced a straight line with very little scatter (Tory and Shannon 1965). Figure 1 of that paper is cited as a typical example. Fitch (1972) remarks that he has never known it to fail. The plotting procedure involves an initial estimate of z_∞ and a revised estimate if the plot curves up or down at the end. Though this is a 'fudge factor', it seriously affects only the last point or two. Furthermore, Tory plots, which correspond to the differential form, also show this straight-line behavior (Fitch 1993) and, within experimental error, extrapolate to the same values of z_∞ and yield the same values of k_1 (Tory and Shannon 1965).

Roberts' equation, which confirms Coe and Clevenger's qualified statement that settling in compression was a function of time, led to many claims which were untrue. Tory and Shannon refuted these by citing their experimental results and those of others. Yoshioka et al. (1955) showed that k_1 is not a true constant (for a given type of slurry), but

$$k_2 = k_1 \varphi_c L \qquad (12.12)$$

is. Tory and Shannon confirmed this result for slurries with a wide variety of initial concentrations and weights of solids. There was some scatter, but a slightly higher value for the lowest weight of solids appeared to be the only systematic variation. Excluding these values, $k = 1.197 \pm 0.102$ (standard deviation). Furthermore, the final concentrations calculated from Roberts' equation were independent of ϕ_0 and depended only on the weight of solids.

We have seen that the highest concentrations form very slowly at the bottom and also propagate very slowly. Roberts' equation does not predict the final concentration profile (only the average concentration), but Tory and Shannon used it to estimate the effects of the delayed emergence of these higher concentrations and the nonlinearity of their rise. They calculated what the profiles would be if Kynch's theory could be applied to these concentrations and compared these profiles to those obtained experimentally. The profiles were surprisingly close, but there were significant differences as shown in their Figures 4, 6 and 11. Using the methods of Chapter 10, it should be possible to construct solids profiles (such as those in Figure 10.3) and check Roberts' equation. The simplicity of Roberts' equation suggests that there may be an easier method. If so, it is not known to us at the moment.

Bibliography

Adorján, L.A., 'A theory of sediment compression', XI International Mineral Processing Congress, Cagliari, Italy, Paper 11:1–22 (1975).

Adorján, L.A., 'Determination of thickener dimensions from sediment compression and permeability test results', *Trans. Inst. Min. Met.* **85**, C 157–163 (1977).

Agricola, G., *De re metallica*, translated from the first Latin edition of 1556 by H.C. Hoover, Dover Publications, Inc., New York 1950.

Ahmadi, G., 'A continuum theory for two phase media', *Acta Mech.* **44**, 299–317 (1982).

Ahmadi, G., 'On the mechanics of incompressible multiphase suspensions', *Adv. Water Res.* **10**, 32–43 (1987).

Ardaillon, E., *Les Mines du Laurion dans l'Antiquité*, Paris 1897.

Arnold, J.H. Jr. and Chopey, N.P., 'New ideas refresh alumina process', *Chem. Engrg.* **67** (24), 108–111, Nov. 28, 1960.

Ashley, H.H., 'Theory of the settlement of slimes', *Min. Scient. Press*, 831–832, June 12, 1909.

Auzerais, F.M., Jackson, R. and Russel, W.B., 'The resolution of shocks and effect of compressible sediments in transient settling', *J. Fluid Mech.* **195**, 437–462 (1988).

Ballou, D., 'Solution to nonlinear hyperbolic Cauchy problems without convexity conditions', *Trans. Amer. Math. Soc.* **152**, 441–460 (1970).

Bardos, C., Le Roux, A.Y. and Nedelec, J.C., 'First order quasilinear equations with boundary conditions', *Comm. Partial Diff. Eqns.* **4**, 1017–1034 (1979).

Barton, N.G., Li, C.H. and Spencer, S.J., 'Control of a surface of discontinuity in continuous thickeners', *J. Austral. Math. Soc. Ser. B* **33**, 269–280 (1992).

Bascur, O., *Modelo fenomenológico de suspensiones en sedimentación*, B.Sc. thesis, University of Concepción 1976.

Batchelor, G.K., 'Sedimentation in a dilute dispersion of spheres', *J. Fluid Mech.* **52**, 245–268 (1972).

Becker, R., *Espesamiento continuo: diseño y simulación de espesadores*, Engineering thesis, University of Concepción 1982.

Bedford, A. and Drumheller, D.S., 'Theories of immiscible and structured mixtures', *Int. J. Engrg. Sci.* **21**, 821–960 (1983).

Bedford, A. and Hill, C.D., 'A mixture theory formulation for particulate sedimentation', *AIChE J.* **22**, No. 5, 938–940 (1976).

Been, K. and Sills, G.C., 'Self-weight consolidation of soft soils: an experimental and theoretical study', *Géotechnique* **31**, 519–535 (1981).

Beenakker, C.W.J., van Saarloos, W. and Mazur, P., 'Many-sphere hydrodynamic interactions III: Influence of a plane wall', *Physica A* **127**, 451–472 (1984).

Behn, V.C., 'Settling behaviour of waste suspensions', *Proc. Amer. Soc. Civil Engrs.* **83**, SA5, 1423-1/20 (1957).

Behn, V.C. and Liebman, J.C., 'Analysis of thickener operation', *J. San. Eng. Div. Proc. Amer. Soc. Civil Engrs.* **89**, 1–15 (1963).

Bénilan, P. and Gariepy, R., 'Strong solutions in L^1 of degenerate parabolic equations', *J. Diff. Eqns.* **119**, 473–502 (1995).

Bénilan, P. and Touré, H., 'Sur l'équation générale $u_t = a(\cdot, u, \varphi(\cdot, u_x)_x + v$ dans L^1: II. Le problème d'évolution', *Ann. Inst. Henri Poincaré, Analyse non linéaire* **12**, 727–761

253

(1995).

Bowen, R.M., 'Theory of mixtures', in *Continuum Physics*, vol. **3**, Ed. A.C. Eringen, Academic Press, New York 1976, pp. 1–127.

Bürger, R., *Ein Anfangs-Randwertproblem einer quasilinearen entarteten parabolischen Gleichung in der Theorie der Sedimentation mit Kompression.* Doctoral thesis, University of Stuttgart 1996.

Bürger, R., 'A mathematical model for sedimentation with compression', in *Analysis, Numerics and Applications of Differential and Integral Equations*, Eds. M. Bach, C. Constanda, G.C. Hsiao, A.M. Sändig and P. Werner, Pitman Research Notes in Mathematics No. 379, Addison Wesley Longman Ltd., Harlow, UK, 41–46 (1998).

Bürger, R., Bustos, M.C. and Concha, F., 'Settling velocities of particulate systems: 9. Phenomenological theory of sedimentation processes: Numerical simulation of the transient behaviour of flocculated suspensions in an ideal batch or continuous thickener', *Int. J. Mineral Process.* **55**, 267–282 (1999a).

Bürger, R. and Concha, F., 'Simulation of the transient behaviour of flocculated suspensions in a continuous thickener', in *Proceedings of the 20th International Mineral Processing Congress (XX IMPC), Aachen, Germany, September 21–26, 1997*, Eds. H. Hoberg and H. v. Blottnitz, GDMB, Clausthal-Zellerfeld, vol. **4**, 91–101 (1997a).

Bürger, R. and Concha, F., 'Numerical simulation of continuous sedimentation of flocculated suspensions', in *Proceedings of the Third Summer Conference "Numerical Modelling in Continuum Mechanics", Prague, Czech Republic, September 8–11, 1997*, Eds. M. Feistauer, R. Rannacher and K. Kozel, Matfyzpress, Charles University, Prague, part 1, 180–186 (1997b).

Bürger, R. and Concha, F., 'Mathematical model and numerical simulation of the settling of flocculated suspensions', *Int. J. Multiphase Flow* **24**, 1005–1023 (1998).

Bürger, R., Concha, F., Fjelde, K.-K. and Karlsen, K.H., 'Numerical simulation of the settling of polydisperse suspensions of spheres', Preprint 99/02, Sonderforschungsbereich 404, University of Stuttgart; submitted to *Powder Technol.* (1999b).

Bürger, R., Concha, F. and Tiller, F.M., 'Applications of the phenomenological theory to several published experimental cases of sedimentation processes', presented at the *UEF Conference on Solid-Liquid Separation Systems*, Kahuku, Oahu, HI, USA, April 18–23, 1999; in review for publication in a special issue of *Separ. Purif. Technol.* (1999c).

Bürger, R., Evje, S. and Karlsen, K.H., 'On some degenerate convection-diffusion initial-boundary value problems with applications to sedimentation-consolidation processes and two-phase flow in porous media', in preparation (1999d).

Bürger, R., Evje, S. and Karlsen, K.H., 'Viscous splitting methods for degenerate convection-diffusion equations with boundary conditions', in preparation (1999e).

Bürger, R., Evje, S., Karlsen, K.H. and Lie, K.-A., 'Numerical methods for the simulation of the settling of flocculated suspensions', Preprint 99/03, Sonderforschungsbereich 404, University of Stuttgart. Presented at the *UEF Conference on Solid-Liquid Separation Systems*, Kahuku, Oahu, HI, USA, April 18–23, 1999; in review for publication in a special issue of *Separ. Purif. Technol.* (1999f).

Bürger, R. and Tory, E.M., 'On upper rarefaction waves in batch settling', in preparation (1999).

Bürger, R. and Wendland, W.L., 'Mathematical problems in sedimentation', in *Proceedings of the Third Summer Conference "Numerical Modelling in Continuum Mechanics", Prague, Czech Republic, September 8–11, 1997*, Eds. M. Feistauer, R. Rannacher and K. Kozel, Matfyzpress, Charles University, Prague, part 1, 18–31 (1997).

Bürger, R. and Wendland, W.L., 'Existence, uniqueness, and stability of generalized solutions of an initial-boundary value problem for a degenerating quasilinear parabolic equation', *J. Math. Anal. Appl.* **218**, 207–239 (1998a).

Bürger, R. and Wendland, W.L., 'Entropy boundary and jump conditions in the theory of sedimentation with compression', *Math. Meth. Appl. Sci.* **21**, 865–882 (1998b).

Bürger, R. and Wendland, W.L., 'Mathematical model and viscous splitting approximation for sedimentation-consolidation processes', to appear in the proceedings of the *4th International Conference on Numerical Methods and Applications NM &A-O(h^4)*, Sofia,

Bulgaria, August 19–23, 1998 (1998c).

Bürger, R., Wendland, W.L. and Concha, F., 'Model equations for gravitational sedimenta-tion-consolidation processes', submitted to *Z. Ang. Math. Mech.* (1998c).

Buscall, R. and White, L.R., 'The consolidation of concentrated suspensions', *J. Chem. Soc. Faraday Trans. I*, **83**, 873–891 (1987).

Bustos, M.C., *On the existence and determination of discontinuous solutions to hyperbolic conservation laws in the theory of sedimentation*, Doctoral thesis, TH Darmstadt 1984.

Bustos, M.C. and Concha, F.,'On the construction of global weak solutions in the Kynch theory of sedimentation', *Math. Meth. Appl. Sci.* **10**, 245–264 (1988a).

Bustos, M.C. and Concha F.,'Simulation of batch sedimentation with compression', *AIChE J.* **34**, 859–861 (1988b).

Bustos, M.C. and Concha, F.,'Boundary conditions for the continuous sedimentation of ideal suspensions', *AIChE J.* **7**, 1135–1138 (1992).

Bustos, M.C. and Concha, F., 'Kynch theory of sedimentation', in *Sedimentation of Small Particles in a Viscous Fluid*, Ed. E.M. Tory, Computational Mechanics Publications, Southampton 1996, pp. 7–49.

Bustos, M.C., Concha, F. and Wendland, W.L.,'Global weak solutions to the problem of continuous sedimentation of an ideal suspension', *Math. Meth. Appl. Sci.* **13**, 1–22 (1990).

Bustos, M.C. and Paiva, F.,'Existence of the entropy solution for the simulation of the sedimen-tation in a continuous thickener', *Revista Notas*, Sociedad Matemática de Chile, **12/13** (1), 47–56 (1994).

Bustos, M.C., Paiva, F and Wendland, W.L.,'Control of continuous sedimentation of ideal suspensions as an initial and boundary value problem', *Math. Meth. Appl. Sci.* **12**, 533–548 (1990).

Bustos, M.C., Paiva, F. and Wendland, W.L., 'Entropy boundary conditions in the theory of sedimentation of ideal suspensions', *Math. Meth. Appl. Sci.* **19**, 679–697 (1996).

Camp, T.R., 'Sedimentation and the design of settling tanks', *Proc. ASCE Pap.* **71**, 445–486 (1945).

Camp, T.R., *Trans. ASCE*, **111**, 895–936 (1946).

Camp, T.R.,'Studies of sedimentation basin design', *Sew. Indust. Wastes* **25**, 1–12 (1953).

Cheng, K.S.,'Asymptotic behavior of solutions of a conservation law without convexity condi-tions', *J. Diff. Eqns.* **40**, 343–376 (1981).

Cheng, K.S.,'A regularity theorem for a nonconvex scalar conservation law', *J. Diff. Eqns.* **61**, 79–127 (1986).

Clevenger, G.H., 'The hydrometallurgical treatment of complex gold and silver ores', *Met. Chem. Engrg.* **14**, 203–210 (1916).

Coe, K.S. and Clevenger, G.H., 'Methods of determining the capacity of slime settling tanks', *Trans. AIME*, **55**, 356–385 (1916).

Colman, J.E., 'Countercurrent washing calculations', *Chem. Engrg.* **70**(5), 93–96, March 4 (1963).

Comings, E.W., 'Thickening calcium carbonate slurries', *Ind. Engrg. Chem.* **32**, No. 5, 663–668 (1940).

Comings, E.W., Pruiss, C.E. and De Bord, C., 'Continuous settling and thickening', *Ind. Engrg. Chem. Des. Process Dev.* **46**, 1164–1172 (1954).

Concha, F. and Almendra, E.R., 'Settling velocities of particulate systems. Part 1. Settling velocities of individual spherical particles', *Int. J. Mineral Process.* **5**, 349–367 (1979).

Concha, F. and Barrientos, A., 'Phenomenological theory of thickening', Eng. Foundation Conference, New Hampshire (1980).

Concha, F. and Barrientos, A., 'Settling velocities of particulate systems. Part 4. Settling of nonspherical isometric particles', *Int. J. Mineral Process.* **18**, 297–308 (1986).

Concha, F. and Bascur, O., 'Phenomenological model of sedimentation', XII Int. Mineral Process. Congress, São Paulo, Brazil, PTC 03/77, 1–25 (1977).

Concha, F. and Bürger, R., 'Wave propagation phenomena in the theory of sedimentation', in *Numerical Methods for Wave Propagation*, Eds. E.F. Toro and J.F. Clarke, Kluwer Academic Publishers, Dordrecht, The Netherlands, 173–196 (1998).

Concha, F. and Bustos, M.C., 'Theory of sedimentation of flocculated fine particles', in *Floc-*

culation, *Sedimentation and Consolidation*, Eds. B.M. Moudgil and P. Somasundaran, AIChE 1986, pp. 275–284.

Concha, F. and Bustos, M.C., 'Settling velocities of particulate systems. Part 6. Kynch sedimentation processes: batch settling', *Int. J. Mineral Process.* **32**, 193–212 (1991).

Concha, F. and Bustos, M.C., 'Settling velocities of particulate systems. Part 7. Kynch sedimentation processes: continuous thickening', *Int. J. Mineral Process.* **34**, 33–51 (1992).

Concha, F., Bustos, M.C. and Barrientos, A., 'Phenomenological model of high capacity thickening', Proceedings of the XIX Int. Mineral Process. Congress, San Francisco, USA, November 1995, Chapter 14, 75–70 (1995).

Concha, F., Bustos, M.C. and Barrientos, A., 'Phenomenological theory of sedimentation', in *Sedimentation of Small Particles in a Viscous Fluid*, Ed. E. M. Tory., Computational Mechanics Publications, Southampton 1996, pp. 51–96.

Concha, F., Bustos, M.C., Oelker, E. and Wendland, W.L., 'Settling velocities of particulate systems 9. Phenomenological theory of sedimentation processes: I batch sedimentation', submitted to *Int. J. Mineral Process.* (1993).

Concha, F. and Christiansen, A., 'Settling velocities of particulate systems. Part 3. Settling velocities of suspensions of particles of arbitrary shape', *Int. J. Mineral Process.* **18**, 309–322 (1986).

Concha, F., Lee, C.H. and Austin, L.G., 'Settling velocities of particulate systems. Part 8. Batch sedimentation of polydispersed suspensions and spheres', *Int. J. Mineral Process.* **35**, 159–175 (1992).

Conway, E. and Smoller, J.A., 'Global solutions of the Cauchy problem for quasi-linear first-order equations in several space variables', *Comm. Pure Appl. Math.* **19**, 95–105 (1966).

Coulson, J.M. and Richardson, J.F., *Chemical Engineering*, Pergamon Press, Oxford, 1962.

Dafermos, C.M., 'Polygonal approximations of solutions of the initial value problem for a conservation law', *J. Math. Anal. Appl.* **38**, 33–41 (1972).

Damasceno, J.J.R., *Uma contribuição ao estudo do espessamento continuo*, Doctoral thesis, COPPE, Rio de Janeiro, Brazil, 1992.

D'Avila, J., 'Análisis da teoria de Kynch', *Revista Brasileira de Tecnologia*, 447–453 (1976).

D'Avila, J.,*Um modelo matemático para a sedimentação.* Doctoral thesis, Federal University of Rio de Janeiro, Brazil, COPPE/UFRJ, PTS 04, 1978.

D'Avila, J., 'Sedimentação, tópicos especiais de sistemas particulados', X ENEMP, Brazil, 167–216 (1982).

D'Avila, J.S., Concha, F. and Telles, A.S., 'Um modelo fenomenológico para sedimentação bidimensional continua', Anais VI Encontro sobre Escoamento em Meios Porosos, Vol. VII, Rio Claro, Brazil, III-1-1–III-1-17 (1978).

D'Avila, J. and Sampaio, R., 'Algumas consideraçãoes sobre o problema da sedimentação I', Anais do VI ENEMP, **1**, Brazil (1978).

Davis, R.H., 'Velocities of sedimenting particles in suspensions', in *Sedimentation of small Particles in a Viscous Fluid*, Ed. E.M. Tory, Computational Mechanics Publications, Southampton 1996, pp. 161–198.

Davis, R.H. and Gecol, H., 'Hindered settling function with no adjustable parameters for polydisperse suspensions', *AIChE J.* **40**, 570–575 (1994).

Deane, W.A., 'Settling problems', *Trans. Amer. Electrochem. Soc.* **37**, 71–102 (1920).

Deerr, N., 'On the settling of precipitates in general and of cane juice precipitates in particular', *Int. Sugar J.* **22**, 618 (1920).

Di Benedetto, E. and Hoff, D., 'An interface tracking algorithm for the porous medium equation', *Trans. Amer. Math. Soc.* **284**, 463–500 (1984).

Diehl, S., 'On scalar conservation laws with point source and discontinuous flux function', *SIAM J. Math. Anal.* **26**, 1425–1451 (1995).

Diehl, S., 'A conservation law with point source and discontinuous flux function modelling continuous sedimentation', *SIAM J. Appl. Math.* **56**, 388–419 (1996).

Diehl, S., 'Dynamic and steady-state behaviour of continuous sedimentation', *SIAM J. Appl. Math.* **57**, 991–1018 (1997).

Diehl, S., 'On boundary conditions and solutions for ideal thickener-clarifier units', Presented at the *UEF Conference on Solid-Liquid Separation Systems*, Kahuku, Oahu, HI, USA,

April 18–23, 1999; in review for publication in a special issue of *Separ. Purif. Technol.* (1999).

Dixon, D.C., 'Momentum balance aspects of free settling theory, part II: Continuous steady state thickening', *Sep. Science* **12**, 193–201 (1979).

Dobbins, W.E., 'Effect of turbulence on sedimentation', *Proc. ASCE* **69** (2), 235–262 (1943).

Dobran, F., 'Theory of multiphase mixtures. A thermomechanical formulation', *Int. J. Multiphase Flow* **11**, 1–30 (1985).

Dorr, J.V.N., 'The use of hydrometallurgical apparatus in chemical engineering', *J. Ind. Engrg. Chem.* **7**, 119–130 (1915).

Dorr, J.V.N., *Cyanidation and Concentration of Gold and Silver Ores*. McGraw-Hill Book Co. Inc., New York 1936.

Drew, D.A., 'Mathematical modeling of two-phase flow', *Ann. Rev. Fluid Mech.* **15**, 261–291 (1983).

Drew, D.A. and Lahey, T. Jr., 'Application of general constitutive principles to the derivation of multidimensional two-phase flow equations', *Int. J. Multiphase Flow* **5**, 243–264 (1979).

Drew, D.A. and Passman, S.L., *Theory of Multicomponent Fluids*, Springer Verlag, New York 1999.

Drew, D.A. and Segel, L.A., 'Averaged equations for two phase flows', *Studies in Appl. Math.* **1**, 205–231 (1971).

Dubois, F. and Le Floch, P., 'Boundary conditions for nonlinear hyperbolic systems of conservations laws', *J. Diff. Eqns.* **71**, 93–122 (1988).

Eklund, L.G. and Jernqvist A., 'Experimental study of the dynamics of vertical continuous thickener I', *Chem. Engrg. Sci.* **30**, 597–605 (1975).

Encyclopedia Britannica, **20**, 155–157.

Espedal, M.S. and Karlsen, K.H., 'Numerical solution of reservoir flow models based on large time step operator splitting algorithms', to be published in Lecture Notes in Mathematics, Springer Verlag 1999.

Evje, S. and Karlsen, K.H., 'Viscous splitting approximation of mixed hyperbolic-parabolic convection-diffusion equations', Report, Institut Mittag-Leffler, Stockholm, Sweden (1997a), to appear in *Numer. Math.*

Evje, S. and Karlsen, K.H., 'A note on viscous splitting of degenerate convection-diffusion equations', IMA Preprint Series #1490, Institute of Mathematics and its Applications, University of Minnesota, Minneapolis, Minnesota, MN, USA (1997b).

Evje, S. and Karlsen, K.H., 'Convergence of MUSCL type schemes for strongly degenerate convection-diffusion type equations', in preparation (1999a).

Evje, S. and Karlsen, K.H., 'Monotone difference approximations of BV solutions to degenerate convection-diffusion equations', *SIAM J. Numer. Anal.*, to appear (1999b).

Evje, S. and Karlsen, K.H., 'Degenerate convection-diffusion equations and implicit monotone difference schemes', to appear in *Proceedings of the Seventh International Conference on Hyperbolic Problems, Zurich, Switzerland, February 1998* (1999c).

Fick, A., 'Über Diffusion', *Ann. der Phys.* **94**, 59–86 (1855).

Fitch, E.B., 'Current theory and thickener design', *Ind. Engrg. Chem.* **58**, 18–28 (1966).

Fitch, E.B., 'Unresolved problems in thickener design theory', *29th Research Conf. Filtration and Separation*, Soc. Chem. Engrs. Japan, 1972.

Fitch, B., 'Kynch theory and compression zones', *AIChE J.* **29**, 940–947 (1983).

Fitch, E.B., 'A two-dimensional model for the free-settling regime in continuous thickening', *AIChE J.* **36**, No. 10, 1545-1554 (1990).

Fitch, E.B., 'Thickening theories—an analysis', *AIChE J.* **39**, 27–36 (1993).

Fitch, E.B. and Stevenson, D.G., 'Gravity separation equipment', in *Solid Separation Equipment Scale-up*, Ed. D.B. Purchas, Uplands Press Ltd., Croydon, England 1977.

Forbes, D.L.H., 'The settling of mill slimes', *Engrg. Min. J.* **93**, 411–415 (1912).

Font, R., 'Compression zone effect in batch sedimentation', *AIChE J.* **34**, 229–238 (1988).

Free, E.E., 'Rate of slimes settling', *Engrg. Min. J.* **101**, 681–686 (1916).

Gaudin, A.M. and Fuerstenau, M.C., 'The transviewer-X rays to measure suspended solids concentration', *Engrg. Min. J.* **159** (9), 110–112 (1958).

Gaudin, A.M. and Fuerstenau, M.C., 'Experimental and mathematical model of thickening',

Trans. AIME **223**, 122–129 (1962).

Glasrud, G.G., Navarrete, R.C., Scriven, L.E. and Macosko, C.W., 'Settling behaviors of iron oxide suspensions', *AIChE J.* **39**, 560–568 (1993).

Glimm, J., 'Solution in the large for nonlinear hyperbolic systems of equations', *Comm. Pure Appl. Math.* **18**, 697–715 (1965).

Godlewski, E. and Raviart, P.A., *Hyperbolic Systems of Conservation Laws*, Ellipses, Paris 1991.

Godlewski, E. and Raviart, P.A., *Numerical Approximation of Hyperbolic Systems of Conservation Laws*, Springer Verlag, New York-Berlin-Heidelberg 1996.

Gösele, W. and Wambsganss, R., 'Thickening process in gravity thickeners', *Ger. Chem. Engrg.* **6**, 351–355 (1983).

Green, M.D., Eberl, M. and Landman, K.A., 'Compressive yield stress of flocculated suspensions: Determination via experiment', *AIChE J.* **42**, No.8, 2308–2318 (1996).

Gurtin, M.E., *An Introduction to Continuum Mechanics*, Academic Press, New York, 1981.

Halmos, P.R., *Measure Theory*, D. Van Nostrand, New York 1950.

Happel, J. and Brenner, H., *Low Reynolds Number Hydrodynamics*, Nijhoff, Dordrecht, The Netherlands, 1983.

Harris, C.C., Somasundaran, P. and Jensen, R.R., 'Sedimentation of compressible materials: analysis of batch sedimentation curves', *Powder Technol.* **11**, 75–84 (1975).

Harten, A., Hyman, J. and Lax, P.D., 'On finite-difference approximations and entropy conditions for shocks', *Comm. Pure Appl. Math.* **29**, 297–322 (1976).

Harten, A. and Lax, P.D., 'A random choice finite difference scheme for hyperbolic conservation laws', *SIAM J. Numer. Anal.* **18**, 289–315 (1981).

Harten, A., Lax, P.D. and van Leer, B., 'On upstream differencing and Godunov-type schemes for hyperbolic conservation laws', *SIAM Rev.* **25**, 35–61 (1983).

Hassett, N.J., 'Design and operation of continuous thickeners', *Ind. Chemist* **34**, 116–120, 169–172, 489–494 (1958).

Hassett, N.J., 'Mechanism of thickening and thickener design', *Trans. Inst. Min. Metall.* **75**, 627 (1965).

Hassett, N.J., 'Thickening, theory and practice', *Min. Sci.* **1**, 24–40 (1968).

Hazen, A., 'On sedimentation', *Trans. ASCE* **53**, 45–71 (1904).

Hesse, C.H. and Tory, E.M., 'The stochastics of sedimentation', in *Sedimentation of Small Particles in a Viscous Fluid*, Ed. E.M. Tory, Computational Mechanics Publications, Southampton 1996, pp. 199–239.

Hill, C.D. and Bedford, A., 'Stability of the equations for particulate sedimentation', *Phys. Fluids* **22**, No. 7, 1252–1254 (1979).

Hill, C.D., Bedford, A. and Drumheller, D.S., 'An application of mixture theory to particulate sedimentation', *J. Appl. Mech.* **47**, 261–265 (1980).

Hinch, E.J. and Leal, L.G., 'Constitutive equations in suspension mechanics. Part 1. General formulation', *J. Fluid Mech.* **71**, 481–495 (1975).

Hoff, D., 'A linearly implicit finite-difference scheme for the one dimensional porous medium equation', *Math. Comp.* **45**, 23–33 (1985).

Holden, H. and Holden, L., 'On scalar conservation laws in one dimension', in *Ideas and Methods in Mathematics and Physics*, Eds. S. Albeverio, J.E. Fenstad, H. Holden and T. Lindstrøm, Cambridge University Press, Cambridge, UK, pp. 480–509 (1988).

Holden, H., Holden, L. and Høegh-Krohn, R., 'A numerical method for first order nonlinear scalar conservation laws in one dimension', *Comput. Math. Appl.* **15**, 595–602 (1988).

Holden, H. and Risebro, N.H., *Front Tracking for Conservation Laws*, Lecture Notes, Department of Mathematics, Norwegian University of Science and Technology, Trondheim, Norway, 1997.

Hopf, E., 'On the right weak solution of the Cauchy problem for a quasilinear equation of the first order', *J. Math. Mech.* **19**, 483–487 (1969).

Jeffrey, D.J., 'Some basic principles in interaction calculations', in *Sedimentation of Small Particles in a Viscous Fluid*, Ed. E.M. Tory, Computational Mechanics Publications, Southampton 1996, pp. 97–124.

Jodrey, W.S. and Tory, E.M., 'Computer simulation of close random packing of equal spheres',

Phys. Rev. A **32**, 2347–2351 (1985).

Kamel, M.T. and Tory, E.M., 'Sedimentation of clusters of identical spheres. I. Comparison methods for computing velocities', *Powder Technol.* **59**, 227–240 (1989).

Kamel, M.T., Tory, E.M. and Jodrey, W.S., 'The distribution of the kst nearest neighbours and its application to cluster settling in dispersions of equal spheres', *Powder Technol.* **24**, 19–34 (1979).

Karlsen, K.H., Brusdal, K., Dahle, H.K., Evje, S. and Lie, K.-A., 'The corrected operator splitting approach applied to a nonlinear advection-diffusion problem', *Comp. Meth. Appl. Mech. Engrg.*, to appear (1999).

Karlsen, K.H. and Lie, K.-A., 'An unconditionally stable splitting scheme for a class of nonlinear parabolic equations', *IMA J. Numer. Anal.*, to appear (1999).

Karlsen, K.H., Lie, K.-A., Risebro, N.H. and Frøyen, J., 'A front-tracking approach to a two-phase fluid-flow model with capillary forces', *In Situ* **22**, 59–89 (1998).

Karlsen, K.H. and Risebro, N.H., 'Corrected operator splitting for nonlinear parabolic equations', *SIAM J. Numer. Anal.*, to appear (1999).

Karlsen, K.H. and Risebro, N.H., 'An operator splitting method for convection-diffusion equations', *Numer. Math.* **77**, 365–382 (1997).

Keyfitz, B.L., 'Solutions with shocks: an example of an L_1 contractive semigroup', *Comm. Pure Appl. Math.* **24**, 125–132 (1971).

Kim, S. and Karrila, S.J., *Microhydrodynamics: Principles and Selected Applications*, Butterworth-Heinemann, Stoneham, 1991.

Kluwick, A., 'Kinematische Wellen', *Acta Mechanica* **26**, 15–46 (1977).

Koglin, B., 'Zum Mechanismus der Sinkgeschwindigkeit in niedrig konzentrierten Suspensionen', in *27 Vorträge des 1. Europäischen Symposions 'Partikelmeßtechnik' in Nürnberg, 17.-19. September 1975*, Dechema-Monographien 1589–1615, vol. **79b**, 235–250 (1976).

Kos, P., 'Gravity thickening of water treatment plant sludges', *J. AWWA*, 272–282 (1977a).

Kos, P., 'Fundamentals of gravity thickening', *Chem. Engrg. Prog.* **73**, 99–105 (1977b).

Kos, P., *Review of sedimentation and thickening.* Fine Particle Processing **2**, 1594–1618, AIME, New York, 1980.

Kos, P., 'Sedimentation and thickening, general overview', in *Flocculation, Sedimentation and Consolidation*, Eds. B.M. Moudgil and P. Somasundaran, AIChE 1986, pp. 39–56.

Kröner, D., *Numerical Schemes for Conservation Laws*, Wiley-Teubner, Chichester & Stuttgart 1997.

Kružkov, S.N., 'Results on the character of continuity of solutions of parabolic equations and some of their applications', *Math. Zametky*, **6**, 97–108 (1969a).

Kružkov, S.N., 'Generalized solutions of the Cauchy problem in the large for nonlinear equations of first order', *Soviet Math. Dokl.* **10**, 785–788 (1969b).

Kružkov, S.N., 'First order quasilinear equations in several independent variables', *Math. USSR Sbornik* **10**, 217–243 (1970).

Kufner, A., John, O. and Fučik, S., *Function Spaces*, Academia, Prague, Czech Republic, 1977.

Kunik, M., 'A solution formula for a non-convex scalar hyperbolic conservation law with monotone initial data', *Math. Meth. Appl. Sci.* **16**, 895–902 (1993).

Kynch, G.J., 'A theory of sedimentation', *Trans. Faraday Soc.* **48**, 166–176 (1952).

Ladd, A.J.C, 'Dynamical simulations of sedimenting spheres', *Phys. Fluids A* **5**, 299–310 (1993).

Ladd, A.J.C., 'Hydrodynamic screening in sedimenting suspensions of non-Brownian spheres', *Phys. Rev. Lett.* **76**, 1392–1395 (1996).

Ladyženskaja, V.A., Solonnikov, V.A. and Ural'ceva, N.N., *Linear and quasilinear equations of parabolic type*, AMS Translations of Mathematical Monographs **23**, Providence, Rhode Island (1968).

Landman, K.A. and White, L.R., 'Solid/liquid separation of flocculated suspensions', *Adv. Colloid Interface Science* **51**, 175–246 (1994).

Landman, K.A., White, L.R. and Eberl, M., 'Pressure filtration of flocculated suspensions', *AIChE J.* **41**, No. 7, 1687–1700 (1995).

Langseth, J.O., 'On an implementation of a front tracking method for hyperbolic conservation laws', *Adv. in Engrg. Software* **26**, 45–63 (1996).

Lax, P.D., 'Weak solutions of nonlinear hyperbolic equations and their numerical computation', *Comm. Pure Appl. Math.* **7**, 159–193 (1954).

Lax, P.D., 'Hyperbolic systems of conservation laws II', *Comm. Pure Appl. Math.* **10**, 537–566 (1957).

Lax, P.D., *Shock waves and entropy.* Contributions to Nonlinear Functional Analysis, Ed. A. Zarantonello, Academic Press, New York 1971, pp. 603–634.

Lax, P.D., *Hyperbolic Systems of Conservation Laws and the Mathematical Theory of Shock Waves.* SIAM Regional Conf. Series Lectures in Appl. Math., No. 11. SIAM, Philadelphia, PA, 1973.

Le Roux, A.Y., 'Etude du problème mixte pour une équation quasi-linéaire du premier ordre', *C. R. Acad. Sci. Paris*, **285**, Série A, 351–354 (1977).

Le Roux, A.Y., *Approximation de quelques problèmes hyperboliques non linéaires.* Thèse d' Etat, Rennes, France, 1979a.

Le Roux, A.Y., *Vanishing viscosity method for a quasilinear first order equation with boundary conditions.* Numerical Analysis of Singular Perturbation Problems, Eds. Hemker and Miller, Academic Press, New York, 493–499 (1979b).

Le Roux, A.Y., 'Convergence of an accurate scheme for first order quasilinear equations', *R.A.I.R.O. Analyse Numérique*, **15**, 151–170 (1981).

Le Veque, R.J., *Numerical Methods for Conservation Laws*, Birkhäuser Verlag, Basel 1992.

Liu, I.S., 'On fluid pressure and buoyancy force in porous media', *Revista Brasileira de Tecnologia* **11**, 35–43 (1980).

Liu, T.P., 'Invariants and asymptotic behavior of solutions of a conservation law', *Proc. Amer. Math. Soc.* **71**, 227–231 (1978).

Málek, J., Nečas, J., Rokyta, M. and Ružička, M., *Weak and Measure-Valued Solutions to Evolutionary PDEs*, Chapman & Hall, London 1996.

Mallareddy, V., *Settling of slurries of praesodymium oxalate*, M. Eng. thesis, McMaster University, Hamilton, Ontario, Canada, 1963.

Marle, C.M., 'Ecoulements monophasiques en milieu poreux', *Rev. Inst. Français du Petrole* **22**, 1471–1509 (1967).

Maude, A.D. and Whitmore, R.L., 'A generalized theory of sedimentation', *Brit. J. Appl. Phys.* **9**, 477–482 (1958).

Michaels, A.S. and Bolger, J.C., 'Settling rates and sediment volumes of flocculated kaolin suspensions', *I&EC Fund.* **1**, 24–33 (1962).

Mishler, R.T., 'Settling slimes at the Tigre Mill', *Eng. Mining J.* **94**, 643–646 (1912).

Nessyahu, H. and Tadmor, E., 'Non-oscillatory central differencing for hyperbolic conservation laws', *J. Comp. Phys.* **87**, 408–463 (1990).

Nichols, H.G., 'Theory of the settlement of slimes', *Min. Scient. Press* 54–56, July 11, 1908.

Nichols, H.G., 'A method of settling slimes, as applied to their separation from solution in cyanide treatment', *Trans. Inst. Min. Met.* **17**, 293–329 (1908).

Nicolai, H. and Guazzelli, E., 'Effect of vessel size on the hydrodynamic diffusion of sedimenting spheres', *Phys. Fluids* **7**, 3–5 (1995).

Oleinik, O.A., 'Discontinuous solutions of nonlinear differential equations', *Amer. Math. Soc. Trans. Serv.* **26**, 95–172 (1957).

Oleinik, O.A., 'Construction of a generalized solution of the Cauchy problem for a quasilinear equation of the first order by the introduction of 'vanising viscosity'', *Amer. Math. Soc. Trans. Serv.*, **33**, 277–283 (1963a).

Oleinik, O.A., 'Uniqueness and stability of the generalized solution of the Cauchy problem for a quasilinear equation', *Amer. Math. Soc. Trans. Serv.* **33**, 285–290 (1963b).

Oleinik, O.A., 'The Cauchy problem for nonlinear equations in a class of discontinuous functions', *Amer. Math. Soc. Trans. Serv.* **42**, 7–12 (1964).

Pane, V. and Schiffman, R.L., 'A note on sedimentation and consolidation', *Géotechnique* **35**, 69–72 (1985).

Petty, C.A., 'Continuous sedimentation of a suspension with a nonconvex flux law', *Chem. Eng. Sci.* **30**, 1451–1458 (1975).

Pickard, D. and Tory, E.M., 'A Markov model for sedimentation', *J. Math. Anal. Appl.* **60**,

349–369 (1977).

Quintard, M. and Whitaker, S., 'Transport in ordered and disordered porous media: volume averaged equations, closure problems and comparison with experiment', *Chem. Eng. Sci.* 48, 2537–2564 (1993).

Rajagopal, K.R. and Tao, L., *Mechanics of Mixtures*, World Scientific, Singapore 1995.

Ralston, O.C., 'The control of ore slimes I', 'The control of ore slimes III', *Engrg. Mining J.* 101, 763–769, 990–994 (1916).

Richards, R.H. and Locke, C.E., *Textbook of Ore Dressing*, McGraw-Hill Book Co., Inc., New York 1909.

Richardson, J.F. and Zaki, W.N., 'Sedimentation and Fluidization I', *Trans. Inst. Chem. Engrs. (London)* 32, 35–53 (1954).

Risebro, N.H., 'A front-tracking alternative to the random choice method', *Proc. Amer. Math. Soc.* 117, 1125–1139 (1993).

Risebro, N.H. and Tveito, A., 'Front tracking applied to a nonstrictly hyperbolic system of conservation laws', *SIAM J. Sci. Stat. Comput.* 12, 1401–1419 (1991).

Risebro, N.H. and Tveito, A., 'A front tracking method for conservation laws in one dimension', *J. Comp. Phys.* 101, 130–139 (1992).

Roberts, E.J., 'Thickening—art or science', *Mining Engrg.* 1, 61–64 (1949).

Saks, S., *Theory of the integral*, Warsaw 1937.

Sampaio, R. and D'Avila, J., 'Condições de salto na sedimentação", Anais do IV ENEMP 1, 1–24, Brazil 1976.

Sampaio, R. and D'Avila, J., 'Algumas considerações sobre o problema da sedimentação II." Anais do VII ENEMP 3, Brazil 1979.

Schiffman, R.L., Pane, V. and Gibson, R.E., 'The theory of one-dimensional consolidation of saturated clays, IV. An overview of nonlinear finite strain sedimentation and consolidation', in: *Sedimentation and Consolidation Models, Predictions and Validations*, Proc. ASCE Symposium, San Francisco (R. Yong, ed.), Young and Townsend Ed., New York, USA 1984.

Schneider, W., 'Kinematic-wave theory of sedimentation beneath inclined walls', *J. Fluid Mech.* 120, 323–346 (1982).

Schneider, W., 'Kinematic wave description of sedimentation and centrifugation processes', in *Flow of Real Fluids*, G.E.A. Meier and F. Obermeier (Eds.), Lecture Notes in Physics, Springer Verlag, Berlin, 326–337 (1985).

Scott, K.J., 'Theory of thickening: factors affecting settling rate of solids in flocculated pulps', *Trans. Inst. Min. Met.* 77, 85–97 (1968a).

Scott, K.J., 'Thickening of calcium carbonate slurries', *I&EC Fund.* 7, 484 (1968b).

Scott, K.J., 'Continuous thickening of flocculated suspensions. Comparison with batch settling and effect of floc compression using pyrophyllite pulp', *I&EC Fund.* 9, No. 3, 422–427 (1970).

Scott, K.J. and Kilgour, D.M., 'The density of random close packing of spheres', *Brit. J. Appl. Phys. (J. Phys. D.)* 2, 863–866 (1969).

Serre, D., *Systèmes de lois de conservation*, vol. I & II, Diderot, Paris 1996.

Shannon, P.T., Dehaas, R.D., Stroupe, E.P. and Tory, E., 'Batch and continuous thickening', *I&EC Fund.* 3, 250–260 (1964).

Shannon, P.T., Stroupe, E.P. and Tory, E., 'Batch and continuous thickening', *I&EC Fund.* 2, 203–211 (1963).

Shannon, P.T. and Tory, E.M., 'Settling of slurries', *Ind. Eng. Chem.* 57, 18–25 (1965).

Shannon, P.T. and Tory, E.M., 'The analysis of continuous thickening', *SME Trans.* 235, 375–382 (1966).

Shirato, M., Kato, H., Kobayashi, K. and Sakazaki, H., 'Analysis of settling of thick slurries due to consolidation', *J. Chem. Engrg. Japan* 3, 98–104 (1970).

Slattery, J.C., 'Flow of viscoelastic fluids through porous media', *AIChE J.* 13, 1066–1071 (1967).

Slattery, J.C., *Momentum, Energy and Mass Transfer in Continua*, Mc Graw-Hill Book Co., New York 1972.

Smirnov, W.I., *Lehrgang der Höheren Mathematik.* Teil V, VEB Deutscher Verlag der Wissenschaften, Berlin 1962.

Smith, G.F., 'On isotropic functions of symmetric tensors, skew-symmetric tensors and vectors', *Int. J. Engrg. Sci.* **9**, 899–916 (1971).

Spencer, S.J., Jenkins, D.R. and Barton, N.G., 'Modelling of batch and continuous sedimentation of suspensions', *Computational Techniques and Applications: CTAC-89*, Eds. W.L. Nogarth and B.J. Noye, Hemisphere, New York, 501–508 (1989).

Stefan, J., 'Über das Gleichgewicht und die Bewegung, insbesondere die Diffusion von Gasmengen', *Sitzungsber. Akad. Wiss. Wien* **63**, No. 2, 63–124 (1871).

Stewart, R.F. and Roberts, E.J., 'The sedimentation of fine particles in liquids. A survey of theory and practice', *Trans. Inst. Chem. Engrg.* **11**, 137 (1933).

Stokes, G.G., 'On the effect of the internal friction on the motion of pendulums', *Trans. Cambridge Phil. Soc.* **9**, No. 2, 8–106 (1851).

Taggart, A.F., *Handbook of Ore Dressing.* John Wiley, Ner York, 1927.

Telles, A., 'The concept of superficial porosity and its implications on flow through porous media', Federal University of Rio de Janeiro, Report COPPE/UFRJ, Brazil (1977).

Talmage, W.P. Private communication (Dorr-Oliver report) (1959), quoted by Fitch (1966).

Talmage, W.P. and Fitch, E B., 'Determining thickener unit areas', *Ind. Engrg. Chem.* **47**, 38–41 (1955).

Thacker, W.C. and Lavelle, J.W., 'Two-phase flow analysis of hindered settling', *Phys. Fluids* **20**, 1577–1579 (1977).

Thacker, W.C. and Lavelle, J.W., 'Stability of settling of suspended sediments', *Phys. Fluids* **21**, No. 2, 291–292 (1978).

Tiller, F., 'Revision of Kynch sedimentation theory', *AIChE J.* **27**, 823–829 (1981).

Tiller, F. and Khatib, Z., 'The theory of sediment volumes of compressible particulate structures', *J. Colloid and Interface Sci.* **100**, 55–67 (1984).

Tiller, F., Hysung, N. B. and Shen, Y. L., 'CATSCAN analysis of sedimentation and constant pressure filtration', Proceedings of the V World Congress on Filtration, Société Française de Filtration, Nice, France, **2**, 80–85 (1991).

Tiller, F. and Tarng, D., 'Try deep thickeners and clarifiers', *Chem. Engrg. Prog.* **91** (3), 75–80 (1995).

Toro, E.F., *Riemann Solvers and Numerical Methods for Fluid Dynamics.* Springer Verlag, Berlin 1997.

Tory, E.M., *Batch and continuous thickening of slurries*, PhD thesis, Purdue University, West Lafayette, Indiana, USA, 1961.

Tory, E.M., 'Absence of concentration gradients in slurries settling at high Reynolds numbers', *I&EC Fund.* **4**, 106–107 (1965).

Tory, E.M., Church, B.H., Tam, M.K. and Ratner, T., 'Simulated random packing of equal spheres', *Canad. J. Chem. Engrg.* **51**, 484–493 (1973).

Tory, E.M. and Hesse, C. H., 'Theoretical and experimental evidence for a Markov model for sedimentation', in *Sedimentation of Small Particles in a Viscous Fluid*, Ed. E.M. Tory, Computational Mechanics Publications, Southampton 1996, pp. 241–281.

Tory, E.M. and Kamel, M.T., 'Mean velocities in polydisperse suspensions', *Powder Tchnol.* **93**, 199–207 (1997).

Tory, E.M., Kamel, M.T. and Chan Man Fong, C.F., 'Sedimentation is container-size dependent', *Powder Technol.* **73**, 219–238 (1992).

Tory, E.M. and Pickard, D.K., 'Extensions and refinements of a mathematical model for sedimentation', *J. Math. Anal. Appl.* **86**, 442–470 (1982).

Tory, E.M. and Shannon, P.J., 'Reappraisal of the concept of settling in compression', *I&EC Fund.* **4**, 194–204 (1965).

Touré, H., *Etude des équations générales* $u_t - \varphi(u)_{xx} + f(u)_x = v$, Doctoral Thesis, University of Besançon, France, 1982.

Truesdell, C., *Rational Thermodynamics.* Second Ed., Springer Verlag, New York 1984.

Truesdell, C. and Noll, W., 'The nonlinear field theories of mechanics', in *Flügge's Handbuch der Physik*, Vol III/3, Springer Verlag, Berlin 1965.

Truesdell, C. and Toupin, R., 'The classical field theories', in *Flügge's Handbuch der Physik*, Vol III/1, Springer Verlag, Berlin 1957.

Turney, M.A., Cheung, M.K., Powell, R.L. and McCarthy, M.J., 'Hindered settling of rod-like particles mesured with magnetic resonance imaging', *AIChE J.* **41**, No. 2, 251–257 (1995).

Ungarish, M., *Hydodynamics of Suspensions*, Springer Verlag, New York 1993.

Verhoeven, J., *Prediction of Batch Settling Behaviour*, B.Eng. thesis, McMaster University, Hamilton, Ontario, Canada, 1963.

Vol'pert, A.I. and Hudjaev, S.I., 'Cauchy's problem for degenerate second order quasilinear parabolic equations', *Math. USSR Sbornik* **7**, 365–387 (1969).

Wallis, G.B., 'A simplified one-dimensional representation of two-component vertical flow and its application to batch sedimentation', in 'Symposium on the Interaction between Fluids & Particles', Inst. Chem. Engrs (London), 1962, pp. 9–16.

Wallis, G B., *One-Dimensional Two-Phase Flow*. McGraw-Hill, New York, 1969.

Wang, C.C., 'A new representation theorem for isotropic functions, parts I & II', *Arch. Rat. Mech. Anal.* **36**, 166–223 (1970).

Whitaker, S., 'Diffusion and dispersion in porous media', *AIChE J.* **13**, 470–477 (1967).

Whitaker, S., 'Advances in the theory of fluid motion in porous media', *Ind. Engrg. Chem.* **12**, 14–78 (1969).

Whitaker, S., 'Flow in porous media I: A theoretical derivation of Darcy's law', *Transport in Porous Media* **1**, 1–30 (1986).

Wilhelm, J.H. and Naide, Y., 'Sizing and operating continuous thickeners', *Mining Engrg.* **33**, 1710–1718 (1981).

Work, L.J. and Kohler, A.S., 'Sedimentation of suspensions', *Ind. Engrg. Chem.* **32**, 1329–1334 (1940).

Wu, Z., 'A note on the first boundary value problem for quasilinear degenerate parabolic equations', *Acta Math. Sc.* **4**, No. 2, 361–373 (1982).

Wu, Z., 'A boundary value problem for quasilinear degenerate parabolic equations', MRC Technical Summary Report #2484, University of Wisconsin, USA.

Wu, Z. and Wang, J., 'Some results on quasilinear degenerate parabolic equations of second order'. Proc. of the 1980 Beijing Symp. of Diff. Geom. and Diff. Eqns., **3**, Gordon & Breach, Science Publishers, New York, 1593–1609 (1982).

Wu, Z. and Yin, J., 'Some properties of functions in BV_x and their applications to the uniqueness of solutions for degenerate quasilinear parabolic equations', *Northeastern Math. J.* **5**, 395–422 (1989).

Yoshioka, N., Hotta, Y. and Tanaka, S., 'Batch settling of homogeneous slurries', *Kagaku Kogaku* **19**, 616–626 (1955).

Yoshioka, N., Hotta, Y., Tanaka, S., Naito, S. and Tsugami, S., 'Continuous thickening of homogeneous flocculated slurries', *Chem. Engrg. Japan* **21**, 66–75 (1957).

Notation Guide

In this list, we provide, where possible, brief definitions of the symbols used in this book, and indicate where they are defined or used first. To make this list as concise as possible, we avoid in this list decriptions which can be looked up in other entries, e.g. we simply write 'B_α' instead of 'the body B_α'.

Some symbols used in the book have an obvious meaning or appear only locally, for example in one theorem and its proof only, and will therefore not be explained here.

Other symbols such as the functions $z_1(t)$, $z_2(t)$ etc. or the times t_1 and t_2 used in Chapters 5, 7 and 8 change their meaning frequently and are therefore not included here either. Some other symbols have different meanings in different parts of this work, which is always indicated.

a	inflection point of f, see p. 76 in Sect. 5.3
a	diffusion coefficient in Ch. 9 and Sect. 11.4, see (9.9)
a (§ 11.3.3)	constant in Wilhelm and Naide's method, see eq. (11.46)
a_1, a_2	endpoints of the domain of f, see p. 52 in Sect. 4.1
$\mathbf{a}_\alpha$	acceleration of p_α, see equation (1.41)
$\bar{\mathbf{a}}_{\alpha q}$	axial vector in equation (1.100)
A (Ch. 4)	set of measure zero introduced in Definition 4.5
A (Sect. 4.4)	maximum of $|f'|$, see Lemma 4.6
A (Ch. 8)	initial height of sediment, see p. 149 in Sect. 8.1
A (Ch. 9)	integrated diffusion coefficient, see (9.9)
$\mathcal{A}$	averaging operator, see (10.6)
$\mathbf{A}_{\alpha q}$	skew tensor, see equation (1.101)
b (Ch. 5)	second inflection point of f, see p. 83 in Sect. 5.4
b (Ch. 11)	constant in Wilhelm and Naide's method, see eq. (11.46)
$\mathbf{b}_\alpha$	body force on B_α, see equation (1.84)
B (Ch. 1)	body, see p. 9 in § 1.2.1
B (Ch. 8)	sediment height given in equation (8.38)
$\mathbf{B}$	basis of ortogonal unit vectors, see equation (1.28)
$\mathbf{B_s}$	left Cauch-Green deformation tensor, see equation (3.19)
B_α	components of a body, see p. 9 in § 1.2.1
$B_{\alpha 0}$, $B_{\alpha c}$	reference configuration of a body, see p. 13 in § 1.2.1
$B_{\alpha t}$	actual configuration of a body, see p. 13 in § 1.2.1

$B_{\alpha t}^+,\ B_{\alpha t}^-$	parts of a body, see equation (1.60)
c (Ch. 4)	constant used in equation (4.5)
c (Ch. 11)	depth of clear water zone, see equation (11.73)
$c_1,\ c_2$	numbers related to geometrical properties of f, see p. 83
C (Ch. 7)	parameter in equation (7.22)
C (Ch. 8)	sediment height given in equation (8.41)
$\mathcal{C}_t$	solution operator for problem (10.2)
$\mathbf{C}_\alpha$	tensor occurring in equation (1.38)
$\mathcal{C}_{\Delta t}^{\Delta z}$	spatial discretization of $\mathcal{C}_{\Delta t}$, see p. 202 in Sect. 10.1
D (Ch. 8)	sediment height given in (8.42)
D (Ch. 11)	Mishler's discharge solid mass flow rate, see eq. (11.1)
$\mathcal{D}$	dilution, see equation (11.3)
$\mathcal{D}_\infty$	dilution at $\varphi = \varphi_\infty$, see p. 251 in Sect. 12.6
$\mathcal{D}_c$	dilution at $\varphi = \varphi_c$, see p. 251 in Sect. 12.6
$\mathcal{D}_D,\ \mathcal{D}_F$	Mishler's discharge/feed dilutions, see eq. (11.2)
$\mathcal{D}_k$	Coe and Clevenger's dilutions, see equation (11.7)
$\mathbf{D}_\alpha$	symmetric part of $\mathbf{L}_\alpha$, see equation (1.48)
$d\mathbf{R}_{\alpha 1,2,3}$	vectors forming $dV_{\alpha\kappa}$, see equation (1.33)
$dV_{\alpha\kappa}$	element of material volume, see equation (1.33)
$\mathbf{e}_i$	Cartesian unit vectors, see equation (1.16)
$\mathbf{e}_I$	unit normal vector to a surface discontinuity, see p. 19
E	sediment height given in (8.44)
E_3	Euclidean three-dimensional space, see p. 9 in § 1.2.1
$E_0,\ E_L$	sets of one-dimensional measure zero, see equation (6.4)
f	flux density function, see equation (2.21)
$f_1,\ f_2$	flux density functions in Sect. 8.5, see equation (8.32)
f_b	one-dimensional drift flux density, see equation (2.12)
f_{bk}	Kynch batch flux density function, see equation (3.61)
f_j'	numerical derivative, see equation (10.10)
f_D	discharge solid flux density, see p. 32 in § 2.2.2
$f_D^{1,2,3}$	values of f_D for steady states in § 10.4.2
f_F	feed solid flux density, see equation (2.22)
$f_F^{1,2,3}$	values of f_F for steady states in § 10.4.2
$f_{F1},\ f_{F2}$	values of f_F considered in Sect. 8.5, see p. 176
$\mathbf{f}$	total flux density of the solid, see equation (2.8)
$\mathbf{f}_b$	drift flux density of the solid, see equation (2.9)
f_k	Kynch flux density function, see equation (3.64)
$\mathbf{f}_\alpha$	deformation function, see equation (1.14)
F (Ch. 4)	numerical entropy flux, see Def. 4.7
F (Ch. 11)	solid mass flow rate of the feed, see eq. (11.1)
F_0	limiting mass feed rate, see equation (11.72)
$\mathbf{F}_s$	solid component rate of deformation tensor, see p. 39
$\mathcal{F}_t$	solution operator for problem (10.4)
$\mathbf{F}_\alpha$	gradient of deformation tensor, see equation (1.27)
$\mathbf{F}_{\alpha 0},\ \mathbf{F}_{\alpha c}$	gradient of deformation tensors, see p. 13 in § 1.2.1

$\mathcal{F}_{\Delta t}^{\Delta z}$	spatial discretization of $\mathcal{F}_{\Delta t}$, see p. 202 in Sect. 10.1
Fr	Froude number, see equation (3.38)
g, $\mathbf{g}$	acceleration of gravity, see p. 39 in Sect. 3.2
g (§ 4.4.1)	numerical flux, see Definition 4.6
g (Ch. 9)	function equivalent to $\widehat{r(\varphi)}\partial\varphi/\partial z$, see Def. 9.1
g_α	mass growth rate, see p. 17 in § 1.2.2
$\bar{g}_\alpha$	rate of mass transfer per unit volume, see p. 17 in § 1.2.2
$\hat{g}_\alpha$	rate of growth per unit mass, see p. 18 in § 1.2.2
$g_{\alpha 1}$	material property G of B_α, see equation (1.17)
$g_{\alpha 2}$	spatial property G of B_α
grad	spatial gradient, see equation (1.20)
G	minimum unit area function, see equation (11.71)
G_α	a property of B_α, see p. 10 in § 1.2.1
$\dot{G}_\alpha$	material time derivative of G_α, see equation (1.21)
Grad	material gradient, see equation (1.19)
h (Ch. 4)	space step, see (4.35)
h (Ch. 11)	depth of hindered settling zone, see equation (11.73)
h (Ch. 12)	compressible sediment height, see p. 246
h_1	inverse of f' restricted to $(-\infty, a)$, see Theorem 5.2
h_2	inverse of f' restricted to (a, ∞), see Theorem 5.2
h_3	inverse of f' restricted to $(-\infty, a)$, see Theorem 5.3
h_4	inverse of f' restricted to (a, b), see Theorem 5.3
h_5	inverse of f' restricted to (b, ∞), see Theorem 5.3
h_i	height interval, see equation (11.14)
H (Ch. 4)	finite difference scheme, see equation 4.52
H (Ch. 9)	Hausdorff measure, see p. 189
H (Ch. 11)	thickener height, see equation (11.73)
H^l, $H^{l,l/2}$	Hölder spaces, see p. 193 in § 9.5.2
$\mathbf{i}$	Cartesian unit vector, see equation (3.16)
$I(a, b)$	interval between a and b, see equation (4.6)
I_0	interval between $\varphi_\infty(t)$ and $\gamma\varphi(0, t)$, see equation (6.5)
I_L	interval between $\varphi_L(t)$ and $\gamma\varphi(L, t)$, see equation (6.7)
$\mathbf{I}$	identity tensor, see equation (1.36)
j	space step index, see (4.35)
$\mathbf{j}$	Cartesian unit vector, see equation (3.16)
$\mathbf{j}_{c\alpha}$	convective flux density, see equation (1.71)
$\mathbf{j}_{D\alpha}$	diffusive flux density, see equation (1.73)
J (Ch. 6)	function defined in equation (6.13)
J (Ch. 10)	number of space intervals, see (10.5)
J_α	dilatation, see equation (1.34)
k	real number in Kružkov's formulation, see Definition 4.5
k (Sect. 4.4)	time step, see (4.35)
k (Sect. 9.7)	coefficient in the solid effective stress equation (9.45)
k (Sect. 12.4)	volume of solids, see p. 242

k_1, k_2	parameter in Roberts' equation, see (12.10) and (12.12)
$\mathbf{k}$	Cartesian unit vector pointing upwards, see eq. (3.16)
K_r	a finite interval used in Definition 4.5
L	initial height of the mixture, see p. 30 in § 2.2.1
L_α	velocity gradient tensor, see p. 15 in § 1.2.1
m (Ch. 1)	mass function of the mixture, see equation (1.3)
m (Ch. 3)	one-dimensional solid-fluid interaction force, see p. 44
m (Sect. 7.5)	index of outer iteration
m (Sect. 7.6)	one plus the number of jump points of φ_I, see p. 141
m_α	mass function of B_α, see equation (1.3)
$\mathbf{m}$	solid-fluid interaction force, see p. 39 in Sect. 3.2
$\mathbf{m}_\alpha$	interaction force, see equation (1.84)
mm	minmod limiter, see equation (10.8)
M	positive constant, see p. 142 in § 7.6.2
M_0, M_1	bounds on v_j^0, see Lemma 4.6
M_1 (Ch. 9)	positive constant given in Lemma 9.5
$M_{2,3,4}$	positive constants given in Lemma 9.5
n	time step index in Sect. 4.4 and Ch. 10, see (4.35), (10.1)
n (Ch. 7)	inner iteration index in Sect. 7.5
n (Sect. 9.7)	exponent in the solid effective stress equation (9.45)
n (Ch. 11)	number of time intervals, see Fig. 11.2
n_z, n_t	components of $\mathbf{n}$ in Ch. 9, see p. 187 in § 9.2.1
$\mathbf{n}$	unit normal vector to ∂R, see p. 39 in Sect. 3.2
$\mathbf{n}$ (Ch. 9)	normal to Γ_φ, see p. 187 in § 9.2.1
N	number of time intervals, see (10.1)
$\mathbb{N}$	set of positive integers, first used on p. 70 in Sect. 4.4
N_τ	number of small time steps per Δt, see p. 202 in Sect. 10.1
O	Mishler's overflow water mass flow rate, see eq. (11.2)
p	pore pressure, see p. 46 in § 3.6.1
p_e	excess pore pressure, see equation (3.56)
p_f	fluid phase pressure, see equation (3.18)
p_f^*	dimensionless fluid pressure, see (3.29)
p_s	solid pressure, see equation (3.21)
p_s^*	dimensionless solid pressure, see (3.29)
p_t	total vertical stress, see equation (3.42)
p_α	particles which are elements of B_α, see p. 9 in § 1.2.1
P_k	parts of the mixture, see equation (1.4)
P_y	compressive yield stress, see p. 244 in Sect. 12.4
q	one-dimensional volume average velocity, see p. 29
$\mathbf{q}$	volume average velocity, see equation (1.83)
q^*	dimensionless volume average velocity, see (3.28)
$q^{1,2,3}$	values of q in § 10.4.2
q_1, q_2	values of q considered in Sect. 8.5, see p. 176
q_α	a particle; see p. 11 in § 1.2.1
r	square root of a, see Def. 9.1

Q_i	constant states of the solution, first used on p. 141
$\mathbb{Q}$	set of rational numbers, first used on p. 70 in Sect. 4.4
$\mathcal{Q}_t$	solution operator for problem (10.3)
Q_D	volumetric discharge flow rate, see p. 32 in § 2.2.2
Q_F	feed volume flux of suspension, see p. 32 in § 2.2.2
Q_O	Mishler's overflow water volume flow rate, see eq. (11.4)
Q_R	overflow volume flux of suspension, see p. 32 in § 2.2.2
Q_T	rectangular computational domain, see p. 95 in Sect. 6.1
$\mathbf{Q}_\alpha$	rotation tensor, see p. 14 in § 1.2.1
$\mathcal{Q}_{\Delta t}^{\Delta z}$	spatial discretization of $\mathcal{Q}_{\Delta t}$, see p. 202 in Sect. 10.1
$\mathbf{r}$	position vector, see equation (1.1)
$\mathbf{r}_q$	position of fixed point in equation (1.100)
R	a fixed, bounded region considered in equations (3.1)–(3.4)
$\mathbb{R}$	set of real numbers, first used in equation (2.15)
$\mathbf{R}_\alpha$	material point, see equation equation (1.12)
S (Ch. 3)	cross-sectional area of settling column, see p. 32 in § 2.2.2
S (Ch. 11)	settling area, see p. 217 in § 11.2.1
S^+, S^-	parts of S_m, see equation (1.60)
S_f	fluid part of porous bed cross-sectional area, see p. 47
S_I	surface of discontinuity, see p. 18 in § 1.2.2
S_m	surface of a body, see equation (1.60)
S_s	surface of discontinuity, see p. 40 in Sect. 3.3
t	time, first used in equation (1.5)
t^* (Ch. 3)	dimensionless time, see (3.28)
t^* (Ch. 11)	time measured in Coe and Clevenger's method, see p. 218
t_0	characteristic time, see p. 45 in Sect. 3.5
t_c	critical time, see p. 113 in Sect. 7.2
t_c^a	approximate value of t_c, see p. 146 in § 7.6.3
t^i	shock intersection times, see p. 142 in § 7.6.1
t_i	points of time considered in Sect. 7.5
(t_k, z_k)	coordinates of point P in Figures 11.3 and 11.4
t_n	points of time considered in Ch. 10
t_x	intersection time of shock $z_3(t)$ with $z = 0$, see eq. (8.40)
t_u	time given in Figure 11.4
t_U	time given in Figures 11.4 and 11.5
t_B	time at which sediment attains height B, see eq. (8.38)
t_D	time at which sediment attains height D, see eq. (8.42)
t_E	time at which sediment attains height E, see eq. (8.45)
T	endpoint of time interval, see p. 60 in § 4.3.5
T (§ 11.3.1)	time given in Figure 11.3 and 11.4
T_D	time given in Figure 11.4
$\mathcal{T}$	time interval, see p. 95 in Sect. 6.1
$\mathbf{T}_f, \mathbf{T}_s$	fluid/solid stress tensor, see p. 39 in Sect. 3.2
$\mathbf{T}_f^E$	fluid extra stress tensor, see equation (3.18)
$\mathbf{T}_I$	inner part of the stress tensor, see equation (1.98)

$\mathbf{T}_\alpha$	partial stress, see equation (1.84)
$\mathbf{T}_\alpha^+, \mathbf{T}_\alpha^-$	limits of $\mathbf{T}_\alpha$ at a jump, see equations (1.89) and (1.90)
u (Ch. 3)	one-dimensional relative solid-fluid velocity, see p. 44
u (Ch. 7)	propagation velocity of a jump, see equation (7.46)
u_∞	settling velocity of a single floc, see p. 45 in Sect. 3.5
$\mathbf{u}$	relative solid-fluid velocity, see equation (2.1)
u^*	dimensionless relative solid-fluid velocity, see (3.29)
$\mathbf{u}_\alpha$	diffusion velocity, see equation (1.95)
$\mathbf{U}_\alpha$	part of the polar decomposition of $\mathbf{F}_\alpha$, see equation (1.37)
UA	unit area, see equation (11.69)
UA_0	basic unit area, see p. 217 in § 11.2.2
v	finite difference approximation function, see eq. (4.51)
v_1, v_2	parameters in equation (7.22)
v_j'	numerical derivative, see equation (10.9)
v_j^n	mesh function, see p. 63 in Sect. 4.4
$\mathbf{v}$	mass average velocity, see p. 20 in § 1.2.2
$\mathbf{v}_I$	velocity of a surface discontinuity, see p. 19 in § 1.2.2
v_f	one-dimensional velocity of the fluid component, see p. 44
$\mathbf{v}_f$	velocity of the fluid component, see p. 27 in Sect. 2.2
v_f^*	dimensionless velocity of the fluid component, see (3.28)
v_s	one-dimensional velocity of the solid component, see p. 44
$\mathbf{v}_s$	velocity of the solid component, see p. 27 in Sect. 2.2
v_s^*	dimensionless velocity of the solid component, see (3.28)
$\mathbf{v}_\alpha$	velocity of p_α, see equation (1.40)
$\mathbf{v}_\alpha^+, \mathbf{v}_\alpha^-$	limits of $\mathbf{v}_\alpha$ at a jump, see equations (1.89) and (1.90)
$V(\chi(B))$	material volume of $\chi(B)$, see p. 8 in § 1.2.1
V^+, V^-	partial volumes, see equation (1.60)
V_0, V_c	material volumes of χ_0 and χ_c, see p. 10 in § 1.2.1
V_i	pulp volume in equation (11.14)
V_m	material volume, see p. 8 in § 1.2.1
V^n	vector of approximate solution values, see equation (10.7)
$\mathbf{V}_\alpha$	part of the polar decomposition of $\mathbf{F}_\alpha$, see equation (1.37)
w	test function, first used in Definition 4.1
W_0	total solid volume per unit area, see p. 220 in § 11.3.1
$\mathbf{W}_\alpha$	skew part of $\mathbf{L}_\alpha$, see equation (1.48)
x_i	spatial coordinates of p_α, see equation (1.16)
X_α	points as elements of the configuration of B_α, see p. 9
$X_{\alpha i}$	material coordinates of p_α, see equation (1.16)
z	upwards-pointing vertical space coordinate, see p. 29
z^*	dimensionless height, see (3.28)
z_0 (Ch. 4)	initial value of characteristic, see equation (4.4)
z_0 (Ch. 11)	initial water/suspension interface height, see Fig. 11.2
z_∞	final sediment height, see p. 251 in Sect. 12.6
z_c	critical height, see equation (7.3)
z_c^a	approximate value of z_c, see p. 146 in § 7.6.3

z_i, z_i^m	iterated values of $z(t)$, see Sect. 7.5
z_j	points of space, see (10.5)
z_k	jump points of φ_I, see p. 141 in § 7.6.1 and eq. (7.54)
z_D	final water/suspension interface height, see Fig. 11.2
z_I	water/suspension interface height, see Fig. 11.2
Z (§ 11.3.1)	height given in Figures 11.3 and 11.4
Z_D	height given in Figure 11.4 and 11.5
$\mathbb{Z}$	set of integers, first used in (4.36)

Greek symbols

α (Ch. 3)	resistance coefficient, see p. 37 in Sect. 3.1
α (Ch. 4)	entropy, see equation (4.14)
α (Ch. 5), $\tilde{\alpha}$	numbers given by geometrical properties of f, see p. 84
α (Ch. 9)	coefficient in solid effective stress equation (9.44)
$\alpha_{1,2}$ (Ch. 5)	numbers given by geometrical properties of f, see p. 86
α_k (Ch. 7)	initial states, see equations (7.42) and (7.53)
α_2^* (Ch. 7)	determined from α_2 by Def. 5.2, see p. 133
β (Ch. 3)	a constant in eq. (3.21); determined in eq. (3.52)
β (Ch. 4)	entropy flux, see equation (4.14)
β (Ch. 5), β^{**}	numbers given by geometrical properties of f, see Def. 5.3
β (Ch. 9)	coefficient in solid effective stress equation (9.44)
γ (Ch. 5)	function given by geometrical properties of f, see Def. 5.5
γ (Ch. 6)	trace operator, see equation (6.4) and Lemma 9.2
γ (Ch. 10)	parameter in numnerical algorithm, see p. 203
γ_1, γ_2	functions defined in Definition (7.3)
Γ_φ	set of jump points of φ, see p. 187 in § 9.2.1
δ (Ch. 3)	degree of compression parameter, see p. 41 in § 3.4.1
δ (Sect. 4.4)	mesh size ratio bound introduced in Lemma 4.7
δ (Ch. 5), $\delta^{\ddagger}$	numbers given by geometrical properties of f, see Def. 5.4
δ (Ch. 6)	a small parameter, first used in equation (6.14)
$\Delta\rho$	solid-fluid density difference, see equation (3.55)
Δt	time interval defined by eq. (7.47) resp. (10.1)
Δt_i	time intervals in Coe and Clevenger's method, see (11.13)
z_I	water/suspension interface height, see Fig. 11.2
Δz	space interval, see (10.5)
ϵ (Ch. 3)	superficial porosity, see p. 47 in § 3.6.1
ϵ (Ch. 6, 9)	viscosity parameter, see equations (6.21) and (9.26)
ϵ (Ch. 7)	a priori error estimate, see equation (7.56)
ϵ_0	set of admissible states at $z = 0$, see Definition 6.2
ϵ_L	set of admissible states at $z = L$, see Definition 6.2
ε	small parameter, see equation (10.11)
η (Ch. 5), η^*	numbers given by geometrical properties of f, see Def. 5.2

η (Ch. 7)	a function defined in equation (7.16)
θ	numerical derivative of A, see equation (10.11)
θ_j^n	a value between v_{j+1}^n and v_{j-1}^n, see Lemma 4.6
κ	dynamic compressibility, see p. 244 in Sect. 12.4
λ (Ch. 4)	mesh size ratio, see equation (4.39)
λ (Ch. 9)	sediment height, see equation (9.43)
λ (Ch. 10)	mesh size ratio, see p. 202 in Sect. 10.1
λ (§ 10.4.2)	sediment height, see p. 211
λ (§ 11.4.3)	loading factor, see equation (11.72)
μ (Ch. 4)	viscosity parameter, see equation (4.24a)
μ (Ch. 5), μ^*	numbers given by geometrical properties of f, see Def. 5.2
μ_δ	a test function, see equation (6.14)
ν_δ	a test function, see equation (6.15)
φ	local solid concentration, see p. 28 in Sect. 2.2
φ^+, φ^-	limiting values of φ at a discontinuity, see p. 53 in Sect. 4.1
φ_+, φ_+	initial states of the Riemann problem, see equation (5.1b)
φ^*	dimensionless concentration, see (3.28)
φ_+^*	determined from φ_+ by Def. 5.2, first used on p. 79
$\overline{\varphi}$	mean value of φ^+ and φ^-, see equation (9.14)
$\tilde{\varphi}$	mean value of φ^l and φ^r, see equation (9.21)
φ_∞	maximum concentration, boundary datum, see eq. (2.13b)
φ_∞^*	determined from φ_∞ by Def. 5.2, first used on p. 119
φ_∞^{**}	determined from φ_∞ by Def. 5.3, first used on p. 113
$\varphi_{\infty 1}^{**}, \varphi_{\infty 2}^{**}$	values of φ_∞^{**} in Sect. 8.5, see equation (8.35)
φ_0	prescribed boundary value, see equation (6.1c)
φ_c	critical concentration, see p. 43 in § 3.4.3
φ_i, φ_i^m	iterated values of φ in Sect. 7.5
φ_i	concentration values in § 11.2.2, first used on p. 218
$\overline{\varphi}_i$	average concentration, see equation (11.15)
φ_j (Sect. 7.6)	abscissas of vertices of Π, see p. 142
φ_k (§ 11.2.2)	concentrations corresponding to $\mathcal{D}_k$, see eq. (11.9)
φ_k (Sect.11.3)	concentration value considered in Figures 11.3 and 11.4
φ^l, φ^r	left and right approximate limits of $\varphi(.,t)$, see p. 189
φ_m	point where f attains a minimum, see p. 154 in § 8.3.1
φ_{min}	point where f attains a minimum, see p. 143 in § 7.6.3
φ_s	concentration value given in eq. (8.9); redefined in eq. 8.22
$\varphi_{s1}, \varphi_{s2}$	values of φ_s in Sect. 8.5, see equation (8.36)
φ_t	tangential point, see p. 88 in § 5.4.2
$\tilde{\varphi}_t$	second tangential point, see p. 119 in § 7.3.2
φ_t^{**}	determined from φ_t by Def. 5.3, first used on p. 88
$\tilde{\varphi}_t^{**}$	determined from $\tilde{\varphi}_t$ by Def. 5.3, first used on p. 121
φ_D	discharge concentration, see p. 32 in § 2.2.2
φ_F	feed concentration, see p. 32 in § 2.2.2
φ_I	initial data, see equation (2.16)
φ_L	concentration prescribed at $z = L$, see equation (2.13c)

$\overline{\varphi_L}$	concentration value given by equation (9.41)
$\underline{\varphi_L}$	concentration value given by equation (9.42)
$\widetilde{\varphi_L}$	function from which φ_L^ϵ is calculated, see equation (9.28)
$\varphi_{L0,1,2}$	values of φ_L considered in Sect. 8.5, see equation (8.27)
φ_L', φ_L''	possible values of φ_L in equation (2.28)
$\varphi_L^{(1),(2),(3)}$	possible solutions of equation (9.40)
φ_{L0}	concentration value defined in equation (8.14)
φ_L^*	determined from φ_L by Def. 5.2, first used on p. 116
φ_L^ϵ	approximation of φ_L, see (9.31)
φ_M	point where f attains a maximum, see p. 156 in § 8.3.2
φ_M^*	determined from φ_M by Def. 5.2, first used on p. 196
φ_M^{**}	determined from φ_M by Def. 5.3, first used on p. 156
φ_R	overflow concentration, see equation (2.26)
φ_α	volume fraction of B_α, see equation (1.7)
φ^ϵ	solution of problem (6.21)
φ_μ	solution of problem (4.24)
φ_χ	solid volume fraction at reference configuration, see p. 39
ϕ	test funtion, first used in Def. 9.1
ϕ_0 (Ch. 6)	initial datum, see equation (6.1b)
ϕ_0 (Ch. 9)	initial datum, see equation (9.1)
ϕ_0	initial concentration, first used on p. 29
Φ	steady state concentration profile, see p. 209 in § 10.4.2
$\Phi^{1,2,3}$	calculated steady states, see p. 211 in § 10.4.2
Φ_D	steady state discharge concentration, see p. 210 in § 10.4.2
$\Phi_D^{1,2,3}$	values of Φ_D for steady states in § 10.4.2
$\Phi_{D\max}$	maximum steady state discharge concentration, see p. 211
Φ_L	value of φ_L at steady state, see p. 210 in § 10.4.2
Φ_L^{1*}, Φ_L^{2*}	values of φ_L in § 10.4.2, see equations (10.18) and (10.19)
$\widetilde{\phi_0}$	function from which ϕ_0^ϵ is calculated, see equation (9.27)
ϕ_0^ϵ	approximation of ϕ_0, see (9.31)
ϕ_0^*	determined from ϕ_0 by Def. 5.2, first used on p. 116
Π	real polygon, see equation (7.51a)
$\hat{\Pi}$	concave hull of Π, see p. 140 in § 7.6.1
Π_-	polygonal function given by equation (7.52)
$\hat{\Pi}_-$	concave hull of Π_-, see p. 141 in § 7.6.1
$\overline{\Pi}_{1,2,3,4}$	line segments of $\hat{\Pi}_-$, see p. 141 in § 7.6.1
ρ	density of the mixture, see p. 20 in § 1.2.2
ρ_f, ρ_s	fluid/solid material densities, see p. 39 in Sect. 3.2
$\overline{\rho_\alpha}$	apparent density, see equation (1.4)
$(\overline{\rho}_\alpha)^+, (\overline{\rho}_\alpha)^-$	limits of $\overline{\rho}_\alpha$ at a jump, see equations (1.89) and (1.90)
$\rho_{\alpha\kappa}, \overline{\rho}_{\alpha 0}, \overline{\rho}_{\alpha c}$	densities in reference configurations, see pp. 9 and 10
ρ_δ	test function, see equation (6.16)
σ	displacement velocity of a discontinuity, see eq. (1.65)
σ_1, σ_2	displacement velocities in Sect. 8.5, see equation (8.34)

σ_e	solid effective stress, see p. 37 in Sect. 3.1
σ_I	descent rate of water-suspension interface, see p. 217
τ	small time step, see p. 202 in Sect. 10.1
$\chi(B)$	configuration of the mixture, see equation (1.2)
χ_α	position function, see equation (1.1)
$\chi_{\alpha\kappa}, \chi_0, \chi_c$	reference configurations, see pp. 9 and 10 in § 1.2.1
ψ (Ch. 4)	a (second) solution of problem (4.1), see Theorem 4.1
ψ (Ch. 9)	second solution considered in § 9.5.4
ψ_0	initial datum of second solution considered in § 9.5.4
ω, ω_ϵ	mollifier functions, see equations (9.29) and (9.30)
Ω (Ch. 4)	computational domain (half plane), see p. 52 in Sect. 4.1
Ω (Ch. 6)	space interval, see p. 95 in Sect. 6.1
Ω_T	computational domain (strip), see p. 60 in § 4.3.5

Other symbols

∂R	boundary of R, see equations (3.1)–(3.4)
$\partial/\partial z^*$	dimensionless spatial derivative, see (3.29)
$\partial/\partial t^*$	dimensionless time derivative, see (3.29)
$\widehat{f(u)}$	functional superposition, see equation (9.15)

Subject Index

Author Index